HEALTHCARE TECHNOLOGIES SERIES 49

Machine Learning in Medical Imaging and Computer Vision

IET Book Series on e-Health Technologies

Book Series Editor: Professor Joel J.P.C. Rodrigues, College of Computer Science and Technology, China University of Petroleum (East China), Qingdao, China; Senac Faculty of Ceará, Fortaleza-CE, Brazil and Instituto de Telecomunicações, Portugal

Book Series Advisor: Professor Pranjal Chandra, School of Biochemical Engineering, Indian Institute of Technology (BHU), Varanasi, India

While the demographic shifts in populations display significant socio-economic challenges, they trigger opportunities for innovations in e-Health, m-Health, precision and personalized medicine, robotics, sensing, the Internet of things, cloud computing, big data, software defined networks, and network function virtualization. Their integration is however associated with many technological, ethical, legal, social, and security issues. This book series aims to disseminate recent advances for e-health technologies to improve healthcare and people's wellbeing.

Could you be our next author?

Topics considered include intelligent e-Health systems, electronic health records, ICT-enabled personal health systems, mobile and cloud computing for e-Health, health monitoring, precision and personalized health, robotics for e-Health, security and privacy in e-Health, ambient assisted living, telemedicine, big data and IoT for e-Health, and more.

Proposals for coherently integrated international multi-authored edited or co-authored handbooks and research monographs will be considered for this book series. Each proposal will be reviewed by the book Series Editor with additional external reviews from independent reviewers.

To download our proposal form or find out more information about publishing with us, please visit https://www.theiet.org/publishing/publishing-with-iet-books/.

Please email your completed book proposal for the IET Book Series on e-Health Technologies to: Amber Thomas at athomas@theiet.org or author_support@theiet.org.

The Institution of
Engineering and Technology

Machine Learning in Medical Imaging and Computer Vision

Edited by
Amita Nandal, Liang Zhou, Arvind Dhaka,
Todor Ganchev and Farid Nait-Abdesselam

The Institution of Engineering and Technology

Published by The Institution of Engineering and Technology, London, United Kingdom

The Institution of Engineering and Technology is registered as a Charity in England & Wales (no. 211014) and Scotland (no. SC038698).

First published 2023

The Institution of Engineering and Technology
Futures Place
Kings Way, Stevenage
Hertfordshire SG1 2UA, United Kingdom

www.theiet.org

British Library Cataloguing in Publication Data
A catalogue record for this product is available from the British Library

ISBN 978-1-83953-593-2 (hardback)
ISBN 978-1-83953-594-9 (PDF)

Typeset in India by MPS Limited

Cover Image: Tom Werner/DigitalVision via Getty Images

Contents

About the editors

Amita Nandal is an Associate Professor in the Department of Computer and Communication Engineering, Manipal University Jaipur, India. She has authored or co-authored over 40 scientific articles in the area of image processing, wireless communication, and parallel architectures. Her research interests include image processing, machine learning, deep learning, and digital signal processing.

Liang Zhou is an Associate Professor at the Shanghai University of Medicine & Health Sciences, China. He received his PhD degree from the Donghua University, Shanghai, China, in 2012. His research interests are focused in the areas of big data analysis and machine learning with applications in the field of medicine and healthcare.

Arvind Dhaka is an Associate Professor in the Department of Computer and Communication Engineering, Manipal University Jaipur, India. He has authored or co-authored over 40 scientific articles in the area of image processing, wireless communication, and network security. His research interests include image processing, machine learning, wireless communication, and wireless sensor networks.

Todor Ganchev is a Professor in the Department of Computer Science and Engineering and the Head of the Artificial Intelligence Laboratory at the Technical University of Varna, Bulgaria. He has authored/co-authored over 180 publications in topics, including biometrics, physiological signal processing, machine learning and its applications. He is a senior member of the Institute of Electrical and Electronics Engineers (IEEE).

Farid Nait-Abdesselam is a Professor in the School of Science and Engineering at the University of Missouri Kansas City, USA. His research interests include security and privacy, networking, internet of things, and healthcare systems. He has authored/co-authored over 150 research papers in these areas.

Preface

Medical imaging is increasingly using machine learning and computer vision. Deep learning allows convolutional neural networks (CNNs) to classify, segment, and identify medical images. This has allowed the creation of new tools and apps to aid in illness detection and treatment. This discipline studies deep learning-enabled medical computer vision, machine learning for medical image analysis, and personalized medicine using image processing, computer vision, and machine learning. Machine learning and computer vision may enhance medical imaging and healthcare, and they will likely become more vital.

Chapter 1 presents machine learning algorithms and applications in medical imaging processing. In this chapter authors discuss several applications using a CNN for medical image quality assessment. It shows that machine learning can identify and handle both diagnostic and non-diagnostic images.

Chapter 2 presents a review of deep learning methods for medical segmentation tasks in brain tumors. Early detection increases patient survival for this widespread medical issue. Deep learning can separate medical images by finding hidden patterns. This chapter discusses brain tumor datasets, fully supervised and non-supervised segmentation methods, constrained sample counts and their solutions, and model interpretability from five perspectives. Authors present key concepts, network design, enhancement strategies, and pros and disadvantages. Considering difficulties, we propose segmentation research directions.

Chapter 3 presents optimization algorithms and regularization techniques using deep learning. Deep learning was employed in optimization and regularization in this book chapter. This chapter covers deep learning methods and deep neural network types. Some fundamental optimization algorithms, regularization methods and kinds were also described. This chapter compares all deep learning methods and deep neural networks.

Chapter 4 presents computer-aided diagnosis in maritime healthcare: review of spinal hernia. This chapter examined marine healthcare and major hernia CAD methods. Seafarers need precise maritime healthcare. Seafarers' medical facilities have limited CAD capability. Shipside physicians favor remote survey and telemedicine for non-emergency situations. Hernia is a serious marine illness. Due the restricted space, seafarers experience several orthopedic diseases. Due to extensive service and restricted repatriation, medical information on hernia severity and progression is scarce. In this chapter, authors examined important herniated disc CAD tools and deep learning medical imaging methods.

Chapter 5 presents diabetic retinopathy detection using AI. In this chapter, authors employed many models to increase the resilience or error-proneness of the diabetic retinopathy (DR) detection process while categorizing data using majority voting in the early phases of the thesis. DR detection, evaluation, feature extraction, categorization, and photo pre-processing are included.

Chapter 6 presents a survey image classification using CNN in deep learning. This chapter details the deep learning CNN-based image categorization system. This chapter explores convolutional brain network models for image grouping and explains image handling and arrangement techniques.

Chapter 7 presents text recognition using convolutional recurrent neural network (CRNN) models based on temporal classification and interpolation methods. Handwritten text digitization and keyboard reduction are in high demand nowadays. Digitalizing handwritten text allows for data storage, searching, alteration, and sharing, while the ability to detect and read data written in the air supports technological advances like augmented reality and virtual reality. When there is a lot of data or no normalized structure, it is tough to manually meet these objectives. This chapter presents a framework to recognize handwritten and air-written text using CRNN, connectionist temporal classification, and interpolation approaches.

Chapter 8 presents microscopic plasmodium classification using robust deep learning strategies for malaria detection. Microscopic pathogen identification is challenging and time-consuming due to similar features. Plasmodium discovery in human RBC will be accelerated by machine-driven microorganism detection. Deep learning classifies, segments, and identifies for near-perfect identification. CNN models classified and identified variant plasmodium species in one slide for closed approximation inference. Classification models such as SE ResNet, ResNeXt, MobileNet, and XceptionNet are fully explored and deployed on a data set after data preparation, augmentation, and regularization, including state-of-the-art comparison.

Chapter 9 presents medical image classification and retrieval using deep learning. In this chapter authors recognize new medical images and forecast after training our model with available data. This method manages medical diagnostics and treatments. Imaging predictions are difficult. Data availability, effectiveness, and integrity hinder deep learning data model prediction. It has been shown that deep learning improves medical imaging and prediction models.

Chapter 10 presents game theory, optimization algorithms, and regularization techniques using deep learning in medical imaging. This chapter discusses MI's deep learning-based mathematical models. Optimization, MI regularization, and game theoretical platforms are discussed. Most deep learning classification tasks are competitive or cooperative games in MI. Optimization algorithms determine the best decision-making approach for these games. Authors also examine several meta-heuristic optimization strategies recently used in MI, which profit from their adaptability, simplicity, and independence from particular conditions.

Chapter 11 presents data preparation for artificial intelligence in federated learning: the influence of artifacts on the composition of the mammography database. Image artifacts impact mammograms while building an artificial intelligence model using federated learning. Authors of this chapter tested federated learning scenarios for a database split among three clients using motion and dust artifacts. Transfer learning classifies pictures into benign, suspicious, and no microcalcifications.

Chapter 12 presents spatial cognition by the visually impaired: image processing with SIFT/BRISK-like detector and two-keypoint descriptor on Android CameraX. This chapter describes a multi-threaded Java Android app that was created using CameraX to improve spatial cognition in visually impaired and blind people. Android phones aid visually impaired spatial cognition. The BRISK algorithm constructs a two-keypoint binary descriptor of defined form.

Chapter 13 presents feature extraction process through hypergraph learning with the concept of rough set classification. Data scientists may extract the most consistent characteristics using this learning approach. Feature extraction uses hypergraph clustering extensively. This chapter applies rough set dimensionality reduction to hypergraph. Rough set defines knowledge base subsets using the unique term reduct. This chapter proposes rough hypergraph-based reduct generation for ASD data set.

Chapter 14 presents machine learning for neurodegenerative disease diagnosis with a focus on amyotrophic lateral sclerosis (ALS). ALS preclinical models, genetics, pathology, biomarkers, imaging, and clinical readouts have improved during the past 10–15 years. Innovative therapies cure neurodegenerative diseases and other medical requirements. This chapter suggests that MRI and PET may detect neurodegenerative disorders early. These methods improve differential diagnosis, therapy innovation, and sickness knowledge.

Chapter 15 presents deep/machine learning methods to identify patterns and detecting misinformation for pandemics in the post-COVID-19 era. This chapter examines post-COVID-19 deep/machine learning methods for pattern discovery and false information detection. CNN extracts text features first. Next, RNN captures text sequence. Deep learning using CNN and RNN models detects erroneous information.

Chapter 16 presents integrating medical imaging using analytic modules and applications. The abstract describes the analytical module structure for medical imaging modalities. The framework improves clinical workflow, treatment planning, and diagnostic accuracy. Machine learning, image processing, and data analytics are used to extract meaningful information from medical pictures and provide a diagnostic report. The framework is adaptable, scalable, and interoperable with clinical systems. The suggested paradigm might greatly improve medical imaging and diagnostics, improving patient outcomes.

This edited book provides a state-of-the-art research on the integration of new and emerging technologies for the medical imaging processing and analysis fields. This book provides future directions to increase the efficiency of conventional imaging models to achieve better performance in diagnoses as well as in the characterization of complex pathological conditions. The book is aimed at a readership of researchers and scientists in both academia and industry in Computer Science and Engineering, Machine Learning, Image Processing, and Healthcare Technologies and those in related fields.

Amita Nandal, Liang Zhou, Arvind Dhaka, Todor Ganchev and Farid Nait-Abdesselam
Editors

Chapter 1

Machine learning algorithms and applications in medical imaging processing

Mukesh Kumar Sharma[1], *Jaimala*[1], *Shubham Kumar*[1] *and Nitesh Dhiman*[1]

Medical images suffer from numerous artifacts, the utmost common of these are blurry artifacts. These kinds of artifacts often yield pictures that are in the form of non-diagnostic quality. To identify blurry artifacts, pictures are prospectively estimated by medical experts to improve image quality. Machine learning can have this ability to develop an automated model to deal with such images and manage these for detection level of these non-diagnostic images. In this work, we developed a convolutional Sugeno's approach-based fuzzy neural network for quality assessment of medical images to explore several applications therein. In this work, we considered the CT scan images of the chest and brain. We also developed machine learning algorithms to enhance these images. We divided each image into several boxes so that we can easily assign a membership value to each cell of the squared box for further investigation in order to make a decision over blurry images.

1.1 Introduction

Artificial intelligence (AI) is an intelligence mechanism for machines, which enables them to initiate human intelligence and mimic their behavior to a certain extent or better. AI was developed keeping in mind to solve complex real-world problems through an approximation of human decision-making capabilities and perform task in more humanistic way. Machine learning is a subset of AI that enables the system to learn from past programming explicitly for each task. This adaptability enables machine learning algorithms to perform operations like prediction or classification. Machine learning makes the data feasible and cost-effective than the manual programming by classification and prediction.

[1]Department of Mathematics, Chaudhary Charan Singh University, India

Machine learning is actually game of data analysis which automates the analytical model building. It is a branch of AI based on the idea that systems can learn from data, identify patterns, and make decisions with minimal human intervention. Alan Turing [1] in 1950 developed the Turing test to excusive if a capture has human-like intelligence 1950s Arthur Samuel coined the phrase machine learning in 1952. Rosenblatt [2] invented "Perceptron" in 1958, an electronic device which worked on principles of biological learning mechanism and simulated human thought process. Rumelhartand and James *et al.* [3] gave the back propagation method for artificial neural networks (ANN). Watkins [4] developed Q-learning that dramatically increases an effectiveness of reinforcement learning. Machine learning is very useful in many fields like medical field, engineering field, and stock analysis. Machine learning in healthcare can be used to improve the quality of patient care and the efficiency of healthcare. Machine learning has carved a niche for itself in the field of medicine. It can be also used for developing a better diagnostic tool to analyze medical images. With the help of medical images, we can find out the location, size, and defects of the body parts.

Many researchers also have used medical images to find out most of the diseases. Fitzpatrick and Maurer [5] first studied the medical image registration. The determination of a one-to-one mapping between the coordinates in one space and those in another is observed, such that points in the two spaces that correspond to the same anatomical point are mapped to each other. Maintz and Viergever [6] tested for various organs and applications, including previous generation ANNs. Pluim *et al.* [7] discussed the medical image survey on mutual-information-based registration. Zheng *et al.* [8] investigated the rough transformations of anatomical structures in medical images. Lempitsky *et al.* [9] studied the echocardiography of heart problems with two-class 3D patch classification. The fundamentals of biomedical images were described in detail by Deserno [10]. Kumar *et al.* [11] discussed the magnetic images or positron emission tomography of non-small-cell lung cancer. Gillies *et al.* [12] studied the process of radiomics and its challenges in the care of patients with cancer. Liu *et al.* [13] discussed the machine learning and convolution network for MR images. Sahiner *et al.* [14] discussed the machine learning in medical images. They analyzed five major areas of application of DL in medical imaging and radiation therapy. Essam *et al.* [15] studied the machine and deep learning techniques for medical images in breast cancer. Laurent *et al.* [16] studied the medical images to determine the potential impact of radiomics on patients. Montero *et al.* [17] reviewed medical imaging with the help of machine learning and AI. Bouhali *et al.* [18] studied the machine learning and AI for veterinary diagnostic images.

The present work is divided into six sections; in the second section, we have discussed some basic concepts related to machine learning, K-mean clustering, and C-mean clustering. In the third section, we have described the Sugeno's approach-based neuro fuzzy system for medical image description. The fourth section of the work contains numerical computations. In the fifth section, we have described a comparative study between the proposed approach and the existing approaches. In the last and final section of this work we have provided conclusion for our proposed system.

1.2 Basic concepts

1.2.1 Machine learning

Machine learning is a subset of AI that enables the system to learn explicitly from past programming for each task. This adaptability enables machine learning algorithms to perform operations like prediction or classification based on the data available as well as make the data feasible and cost-effective than manual programming.

1.2.2 Stages for conducting machine learning

There are generally three stages for conducting machine learning methodology:

1. The first stage is called the training stage. In this stage, the inputs are mapped with the expected outputs. This stage prepares the machine learning model for classification and predictions.
2. The second stage is the validation and testing stage. In this stage, we test the model with same test samples and measure its performance to know how well it has been trained. Measures like errors, accuracy time, and space complexities are observed.
3. The third stage is the final stage called the application stage. In this stage, the model is introduced to the real world to extract the necessary information to solve real-life problems.

1.2.3 Types of machine learning

In 1995, two popular methods were introduced in the field of machine learning, as mentioned below:

1. Random forest decision (Tinkam) [19]
2. Support vector machines (Cortes and Vapnik) [20]

Machine learning uses many types of learning techniques (as shown in Figure 1.1) that are primarily classified as follows.

1.2.3.1 Supervised learning

In this type of learning, the machine learns under supervision. It was first provided by a set of labeled data and training data (Kotsiantis, 2006) [21]. Here, the training data serves as an instructor with inputs coupled with correct outputs. The learning algorithm in its training phase will search for certain patterns and features to relate with labeled outputs. Following the training phase, the supervised learning algorithm will be assigned new inputs and evaluate the label of new inputs are prior training results.

1.2.3.2 Unsupervised learning

Machine learning learns on its own without any guidance from the learning algorithm. The goal is to find the construction and patterns from the input data. It does not require any supervision. It finds patterns from a given data on its own.

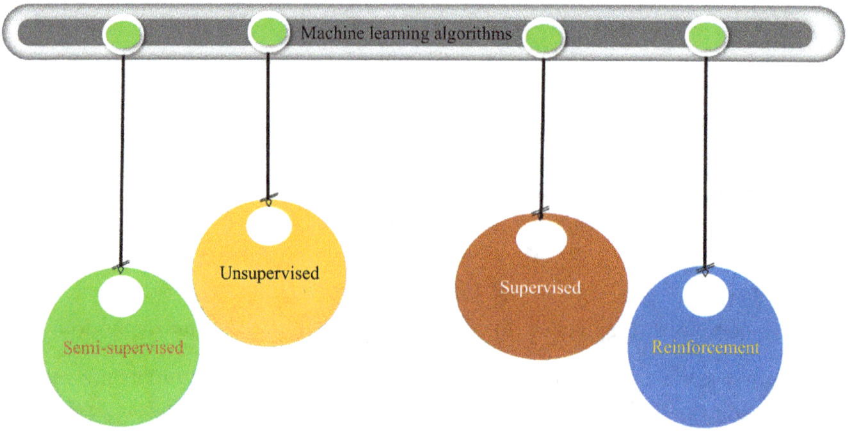

Figure 1.1 Machine learning

1.2.3.3 Semi-supervised learning

Semi-supervised learning is an intermediate state between supervised and unsupervised learning. Supervised learning uses labeled data that is very expensive, whereas unsupervised learning uses unlabeled data for maximizing prediction. Semi-supervised learning is utilized in cases when there are minimum labeled data that uses maximum unlabeled data.

1.2.3.4 Reinforcement learning

This type of learning is based upon an agent (the learning algorithm) and its interaction with a different dynamic environment. The interaction here is mainly trial and error where the agent at a particular state performs an action based on a situation and it gets reloaded for a correct decision and penalized for an incorrect area and based on the positive rewards the system trains itself and becomes ready to predict new data when presented.

Now, we will discuss the algorithms used for different learning approaches.

1.2.3.5 Supervised learning algorithms

Supervised learning can be further categorized into algorithms for regression and classification. Classification algorithms are used for predicting distinct class labels, whereas regression algorithms are used to predict continuous data.

(a) Linear regression: The statistical method is used to predict the dependent variable Y with respect to the independent variable X (as shown in Figure 1.2)

$$Y = f_0 + f_1 X + e$$

where Y is the dependent variable; X is the independent variable; f_0 is the intercept of Y; f_1= slope; and e = error.

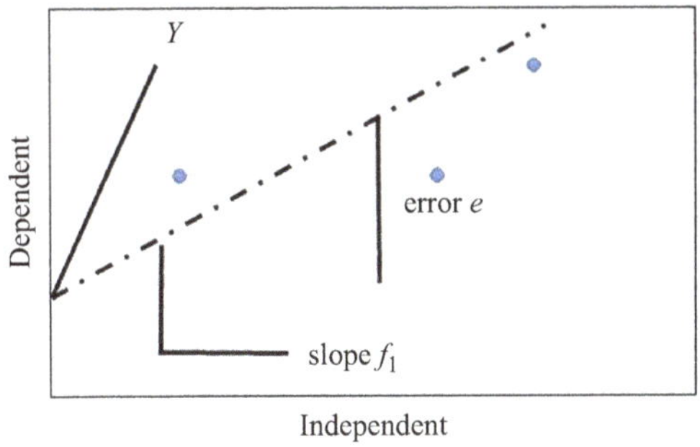

Figure 1.2 Linear regression

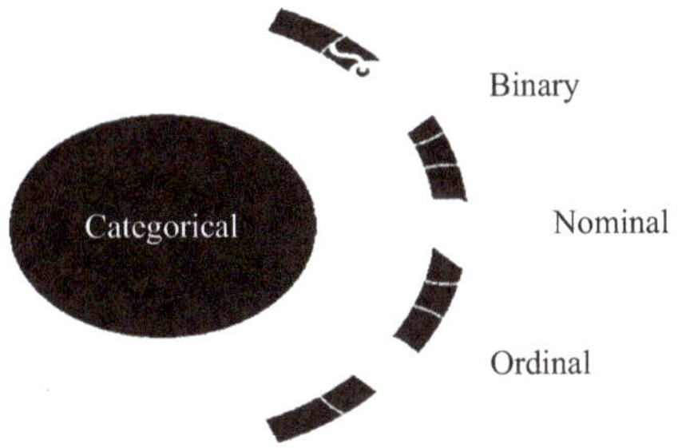

Figure 1.3 Categorical representation

(b) Logistic regression: Classical algorithm is used to predict dependent variable Y for a given independent variable X, where the dependent variables are categorical. This technique uses a cost function called sigmoid function instead of linear function and is majorly used for binary classification, where we need to predict the probability of dependent variables that belong to which class.

Categorical variables are of three types as shown in Figure 1.3.

Categorical variables with two levels for forecasting a consequence variable Y and the basis of a forecaster variable X can be written as

$$Y = b_0 + b_1 X$$

where b_0 and b_1 represent the interact and slope, respectively.

(c) Decision tree: This approach is the most prominent and extensively used technique both for classification problems where the prediction of output is a categorical value and regression problems where the prediction of output is a continuous value. A decision tree structure consists of nodes and branches where the internal nodes denote the attribute conditions. The class labels are represented by the leaf terminal node and the branches denote the value to be considered.

(d) Naive Bayesian model: Naive Bayesian algorithm is used for both discrete and continuous attributes, but mostly it is used for classification problems. This method is famous because it is easy to construct and interpret, which makes it suitable for larger datasets.

Naive Bayes is based upon the fundamental of the Bayesian theorem that operates as a conditional probability.

$$P\left(\frac{Y}{X}\right) = \frac{P\left(\frac{X}{Y}\right).P(Y)}{P(X)}$$

Here, X and Y are two events that are considered.
$P(Y/X)$ is the posterior probability of Y given X has occurred;
$P(Y)$ is the prior probability of Y;
$P(X)$ is the marginal probability of X;
$P(X/Y)$ is the likelihood probability of X given Y has occurred.

Naive Bayesian classifier is an application of Naive Bayesian algorithm that utilizes the Bayesian theorem for classification problems. Here, Y may be any particular record that is to be categorized depending upon X, where $X = (x_1, x_2, \ldots .x_n)$ is a set of independent features. So, we can rewrite the Bayesian approach as

$$P\left(\frac{Y}{(x_1, x_2, \ldots .x_n)}\right) = \frac{P\left(\frac{x_1}{Y}\right).P\left(\frac{x_1}{Y}\right).P\left(\frac{x_2}{Y}\right) \ldots\ldots .P\left(\frac{x_n}{Y}\right).P(Y)}{P(x_1).P(x_2) \ldots\ldots P(x_n)}$$

$$= \frac{P(Y).\prod_{i=1}^{n} P\left(\frac{x_i}{Y}\right)}{P(x_1).P(x_2) \ldots\ldots P(x_n)}$$

Since the denominator remains constant for all inputs, it can be removed

$$P\left(\frac{Y}{(x_1, x_2, \ldots .x_n)}\right) \propto P(Y).\prod_{i=1}^{n} P\left(\frac{x_i}{Y}\right)$$

This class variable Y with the highest likelihood must be calculated using all result of the class variable.

$$P(Y) = argmaxP(Y).\prod_{i=1}^{n} P\left(\frac{x_i}{Y}\right)$$

(e) Support vector machine: Support vector machine works on a hyper plane for separating SVM data points in the working theory of SVM. It also draws two marginal lines along with the hyper plane in such a way that the distance between the two marginal lines is maximum, thereby maximizing the precision of the classification to ensure that the problem is ready linearly separable. The SVM algorithm is popular as it can perform not only linear classification but also non-linear classification through kernel trick. The main goal of SVM kernel is to transform lower dimension into higher dimension so that the hyper plane can be easily categorized as the non-linear data point (as shown in Figure 1.4).

(f) K-nearest neighbours: K-nearest neighbours (KNN) is considered to be the most basic fundamental supervised learning algorithm for classification purposes. This non-parametric technique is a form of lazy learning. Whereas the function is only located, all the calculations are postponed before function analysis is carried out. When a test sample is introduced for the classification process, the algorithm starts. Searching all the instances in the training set is done by finding the distances of the test sample with all the respective training data to determine the nearest neighbours of the test sample easily according to the maximum nearest neighbour with the shortest distance.

(g) Artificial neural network: ANN is a machine learning technique (technology) motivated by human brains neural network system. ANN consists of a series of interconnected nodes that represent artificial neurons similar to the neurons connection in our brain (Yagnanarayna 2009) [22], which allows the transmission of signals from one node to another. Inputs in the form of real numbers are processed in the nodes, and a series of inputs is determined by a non-linear activation function. The ANN learns by changing the weight values. The ANN comprises three classes of nodes as shown in Figure 1.5.

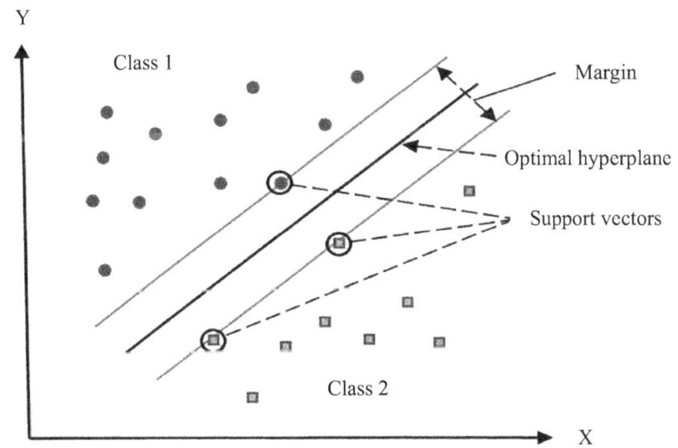

Figure 1.4 Support vector machine

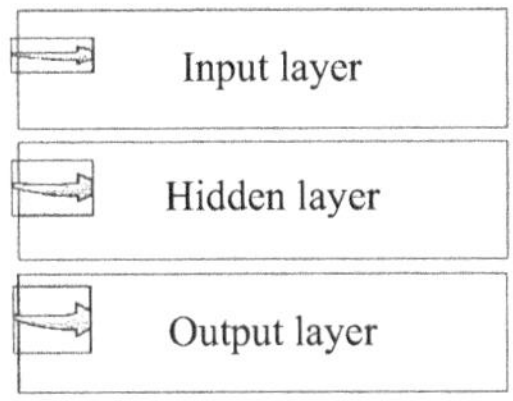

Figure 1.5 Artificial neural network

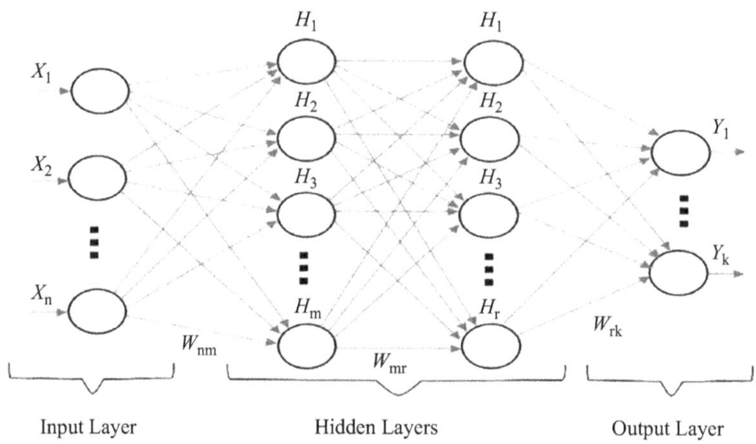

Figure 1.6 Graphical representation of ANN

In the supervised learning of ANN, the output is already known and the predicted output is matched with the actual output. The parameters will be modified depending upon the error which will be then back propagated to adjust the weight and the process will continue till the predicted outputs match the expected output as shown in Figure 1.6.

1.2.3.6 Unsupervised learning algorithm

Two main algorithms are majorly used in unsupervised learning which are discussed as follows.

(a) K-means clustering: The k-means algorithm is a partitioning clustering algorithm which partitions the unlabeled datasets into clusters by finding the optimal centroid in the high-dimensional vector space. In this iterative process, the data points are allocated to a specific cluster by calculating the minimum distance (Euclidean distance or Manhattan distance) from the data points to the centroids.

    The process of clustering is shown in Figure 1.7.

Example 1 Cluster the following eight points (with (x, y) representing the locations) into three clusters.

$A_1(2,10)$, $A_2(2,5)$, $A_3(8,4)$, $A_4(5,8)$, $A_5(7,5)$, $A_6(6,4)$, $A_7(1,2)$, and $A_8(4,9)$. Let the initial cluster centers are as shown in Table 1.1.

The distance function between two points $a = (x_1, y_1)$ and $b = (x_2, y_2)$ is shown as $d(a,b) = |x_2 - y_2| + |x_1 - y_1|$, using k-mean algorithms to find the three clusters after the second iteration.

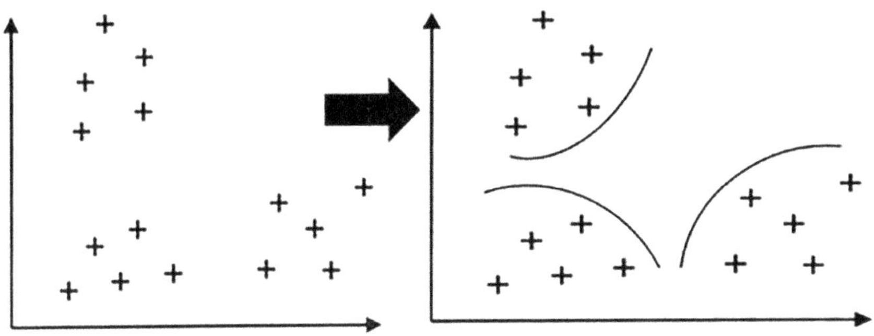

Figure 1.7 Cluster process

Table 1.1 Initial clusters

1	4	7
$A_1(2,10)$,	$A_4(5,8)$,	$A_7(1,2)$,

Table 1.2 Clusters

Clusters	Points	A_1 (2,10) Distance mean-1	A_4 (5,8) Distance mean-2	A_7 (1,2) Distance mean-3	Cluster
A_1	(2,10)	0	5	9	1
A_2	(2,5)	5	6	4	3
A_3	(8,4)	12	7	9	2
A_4	(5,8)	5	0	10	2
A_5	(7,5)	10	5	9	2
A_6	(6,4)	10	5	7	2
A_7	(1,2)	9	10	0	3
A_8	(4,9)	3	2	10	2

Solution: we have data as shown in Table 1.2.

The minimum value in cluster-1 is $A_1(2,10)$, cluster-2 are $A_3(8,4)$, $A_4(5,8)$, $A_5(7,5)$, $A_6(6,4)$, $A_8(4,9)$, and cluster-3 are $A_2(2,5)$ and $A_7(1,2)$.

Next, we need to re-compute the new clusters (means). We do this by taking the means of all points in each cluster.

For cluster-1, we only have one point $A_1(2,10)$ which was old mean, so the cluster will remain same.

For cluster-2, we have points 3, 4, 5, 6, and 8; therefore, the centroid is

$$\left(\frac{8+5+7+6+4}{5}, \frac{4+8+5+9+4}{5}\right) = (6,6)$$

For cluster-3, we have points 2 and 7; therefore, the centroid is

$$\left(\frac{2+1}{2}, \frac{5+2}{2}\right) = (1.5, 3.5)$$

The minimum value in the cluster-1 are $A_1(2,10)$ and $A_8(4,9)$, cluster-2 are- $A_3(8,4)$, $A_4(5,8)$, $A_5(7,5)$, $A_6(6,4)$, and cluster-3 are $A_2(2,5)$ and $A_7(1,2)$.

Next, we need to re-compute the new clusters (means). We do so by taking the means of all points in each cluster.

For cluster-1, we have points 1 and 8; therefore, the centroid is

$$\left(\frac{2+4}{2}, \frac{10+9}{2}\right) = (3, 9.5)$$

For cluster-2, we have points 3, 4, 5, and 6; therefore, the centroid is

$$\left(\frac{8+5+7+6}{4}, \frac{4+8+5+4}{4}\right) = (6.5, 5.25)$$

For cluster-3, we have points 2 and 7; therefore, the centroid is

$$\left(\frac{2+1}{2}, \frac{5+2}{2}\right) = (1.5, 3.5)$$

Tables 1.3 and 1.4 describe the cluster and distance means, respectively. Again, if we can move finally when the clusters come out to be same as in the earlier step then the image can be made as shown in Figure 1.8.

(b) Fuzzy C-means clustering: Unsupervised clustering is driven by the necessity to find interesting grouping in a given dataset. When the data is ambiguous in nature, then we turn toward fuzzy mean clustering.

In case of unsupervised learning, the areas of image processing and pattern recognition are used to prove the importance of unsupervised learning.

Table 1.3 Cluster and distance means

Clusters	**Points**	**(2,10) Distance mean-1**	**(6,6) Distance mean-2**	**(1.5,3.5) Distance mean-3**	**Cluster**
A_1	(2,10)	0	8	7	1
A_2	(2,5)	5	5	2	3
A_3	(8,4)	12	4	7	2
A_4	(5,8)	5	3	8	2
A_5	(7,5)	10	2	7	2
A_6	(6,4)	10	2	5	2
A_7	(1,2)	9	9	2	3
A_8	(4,9)	3	5	8	2

Table 1.4 Distance means

Clusters	Points	(3,9.5) Distance mean-1	(6.5,5.25) Distance mean-2	(1.5,3.5) Distance mean-3	Cluster
A_1	(2,10)	1.5	9.25	7	1
A_2	(2,5)	5.5	4.75	2	3
A_3	(8,4)	10.5	2.75	7	2
A_4	(5,8)	3.5	4.25	8	2
A_5	(7,5)	8.5	0.75	7	2
A_6	(6,4)	8.5	1.75	5	2
A_7	(1,2)	9.5	8.75	2	3
A_8	(4,9)	1.5	6.25	8	2

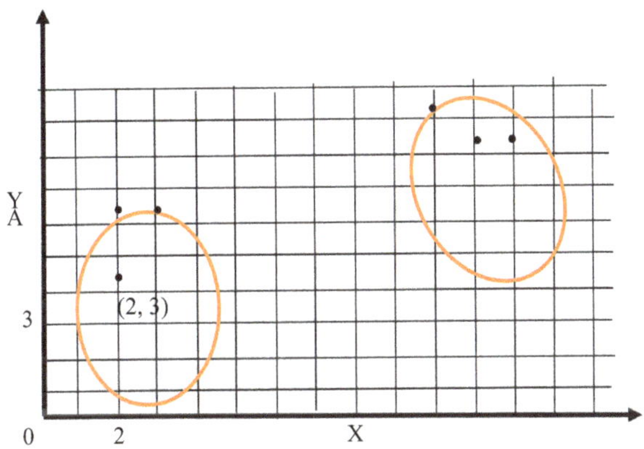

Figure 1.8 Clustered data

Let X be the set of data and x_i be an element of X,
A partition $P = \{c_1, c_2, c_3, \dots c_l\}$ of X is hard if and only if,

(i) $\forall x_i \epsilon X, \exists c_j \epsilon P$ such that $x_i \epsilon c_j$
(ii) $\forall x_i \epsilon X, x_i \epsilon c_j$ then x_i does not belong to c_j where $k \neq j$,and $c_k, c_j \epsilon P$

Let X be the set of data and $x_i \in X$.
Let us consider a partition $P = \{c_1, c_2, c_3, \dots c_l\}$ of X is soft if and only if,

(i) $\forall x_i \epsilon X \forall c_j \epsilon P$ such that $0 \leq \mu_{c_j} \leq 1$,
(ii) $\forall x_i \epsilon X \exists c_j \epsilon P$ such that $\mu_{c_j}(x_i) > 0$.

where $\mu_{c_j}(x_i)$ signifies the degree to which x_i belongs to c_j with $\sum_j \mu_{c_j}(x_i) = 1 \forall\, x_i \epsilon X$.
Partition $P = \{c_1, c_2, c_3, \dots c_k\}$ of the dataset X contains compact separated cluster c_s if any two points (in a cluster) are nearer than the distance between the two points (in different clusters) and vice versa.

Assuming that a dataset contains compact, well-separated clusters, the goal of the hard c-means algorithm is two folds:

1. Calculate the center of clusters.
2. To find the clusters level of each fact in the dataset.

$$J(P, V) = \sum_{j=1}^{c} \sum_{x_i \epsilon X} ||x_i - V_j||^2$$

where v is the vector of cluster centers

$O =$ Minimal value for a given dataset

$j =$ Also function of partition P.

The hard C-mean (HCM) algorithm rescues for the true cluster center by considering the following two steps:

1. Finding the current position base on the existing cluster
2. Adapting the current cluster center by means of a gradient descent technique to minimize the O function.

The "fuzzy c-means (FCM)" algorithm specifies the hard c-means algorithm to permit a point that belongs to numerous clusters. Consequently, it produces a soft divider for an assumed dataset. In detail, it produces a constrained soft divider to do this. The objective function O_1 of hard c means can be generalized in the following ways:

1. The fuzzy truth grade in clusters was combined into the formulation
2. A supplementary parameter was presented as a weight exponent in the fuzzy truth value grade.

The generalized objective function O_m is

$$O_m(P, V) = \sum_{i=1}^{k} \sum_{x_k \epsilon X} (\mu_{c_k}(x_k))^m ||x_i - v_j||^2$$

where the fuzzy divider of dataset X is molded by $c_2, c_3, \ldots c_k$ and the parameter m denotes the weight.

Example 2 Let us consider a dataset of six points, each of which has two features Γ_1 and Γ_2. Assume that we want to use FCM to partition the dataset into two clusters. Consider we get the parameter in FCM at z, and the initial prototype to $v_1 = (5, 5)$ and $v_2 = (10, 10)$.

Dataset: Dataset is partitioned as shown in Table 1.5.

Solution: The initial truth grades of the two clusters are classified as

$$\mu_{c_1}(x_1) = \frac{1}{\sum_{j=1}^{2} \left(\frac{||x_1 - v_1||}{||x_1 - v_j||} \right)^2}$$

Table 1.5 Dataset

	Γ_1	Γ_2
x_1	02	12
x_2	04	09
x_3	07	13
x_4	11	05
x_5	12	07
x_6	14	04

$$||x_1 - v_1||^2 = 3^2 + 7^2 = 58$$

$$||x_2 - v_1||^2 = 1^2 + 4^2 = 17$$

$$||x_1 - v_2||^2 = 8^2 + 2^2 = 68$$

$$||x_2 - v_2||^2 = 6^2 + 1^2 = 37$$

$$\mu_{c_1}(x_1) = \frac{1}{\frac{58}{58} + \frac{58}{68}} = \frac{1}{1.853} = 0.5397; \mu_{c_2}(x_1) = \frac{1}{\frac{68}{58} + \frac{68}{68}} = \frac{1}{2.172} = 0.4604$$

$$\mu_{c_1}(x_2) = \frac{1}{\frac{17}{17} + \frac{17}{37}} = \frac{1}{1.459} = 0.6854; \mu_{c_2}(x_1) = \frac{1}{\frac{37}{17} + \frac{37}{37}} = \frac{1}{3.176} = 0.3148$$

Consequently, using early samples as the two clusters, the truth grade indicates that x_1 and x_2 are supplementary in the first cluster, while outstanding points are more in the second cluster.

(i) Principal component analysis: Primary component analysis (PCA) [23] is a dimension reduction statistical approach frequently used to decrease the dimensionality of large datasets by converting a large set into a smaller variable that still has the majority of information within that large set. Karl Pearson invented this technique in the year 1901 and it was further developed by Harold Hotelling in 1933. This machine learning algorithm can be applied to noise filtration, image compression, and many other tasks.

(ii) A priori algorithm: A breadth-first search method known as the a priori algorithm calculates the support between elements. This support effectively maps data item's reliance on one another, which can help us to understand which data items affect the probability of something happening to the other data items. Bread for instance influenced the customer to purchase milk and eggs, so mapping helps maximize the store profit using this algorithm, which yields rules for its output mapping of such kind that can be learned.

1.2.3.7 Semi-supervised algorithm

(a) Self-training: Self-training is a self-thought algorithm which is generally used in semi-supervised learning; in this process, a classifier is trained first with the limited amount of data which is labeled. After training, the classifier is used for the unlabeled data classification; the training normally contains the most accurate unlabeled points along with their labeled predicted. This process is repeated till an optimal point is obtained. In order to learn itself, the classifier uses its own prediction called the self-training method.

(b) Graph method: This is a graph-based technique: This learning method constructs a graph $G = (V, E)$ from the training data, where V represent the vertices that are labeled and unlabeled data and E denotes the undirected edges connecting the nodes (i, j) with weight (W_i, j). The similarity of two instances is expressed by an edge between two vertices (x_i, x_j). This represents a graph structure where there are same labeled nodes and the majority of unlabeled nodes. All nodes are connected via edges and weights play an important part.

1.2.3.8 Reinforcement learning algorithms

(a) Q-learning: It is a “model-free” reinforcement learning procedure to study the cost of an act in a particular state. Q-learning does not need a situation-based model; hence, we call it as a “model-free” learning, and it can deal with the complications associated with stochastic transitions.
(b) Temporal difference learning: It is based on a class of reinforcement learning methods that study by bootstrapping from the present evaluation of the function value. These approaches sample from the situation, like Monte Carlo methods, and perform updates (dynamic programming).
(c) Multi-agent reinforcement learning: Multi-agent reinforcement learning (MARL) is a sub-field of reinforcement learning. It focuses on learning the performance of multiple learning agents that exist in a communal situation. Each mediator is inspired by its own rewards, and does movements to advance its own interests.
(d) Self-play: It is a method to improve the act of reinforcement learning agents. Naturally, agents study to enhance their act by playing against themselves.

1.3 Proposed algorithm for supervised learning based on neuro-fuzzy system

In this segment, we discussed about the including factors and described the five layered structural frameworks of the proposed Sugeno’s approach-based neuro fuzzy system with the approach of two input and one output fuzzy rules [24–26].

1.3.1 Input factors

Medical image suffers from several artifacts. This quality of image occurs blurring artifacts that are of non-diagnostic quality. To detect such blurring artifacts, let us

Table 1.6 Partition of image

n_1	n_2	n_3	n_4	n_5
n_6	n_7	n_8	n_9	n_{10}
n_{11}	n_{12}	n_{13}	n_{14}	n_{15}
n_{16}	n_{17}	n_{18}	n_{19}	n_{20}
n_{21}	n_{22}	n_{23}	n_{24}	n_{25}

take a frame of a CT scan of the chest showing the lungs. Now divide this image into 25 boxes and assign a membership value to each dataset as shown in Table 1.6 (where n_i; $i = 1$ to 25 represents a membership value).

1.3.2 Output factors

The output factor is categorized into three linguistic categories in terms of quality of medical images (poor, moderate, and excellent quality).

Output	**Bad**	**Moderate**	**Excellent**
y_i; i = 1 to m	[0 0.3]	(0.3 0.7]	(0.7 1]

The various components of the proposed neuro fuzzy system are described below.

Layer 1 (input layer)

Divide the figure into $n \times n = m$ squares in order to tackle 25 input factors, namely I_i; $i = 1$ to m. Categorize each input into three linguistic categories:

Input	**Linguistic range**
I_i; $i = 1$ to m	[0 1]

Layer 2 (fuzzification)

In this layer, we receive an input value and determine the grade at which the element belongs to the fuzzy set. We obtained an fuzzified output in the form of membership value. These membership values will work as an input for layer 3. During the study, triangular membership functions are considered which are defined as follows:

$$\mu(\delta) = \begin{cases} \dfrac{\delta - s}{t - s} & s \leq \delta < t \\ \dfrac{u - \delta}{u - t} & t \leq \delta \leq u \\ 0 \text{ else} \end{cases}$$

with $s < t < u$ on real line.

Layer 3 (fuzzy rules)

Firing strength of the rules is obtained by considering the following expression:

$$R_i\text{: If } I_1 \text{ is } A_1 \text{ and } I_2 \text{ is } A_2 \ldots and \quad I_m \text{ is } A_m \quad then \quad y_i \text{ is } \sum_{i=1}^{m} w_i * I_i$$

Layer 4 (output layer)

We are working with the Sugeno's approach-based neuro fuzzy system with two input and one output-based system. So, the final aggregated output of the layer is given by

$$y = \begin{cases} \sum_{i=1}^{\frac{m}{2}} y_i; \; if \quad m \quad is \quad even; \\ \sum_{i=1}^{\frac{m}{2}+1} y_i; \; if \quad m \quad is \quad odd \end{cases}$$

1.4 Application in medical images (numerical interpretation)

To understand the image proposed model, we observed the quality of the medical image [27] through this model. For this purpose, we took a sample of two medical images (see Figures 1.9–1.12) [28] in order to find the diagnostic class. Given numerical computation will help to demonstrate between various classes of image study.

Example 3:

Layer 1 (input layer)

Divide the figure into 25 squares in order to tackle 25 input factors, namely I_i; $i = 1$ to 25. Categories each input into three linguistic categories.

Layer 2 (fuzzification)

In this layer, we receive an input value and determine the grade at which the element belongs to the fuzzy set. We obtained an fuzzified outputs in the form of membership value. These membership values will work as an input for layer 3. During the study, triangular membership functions are considered which are defined as follows:

$$\mu(\delta) = \begin{cases} \dfrac{\delta - s}{t - s} & s \leq \delta < t \\ \dfrac{u - \delta}{u - t} & t \leq \delta \leq u \\ 0 \text{ else} & \end{cases}$$

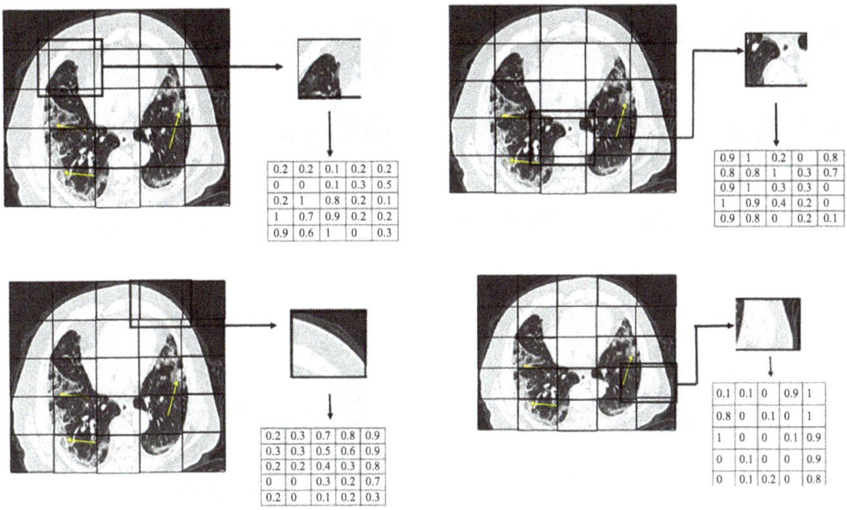

Figure 1.9 Computation 1

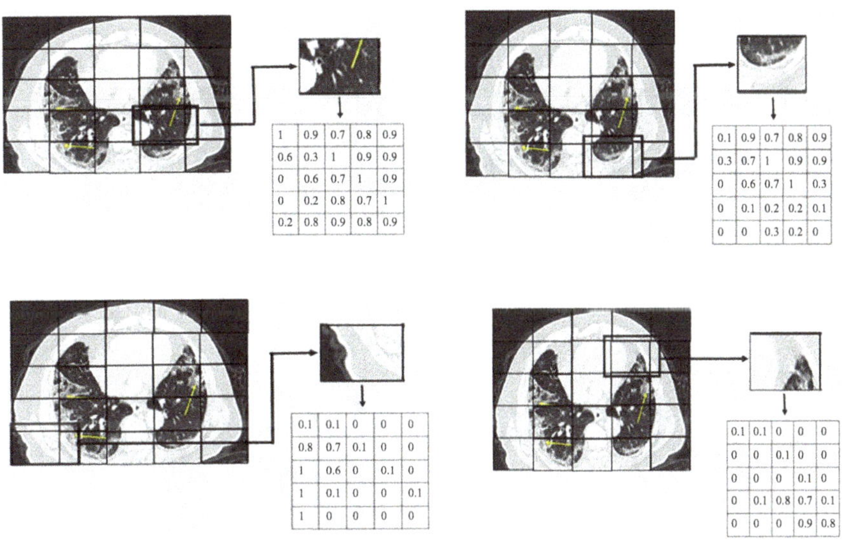

Figure 1.10 Computation 2

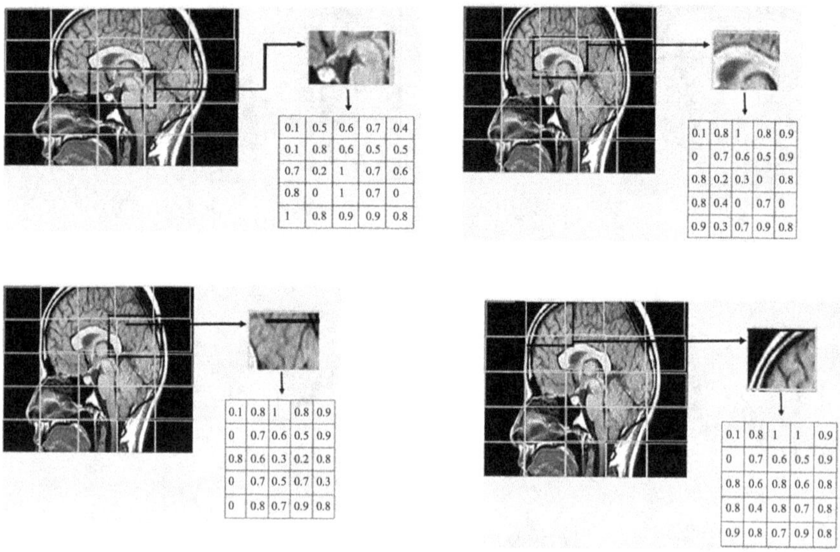

Figure 1.11 Computation 3
[Source: *https://www.aboutkidshealth.ca/article?contentid=1334&language=english*]

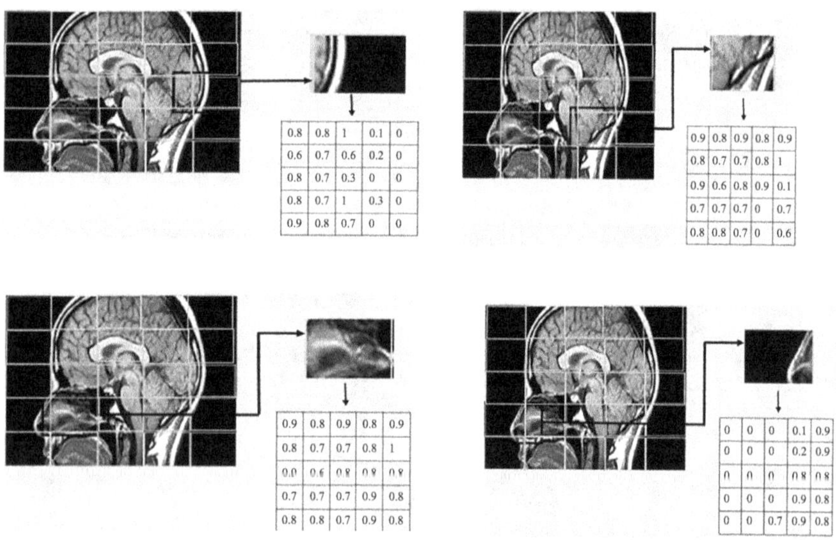

Figure 1.12 Computation 4

with $s < t < u$ on real line.

Input	Triangular fuzzy number	Input	Triangular fuzzy number
I_1	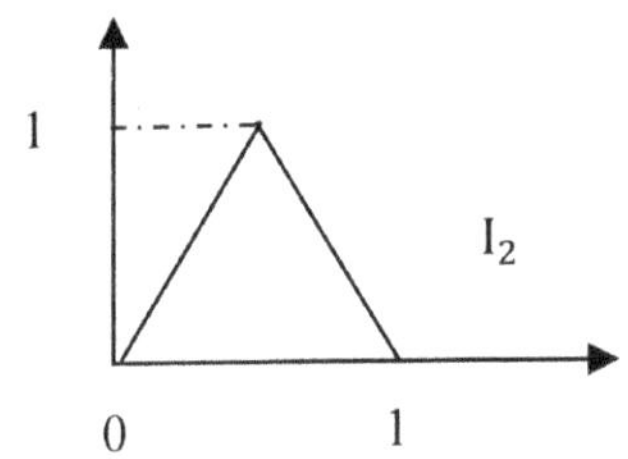	I_2	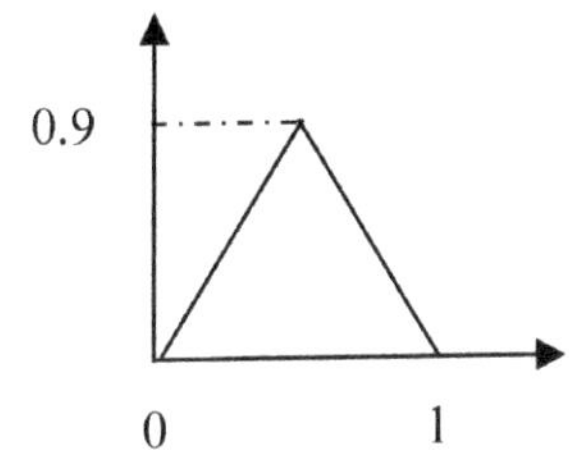

Layer 3 (fuzzy rules)

Firing strength of the rules is obtained by considering the following expression:

Choosing weight $w_i's$ in such a manner so that $\sum_i w_i = 1$, for instance, take each $w_i = 0.04$, we have the following fired rules:

$$R_1\text{: If } I_1 \text{ is 0.6 and } I_2 \text{ is 0.2 then } y_1 = 0.6w_1 + 0.2w_2$$

$$= 0.6(0.04) + 0.2(0.04) = 0.032$$

$$R_2\text{: If } I_3 \text{ is 0.9 and } I_4 \text{ is 0.2 then } y_2 = 0.9w_3 + 0.2w_4$$

$$R_3\text{: If } I_5 \text{ is 0.4 and } I_6 \text{ is 0.2 then } y_3 = 0.4w_5 + 0.2w_6$$

$$R_4\text{: If } I_7 \text{ is 0.5 and } I_8 \text{ is 0.6 then } y_4 = 0.5w_7 + 0.6w_8$$

$$R_5\text{: If } I_9 \text{ is 0.3 and } I_{10} \text{ is 0.2 then } y_5 = 0.3w_9 + 0.2w_{10}$$

$$R_6\text{: If } I_{11} \text{ is 0.1 and } I_{12} \text{ is 0.2 then } y_6 = 0.1w_{11} + 0.2w_{12}$$

$$R_7\text{: If } I_{13} \text{ is 0.5 and } I_{14} \text{ is 0.6 then } y_7 = 0.5w_{13} + 0.6w_{14}$$

$$R_8\text{: If } I_{15} \text{ is 0.1 and } I_{16} \text{ is 0.2 then } y_8 = 0.1w_{15} + 0.2w_{16}$$

$$R_9\text{: If } I_{17} \text{ is 0.6 and } I_{18} \text{ is 0.4 then } y_9 = 0.6w_{17} + 0.4w_{18}$$

$$R_{10}\text{: If } I_{19} \text{ is 0.1 and } I_{20} \text{ is 0.3 then } y_{10} = 0.1w_{19} + 0.3w_{20}$$

$$R_{11}\text{: If } I_{21} \text{ is 0.3 and } I_{22} \text{ is 0.4 then } y_{11} = 0.3w_{21} + 0.4w_{22}$$

$$R_{12}\text{: If } I_{23} \text{ is 0.1 and } I_{24} \text{ is 0.9 then } y_{12} = 0.1w_{23} + 0.9w_{24}$$

$$R_{13}\text{: If } I_{25} \text{ is 0.1 and } I_1 \text{ is 0.9 then } y_{12} = 0.1w_{25} + 0.9w_1$$

Layer 5 (output layer)

As we are adopting Sugeno's approach-based neuro fuzzy system, so the output layer is given by

$$y = \sum_{i=1}^{\frac{m}{2}+1} y_i = 9.9 \times 0.04 = 0.396$$

Example 4:

Layer 3 (Fuzzy rules)

We have the following fired rules:

R_1: If I_1 is 0.2 and I_2 is 0.3 then $y_1 = 0.2w_1 + 0.3w_2$

$= 0.2(0.04) + 0.3(0.04) = 0.02$

R_2: If I_3 is 0.3 and I_4 is 0.4 then $y_2 = 0.3w_3 + 0.4w_4$

R_3: If I_5 is 0.7 and I_6 is 0.8 then $y_3 = 0.7w_5 + 0.8w_6$

R_4: If I_7 is 0.5 and I_8 is 0.6 then $y_4 = 0.5w_7 + 0.6w_8$

R_5: If I_9 is 0.3 and I_{10} is 0.2 then $y_5 = 0.3w_9 + 0.2w_{10}$

R_6: If I_{11} is 0.3 and I_{12} is 0.2 then $y_6 = 0.3w_{11} + 0.2w_{12}$

R_7: If I_{13} is 0.5 and I_{14} is 0.7 then $y_7 = 0.5w_{13} + 0.7w_{14}$

R_8: If I_{15} is 0.1 and I_{16} is 0.2 then $y_8 = 0.1w_{15} + 0.2w_{16}$

R_9: If I_{17} is 0.6 and I_{18} is 0.2 then $y_9 = 0.6w_{17} + 0.2w_{18}$

R_{10}: If I_{19} is 0.5 and I_{20} is 0.3 then $y_{10} = 0.5w_{19} + 0.3w_{20}$

$*R_{11}$: If I_{21} is 0.3 and I_{22} is 0.6 then $y_{11} = 0.3w_{21} + 0.6w_{22}$

R_{12}: If I_{23} is 0.3 and I_{24} is 0.9 then $y_{12} = 0.3w_{23} + 0.9w_{24}$

R_{13}: If I_{25} is 0.2 and I_1 is 0.9 then $y_{12} = 0.2w_{25} + 0.9w_1$

Layer 4 (Output layer)

As we are adopting Sugeno's approach-based neuro fuzzy system, so the output layer is given by

$$y = \sum_{i=1}^{\frac{m}{2}+1} y_i = 11.1 \times 0.04 = 0.444$$

1.5 Comparison of proposed approach with the existing approaches

In this section, we have discussed a comparative study between the existing approach and our proposed approach as shown in Table 1.7.

Table 1.7 Comparative study between the proposed approach and the existing approaches

Algorithms	Advantages	Disadvantages	Applicability
Reinforcement learning	Innovative, goal-oriented, and adaptable	It prefers to solve complex problems not simple ones. It is time taking and it requires high maintenance cost	Game optimization and simulating synthetic environments, self-driving cars, etc.
Unsupervised learning	It is useful in dimensionality reduction. It can find the patterns in data; it can solve the problem by learning the data and without label classification	Less accuracy, time spending, and it cannot give an accurate information about data sorting	Anomaly detection, clustering, dimensionality reduction, visualization, etc.
Semi-supervised learning	It is easy to understand and it is powerful when we have limited label data and overflowing unlabeled data	It cannot handle complex problems and it requires high computations	In search engines, analysis of images and audio, etc.
Proposed supervised learning-based neuro fuzzy model	It has the capability to produce an output from the earlier experience	It is completely dependent on human expertise	Object recognition, speech recognition, bio informatics, spam detection, medical image classification, etc.

1.6 Conclusion

Machine learning algorithms are applied over various statistical models that processer systems use to achieve a specific task. Machine learning techniques have many applications in daily life, including data mining, predictive analytics, and image processing. The basic advantage of using machine learning algorithms is that once a procedure learns what and how to do with data, it can easily do its work automatically. In this work, a brief overview of past applications of various machine learning algorithms has been discussed, including supervised, unsupervised, semi-supervised, and reinforcement learning. Supervised learning can be defined by its use of labeled datasets to train procedures that predict outcomes precisely. The given supervised learning-based neuro fuzzy system adjusts the weights used in it, until the model has been fitted correctly. The proposed model helps medical experts in the classification of medical images.

In this work, we studied several CT scanned images to detect blurry artifacts in medical imaging and classify the quality of non-diagnostic images. The results of numerical computation were 0.396 and 0.444, described as moderate medical images. The proposed model is able to identify the levels between diagnostic and non-diagnostic images. More significantly, we examined and offered several tasks in the classification of medical imaging and the utility of the proposed approach over the consequences on standard performance models used in the study of medical images.

References

[1] Turing A.M. (1950) Computing machinery and intelligence, *Mind*, vol. 59, pp. 433–460.

[2] Rosenblatt F. (1958) The perceptron: a probabilistic model for information storage and organization in the brain, Cornell Aeronautical Laboratory, *Psychol. Rev.*, vol. 65(6), pp. 386–408.

[3] Rumelhart D.E. and James M. (1986) *Parallel Distributed Processing: Explorations in the Microstructure of Cognition*, MIT Press Cambridge, ISBN 978-0-262-63110-5.

[4] Watkins C.J.C.H. (1989) *Learning from Delayed Rewards* (PDF), PhD thesis. University of Cambridge. EThOS uk.bl.ethos.330022.

[5] Fitzpatrick J.M. and Maurer C.R. (1993) A reviews of medical image registration. *Interactive Image-guided Neurosurgery*, vol. 1, pp. 17–44.

[6] Maintz J.B. and Viergever M.A. (1998) A survey of medical image registration, *Med. Image Anal.*, vol. 2, pp. 1–36.

[7] Pluim J.P., Maintz J.A. and Viergever M.A. (2003) Mutual-information-based registration of medical images: a survey. *IEEE Trans. Med. Imaging*, vol. 22, pp. 986–1004.

[8] Zheng Y., Barbu A., Georgescu B., Scheuering M. and Comaniciu D. (2008) Four-chamber heart modeling and automatic segmentation for 3D cardiac CT volumes using marginal space learning and steerable features. *IEEE Trans. Med. Imaging*, vol. 27 (11), pp. 1668–1681.

[9] Lempitsky V., Verhoek M., Noble J.A. and Blake A. (2009) Random forest classification for automatic delineation of myocardium in real-time 3D echocardiography. In: Nicholas Ayache, Herve Delingette, Maxime Sermesant (eds.), *Functional Imaging and Modeling of the Heart*, Springer Berlin Heidelberg, pp. 447–456.

[10] Deserno T.M. (2010) Fundamentals of biomedical image processing. In: Thomas Martin Deserno (ed.), *Biomedical Image Processing*, Springer Berlin, Heidelberg, pp. 1–51.

[11] Kumar V., Gu Y., Basu S., *et al.* (2012) Radiomics: the process and the challenges, *Magn. Reson. Imaging* vol. 30(9), pp. 1234–1248.

[12] Gillies R.J., Kinahan P.E. and Hricak H. (2016) Radiomics: images are more than pictures, they are data, *Radiology*, vol. 278(2) pp. 563–577.

[13] Liu C., Wu X., Yu X., Tang Y., Zhang J. and Zhou J. (2018) Fusing multi-scale information in convolution network for MR image super-resolution reconstruction, *Biomed. Eng. Online*, vol. 17, p. 114.

[14] Sahiner B., Pezeshk A., Hadjiiski L.M, *et al.* (2019) Deep learning in medical imaging and radiation therapy, *Med. Phys.*, vol. 46(1), pp. 1–36.

[15] Essam H.H., Marwa M. E., Abdelmgeid A. A. and Ponnuthurai N.S. (2020) Deep and machine learning techniques for medical imaging-based breast cancer: a comprehensive review, *Expert Syst. Appl.*, vol. 167, pp. 114–161 https://doi.org/10.1016/j.eswa.2020.114161

[16] Laurent D., Theophraste H., Alexandre C.N.P. and Deutschb E. (2020) Reinventing radiation therapy with machine learning and imaging biomarkers (radiomics): state-of-the-art, challenges and perspectives, *Methods*, vol. 188, pp. 44–60. (https://doi.org/10.1016/j.ymeth.2020.07.003)

[17] Montero A.B., Javaid U., Vald´es G., *et al.* (2021) Artificial intelligence and machine learning for medical imaging: a technology review, *Phys. Med.*, vol. 83, pp. 242–256.

[18] Bouhali O., Bensmail H., Sheharyar A., David F. and Johnson J.P. (2022) A review of radiomics and artificial intelligence and their application in veterinary diagnostic imaging, *Vet. Sci.*, vol. 9(12), pp. 620–634.

[19] Ho T. K. (1995) Random decision forests. In *Proceedings of 3rd International Conference on Document Analysis and Recognition, IEEE*, vol. 1, pp. 278–282.

[20] Cortes C. and Vapnik V. (1995) Support-vector networks. *Mach. Learn.* vol. 20, pp. 273–297.

[21] Kotsiantis S.B., Zaharakis I.D. and Pintelas P.E. "Machine learning: a review of classification and combining techniques", *Artif. Intell. Rev.*, vol. 26, pp. 159–190, (2006).

[22] Yegnanarayana B. Artificial Neural Network, PHI Learning Pvt. Ltd., Delhi, 2009.

[23] Sharma A.K., Nandal A., Dhaka A., and Dixit R., Medical image classification techniques and analysis using deep learning networks: a review, In Ripon Patgiri, Anupam Biswas, Pinki Roy (eds.), *Health Informatics: A Computational Perspective in Healthcare*, Springer Singapore, pp. 233–258, 2021.

[24] Zhou L., Chaudhary S., Sharma M. K., Dhaka A., and Nandal A., (2023) Artificial neural network dual hesitant Fermatean fuzzy implementation in transportation of COVID-19 vaccine, *J. Organ. End User Comput. (JOEUC), IGI Global*, vol. 35(2), pp. 1–23.

[25] Sharma M.K., Dhiman N., Mishra V.N., Mishra L. N., Dhaka A., and Kounda D. (2022) Post-symptomatic detection of COVID-2019 grade based mediative fuzzy projection, *Comput. Electr. Eng.*, Elsevier, vol. 101, pp. 1–20.

[26] Nandal A., Blagojevic M., Milosevic D., Dhaka A., and Mishra L. N. (2021) Fuzzy enhancement and deep hash layer based neural network to detect Covid-19, *J. Intell. Fuzzy Syst.*, IOS Press, vol. 41(1), pp. 1341–1351.

[27] Sharma A.K., Nandal A., Dhaka A., and Dixit R. (2020) A survey on machine learning based brain retrieval algorithms in medical image analysis, *Health and Technology*, Springer, vol. 10, pp. 1359–1373.

[28] Khare R.K., Sinha G.R., and Kumar S. (2017) Fuzzy based contrast enhancement method for lung cancer CT images, *International Journal of Engineering and Computer Sciences*, vol. 6, no. 5, pp. 21201–21204.

Chapter 2

Review of deep learning methods for medical segmentation tasks in brain tumors

Jiaqi Li[1] and Yuxin Hou[1]

Detection and segmentation of brain tumors present an arduous and time-consuming challenge in medical image processing. This is a significant medical issue affecting individuals of all ages worldwide, and early detection can substantially improve patient survival rates. Deep learning techniques can extract concealed features from medical images and successfully complete image segmentation tasks, which makes it a highly sought-after area of research for brain tumor segmentation. In this chapter, we provide a detailed summary of brain tumor datasets, fully supervised and non-fully supervised segmentation methods, the problem of small sample sizes, as well as its solutions, and model interpretability from five different perspectives. We focus on the fundamental concepts, network structures, improvement strategies, and an overview of the advantages and disadvantages of each method. Furthermore, taking into account the current challenges faced by segmentation methods, we propose promising avenues for future research in this field.

2.1 Introduction

Brain tumors are abnormal masses of tissue located within the skull in which abnormal cells grow and proliferate uncontrollably [1]. Meningiomas and astrocytomas, such as neuroglioblastomas, are the most common primary brain tumors in adult patients. Gliomas can be categorized into four grades based on the histological and cellular characteristics of glioblastomas. Grades 1 and 2 gliomas are considered low-grade gliomas (LGG), while grades 3 and 4 are high-grade gliomas (HGG). Typically, both grades are treated using a combination of surgery, radiotherapy, and chemotherapy [2]. However, due to its aggressive nature and limited response to concurrent chemotherapy with standard care, the treatment of HGG may not be as effective as LGG, especially when dealing with glioblastomas, a more common and malignant primary brain tumor [3]. Because gliomas have a

[1]School of Health Science and Engineering, University of Shanghai for Science and Technology, China

high death rate, early detection is essential to increasing the possibility that the disease will be treatable.

Magnetic resonance imaging (MRI) is a commonly used non-invasive technique for brain imaging, providing extensive information about the soft tissues of the body without requiring surgery. This imaging method offers high-contrast images without ionizing radiation or contrast agents, providing an objective and specific view of the anatomy and surrounding tissues. MRI enables precise localization and characterization of lesions, making it a crucial tool for the accurate segmentation of gliomas. Each MRI modality, including T1, T2, T1 contrast, and FLAIR images, provides crucial information about the various soft tissues of the human brain. By integrating this data, comprehensive information can be obtained, leading to improved patient prognosis, diagnosis, and follow-up treatment.

Despite the considerable benefits offered by MRI, accurate segmentation of brain tumors in MRI scans remains a challenging task. This is primarily due to the highly heterogeneous appearance and shape of the tumors and their resemblance to healthy brain tissue, which increases the risk of misdiagnosis. In addition, the heavy workload of reviewing MR images and the subsequent slow review times, coupled with the lack of timely feedback to patients, exacerbates the problem. As a result, segmentation of images has become increasingly challenging. However, recent advances in deep learning-based models have achieved significant results in segmenting irregular tumor areas from normal areas in medical tumor images. This has been achieved by leveraging a shallow to deep multi-level network structure, which facilitates the learning of low-level features from the original image and intermediate-level features extracted by different convolutional kernels, to form progressively abstracted high-level semantic features [4,5].

This chapter represents the cutting-edge research in brain tumor segmentation using MRI images, and its structure is illustrated in Figure 2.1. First, it compiles commonly used datasets in the field of brain tumor medical image processing. Second, it provides a comprehensive summary of deep learning-based brain tumor detection and segmentation methods, delving into two different supervised approaches, namely fully supervised and non-fully supervised. It then introduces the small-sample problem and its solutions, as well as model interpretability, before comparing and analyzing the algorithms' performance and characteristics in a comprehensive manner. Finally, it forecasts the challenges faced by deep learning in the field of medical image processing of brain tumors and potential future research directions.

2.2 Brain segmentation dataset

2.2.1 BraTS2012-2021

The BraTS2012-2021 dataset (http://braintumorsegmentation.org/) is derived from the "Multimodal Brain Tumor Segmentation" competition organized by the MICCAI Association. This annual challenge has been running since 2012 and aims to provide up-to-date data for brain tumor segmentation research. The BraTS

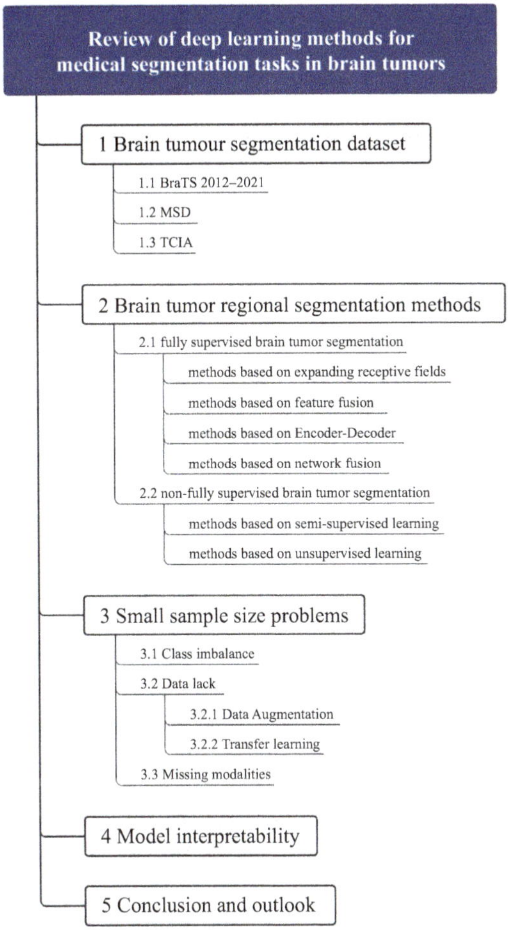

Figure 2.1 The overall framework of this chapter

dataset primarily focuses on glioma segmentation and is widely used in measuring the performance of various brain tumor segmentation methods. It contains pre-operative MRI image data obtained through different clinical protocols and scanners from multiple institutions. Among these datasets, BraTS2013 [6], BraTS2015, BraTS2017, and BraTS2018 are the most commonly used, while BraTS2021 [7] is the most recent. The following section provides a brief description of these five datasets.

The BraTS dataset comprises a training set, a validation set, and a test set, with a slice dimension of $240 \times 240 \times 155$ (where 240×240 denotes the size, and 155 denotes the volume), and is stored in NIfTI file format (.nii.gz). The training set is publicly available, while the test set is not, and is primarily used for online evaluation of the method. The BraTS2013 and BraTS2015 datasets are also referred to as the Leaderboard for the validation set and Challenge for the test set, respectively. The BraTS2015 training set contains 220 cases of HGG and 54 cases of LGG, and each case in both series datasets includes four different image modalities, namely T1, T1c, T2, and FLAIR. Depending on the clinical application, for multi-category brain tumor segmentation tasks, regions such as complete tumor, core tumor, and enhancing tumor are used for automatic image segmentation. The BraTS dataset has undergone several changes since 2017, with the labels transformed from five to four, the necrotic (label 1) and non-enhancing (label 3) unified and merged into the necrotic and non-enhancing tumor core (label 1), and label 3 removed. While BraTS2018 has the same training set data as BraTS2017, consisting of 210 HGG cases and 75 LGG cases, the validation and test sets have been updated. The total number of cases in BraTS2021 is continuously increasing, rising from just over 600 to 2040, and the training set comprises 1251 cases, each including four different imaging modalities: T1, T1Gd, T2, and FLAIR. The segmentation labels provided are 1 for the necrotic tumor core (NCR), 2 for the peritumoral edematous/invaded tissue (ED), 4 for ET, and 0 for other. The three final parts to be segmented are the whole tumor (i.e., WT; label 1+2+4), the tumor core (i.e., TC; label 1+4), and the enhancing tumor (ET). For better application to the fully supervised model training process, each imaging dataset was manually segmented by one to four assessors using the same annotation procedure and validated by experienced neuroradiologists. the Hausdorff distance (95%) and dice score were used to determine the accuracy of segmentation.

2.2.2 MSD

The Medical Segmentation Decathlon (MSD, http://medicaldecathlon.com/) is a compilation of ten medical image segmentation datasets [8] designed to serve as a benchmark for evaluating segmentation algorithms across various domains, including the brain, heart, liver, and prostate. The Brain dataset comprises the same cases as the BraTS 2016 and BraTS 2017 challenges, and the utilized sequences are T1, T1-Gd, T2, and FLAIR. The Brain Tumor dataset consists of 750 images, with 484 case images in the training set and 266 cases in the test set, stored in NIfTI (.nii.gz) format. The corresponding ROIs for segmentation targets are three subregions of tumors: edema, enhancing, and non-enhancing. To enhance model development and training, more accurate data labels are provided. Table 2.1 provides a summary of the datasets (BraTS series and MSD) typically used in the development of brain tumor segmentation algorithms.

2.2.3 TCIA

The Cancer Imaging Archive is a publicly available medical image dataset that is continuously updated and funded by the Cancer Imaging Program, a division of the United States National Cancer Institute. It is managed by the Frederick National Laboratory for Cancer Research. Currently, it comprises a total of 17 subsets of brain tumor data. It is worth noting that the number of LGGs and HGGs in some datasets might change as the dataset is in a state of constant updating. Each dataset

Table 2.1 Commonly datasets used for brain tumor segment

Dataset	Subset	Format	Number	Modality[1]	Label[2]	Available
BraTS2013	Training	NIFTI	20 HGG +10 LGG	T1, T1c, T2 and FLAIR	0, 1, 2, 3, 4	Public
	Leaderboard		21 HGG +4 LGG			
	Challenge		10HGG			
BraTS2015	Training	NIFTI	220 HGG +54 LGG	T1, T1c, T2 and FLAIR	0, 1, 2, 3, 4	Public
	Leaderboard		110 cases			
	Challenge		–			
BraTS2017	Training	NIFTI	210 HGG +75 LGG	T1, T1c, T2 and FLAIR	0, 1*[3], 2, 4	Public
	Validation		46 cases			
	Testing		146 cases			
BraTS2018	Training	NIFTI	210 HGG +75 LGG	T1, T1-Gd, T2 and FLAIR	0, 1*, 2, 4	Public
	Validation		66 cases			
	Testing		191 cases			
BraTS2021	Training	NIFTI	1251 cases	T1, T1-Gd, T2 and FLAIR	NCR(1*), ED(2), 4 and 0	Public
	Validation		219 cases			
	Testing		570 cases			
MSD	Training	NIFTI	484 cases	T1, T1-Gd, T2 and FLAIR	Edema, enhancing and non-enhancing tumor	Public
	Testing		266 cases			

[1]T1 refers to a magnetic resonance imaging (MRI) sequence that highlights certain types of tissue, while T1c indicates T1-weighted images. T1-Gd denotes T1-weighted images acquired after the administration of Gadolinium, a type of contrast agent. T2 is another MRI sequence that highlights different types of tissue, and FLAIR stands for Fluid Attenuated Inversion Recovery, which is another type of MRI sequence.

[2]Label 0 is used for normal tissue, 1 for necrosis, 2 for peritumoral edema, 3 for non-enhancing tumor, and 4 for enhancing tumor.

[3]Label 1* is used to represent the necrotic and non-enhancing tumor core, which distinguishes it from previous labels used in 2013 and 2015.

is provided in various formats, which are summarized in Table 2.2. Other datasets such as The Ischemic Stroke Lesion Segmentation, The Open Access Series of Imaging Studies, and others have also been used. However, as their objectives are less relevant to brain tumor segmentation and they are not used as frequently, they are not described in this context.

2.3 Brain tumor regional segmentation methods

Automated and precise segmentation of brain tumors remains a challenging task due to a multitude of issues such as uncertain tumor locations, significant spatial and structural heterogeneity among brain tumors, and ambiguous tumor borders.

Table 2.2 Commonly datasets used for brain tumor segment

Collection	Cancer type	Subjects	Data types	Supporting data[1]	Access	Status	Updated
CPTAC-GBM	Glioblastoma multiforme	189	CT, MR, pathology	Clinical, genomics, proteomics	Limited	Ongoing	2023-02-24
Meningioma-SEG-CLASS	Meningioma	96	MR, RTSTRUCT	Clinical	Limited	Complete	2023-02-13
UCSF-PDGM	Diffuse glioma	495	MR	Clinical, genomics, image analyses, software/source code	Public	Complete	2023-01-11
UPENN-GBM	Glioblastoma	630	MR	Clinical, image analyses	Public	Complete	2022-10-24
GBM-DSC-MRI-DRO	Phantom	3	MR	–	Public	Complete	2022-07-21
GLIS-RT	Glioblastoma and low-grade glioma	230	CT, MR, REG, RTSTRUCT	–	Limited	Complete	2021-12-20
ACRIN-DSC-MR-Brain (ACRIN 6677/RTOG 0625)	Glioblastoma multiforme	123	MR, CT	Clinical	Limited	Complete	2020-09-09
TCGA-LGG	Low-grade glioma	199	MR, CT, Pathology	Clinical, genomics, image analyses	Limited	Complete	2020-05-29
TCGA-GBM	Glioblastoma multiforme	262	MR, CT, DX, Pathology	Clinical, genomics, image analyses	Limited	Complete	2020-05-29
QIN GBM Treatment Response	Glioblastoma multiforme	54	MR	–	Limited	Complete	2020-04-03
ACRIN-FMISO-Brain (ACRIN 6684)	Glioblastoma multiforme	45	CT, MR, PT	Clinical	Limited	Complete	2019-09-16
QIN-BRAIN-DSC-MRI	Low and high grade glioma	49	MR	–	Limited	Complete	2019-08-28
Brain-Tumor-Progression	Brain cancer	20	MR	Image analyses	Limited	Complete	2019-02-05
LGG-1p19qDeletion	Low-grade glioma	159	MR	Image analyses, genomics	Limited	Complete	2017-09-30
IvyGAP	Glioblastoma	39	MR, Pathology	Clinical, genomics	Limited	Complete	2016-12-30
RIDER Neuro MRI	Brain cancer	19	MR	–	Limited	Complete	2015-09-07
REMBRANDT	Low and high grade glioma	130	MR	Clinical, genomics, image analyses	Limited	Complete	2011-11-03

[1]Source: Above data from www.cancerimagingarchive.net.
If no relevant information is available, "–" is used.

Convolutional neural networks (CNNs) have made significant advancements in the interpretation of medical pictures recently, showing improved segmentation accuracy for 2D natural images and 3D modeling of medical images, and outperforming conventional image segmentation methods by a considerable margin. Deep learning is an exemplary data-driven model, and this section will rely on the number of annotations and the annotation approach, categorizing the segmentation techniques into fully supervised brain tumor segmentation and non-fully supervised brain tumor segmentation.

2.3.1 Fully supervised brain tumor segmentation

Currently, fully supervised deep learning is the most commonly used model architecture for brain tumor segmentation, boasting the highest segmentation quality and broadest impact scale. This approach maximizes the utilization of annotated samples and extracts local features and details, thereby improving training efficacy and increasing segmentation accuracy to a certain extent. The supervised deep learning method transforms the problem of image segmentation into that of tumor pixel classification. This allows the model to output pixels corresponding to a specific target class and ultimately produce a segmentation map that corresponds to the input. Unlike traditional algorithms, this CNN model can directly use images as network inputs, and pooling layer is used to increase the effective range of the perceptual field and to perform feature fusion, effectively improving the feature extraction capability of the model. However, continuous downsampling and pooling layers in this network result in the loss of critical information in many local regions, ignoring the correlation between the local and the global, while also limiting network interpretability due to poor feature encapsulation. To address these issues, researchers have proposed several new approaches based on CNNs, categorized according to the improved features of their methods, including those based on expanding receptive fields, feature fusion, encoder–decoder, and network fusion.

2.3.1.1 Methods based on expanding receptive fields

Currently, the most widely used and impactful model architecture for brain tumor segmentation is fully supervised deep learning. This framework utilizes annotated samples to extract local features and details, resulting in improved training and increased accuracy. By transforming the image segmentation problem into a tumor pixel classification problem, the model can output the pixels of a target class and generate a corresponding segmentation map. There is no requirement for difficult feature extraction or data reconstruction as observed in conventional methods when using a CNN model that enables the direct usage of images as network inputs. However, the use of continuous downsampling and pooling layers can lead to the loss of effective information in local regions and poor network interpretability. To address these issues, researchers have proposed various new approaches based on CNNs, such as those based on expanding receptive fields, feature fusion, encoder–decoder, and network fusion.

The region of the feature map for each layer of the CNN that corresponds to the pixel points on the original picture is referred to as the "receptive field" in the context of deep learning. This allows the network to capture a neuron's scope of perception for different locations in the image, such as for the purpose of identifying brain tumors. A larger receptive field size permits access to a greater image area, providing access to higher-level features like tumor size, shape, and location. Conversely, a smaller receptive field size tends to capture more local information such as tumor texture and boundaries. In the field of brain tumor segmentation, improving the representational capabilities of the model can be achieved by expanding the visual perceptual field to capture richer feature information, leading to better segmentation accuracy. To this end, Pereira *et al.* [9] developed a CNN-based automatic segmentation method using small 3×3 convolutional kernels, which not only increased the effective range of the receptive field but also allowed for deeper network layers. By incorporating intensity normalization and data enhancement techniques, this model proved highly effective for segmenting brain tumors in MRI images, achieving first place in the BraTS 2013 challenge.

The conventional technique for automatic brain tumor segmentation using deep convolutional neural networks (DNN) involves a pooling layer that gradually expands the effective range of the receptive field and merges background information. However, in brain tumor images, there are usually more healthy regions than non-healthy ones, leading to weak recognition of details and increased learning of extraneous background information, resulting in poor segmentation outcomes during training. To mitigate this issue, Shaikh *et al.* [10] utilized DenseNet [11], which repeatedly employs dense blocks to effectively utilize features, address the gradient disappearance problem, enable deep-layer learning, improve back-propagation of gradients, and facilitate easier training of the network while decreasing the number of parameters. However, this method may result in disadvantages such as the loss of spatial information, inadequate handling of multi-scale targets, and insufficient feature representation of low-level features, as well as weak segmentation of smaller tumors for higher-level features. To address these issues, dilated convolution (also known as atrous convolution) is a convolutional method that tackles the problem of image semantic segmentation, where down-sampling reduces the image resolution and consequently results in information loss. Havaei *et al.* [12] developed a dual-path 2D CNN brain tumor segmentation network that comprises both local and global information, with each path employing different-sized convolutional kernels to extract feature information from distinct contexts. One path employs two Maxout convolutional layers to concentrate on small local pixel regions, while the other path utilizes fully connected Maxout layers to focus on larger perceptual fields. The last layer of network makes use of the convolution of fully connected layers to increase operating speed. The implementation of dilated convolution can resolve the inherent conflict between feature resolution and perceptual field. The use of multiple dilated convolutions can extract richer detail information and refine segmentation outcomes. However, Moreno Lopez and Ventura [13] introduced residuals to the network, which resulted in the dilated residual network. This model ensured that both range and visual perceptual fields were covered while maintaining image resolution during downsampling, but it was unable to capture the 3D features of MRI images. HighRes3DNet [14], a 3D CNN structured model, was proposed to address this issue, which maintained high-resolution multi-scale features through detailed convolution and residual concatenation. It acquired more comprehensive brain tumor features while enhancing feature representation and refining segmentation outcomes. It is important to note that increasing the perceptual field aimlessly can lead to information redundancy and ineffective detailed convolution operations, which can degrade network performance. Therefore, Sun *et al.* [15] used a multi-path architecture for feature extraction while employing 3D detailed convolution in each path. With the use of this method, features from multimodal MRI images with various receptive fields may be successfully extracted, improving the network's ability to represent information.

From the above analysis, it is evident that the approach of expanding the perceptual field provides several benefits:

1. By using small convolutional kernels of size 3×3 or smaller (1×1), the network depth can be increased while reducing the number of model parameters and computational complexity. Furthermore, using more non-linear activation functions allows for a stronger semantic representation of the network.
2. The use of residual connectivity [16] and DenseNet can resolve network degradation problems and gradient disappearance explosion problems, significantly enhancing the depth and performance of networks that can be efficiently trained.
3. The use of detailed convolution enables faster sampling rates, extending the effective area of the receptive field while preserving the spatial dimensions of the image. This results in faster extraction of tumor features and more accurate segmentation results for tasks that require maintaining the spatial dimension of the target as the network deepens.

However, this method also has the following problems that need to be solved:

1. In practice, the use of smaller convolutional kernels can make the receptive field too small to represent its features, while the relatively fixed shape of the convolutional kernels for detailed results in a network that is relatively poor at adapting to changes in image size and extracting features from irregular brain tumor regions.
2. Too large a sampling rate tends to invalidate the detailed convolution operation, resulting in local information loss. Repeated application of detailed convolution can cause a checkerboard effect and also cause some brain tumor features to be lost or feature information acquired at a distance to be irrelevant, occupying running space and consuming large amounts of memory.

2.3.1.2 Methods based on feature fusion

In neural networks, the lower layer features contain more detailed and localized information regarding the tumor location, but are relatively less semantic and noisier as they undergo less convolution. As the network structure becomes deeper, DNNs tend to lose shallow details and are unable to leverage multi-scale feature information efficiently. To overcome this issue and prevent the neglection of feature information correlation between different blocks within a larger perceptual field and the checkerboard effect caused by recycling detailed convolution, the feature representation capabilities of the network have been improved by the use of feature fusion techniques. By combining features from various regions and levels to obtain implicit contextual information in the image, these techniques can effectively improve the segmentation rate and performance of the network, as well as significantly reduce the operational consumption, thereby circumventing the issues associated with methods based on expanding the receptive field.

In the realm of multimodal data analysis, data-level fusion, also known as early fusion, has long been the traditional method of combining multiple data sources. Unfortunately, this method frequently results in the input of redundant vectors and has limitations in terms of illustrating the complementary nature between various modalities. As such, late fusion has become the more popular approach, which involves analyzing data sources independently, training each modality separately in the early stage, and fusing the model outputs in the later decision stage. In recent years, researchers have utilized feature pyramid networks (FPNs) and atrous convolutions to further enhance the performance of late fusion. For example, Zhou *et al.* [17] implemented a 3D atrous-convolution feature pyramid network (AFPNet) to fuse features before predictive segmentation, which significantly improved the model's ability to discriminate between tumors of different sizes. Additionally, a 3D fully connected conditional random field (CRF) was constructed to improve appearance and spatial consistency of the structural segmentation. Alternatively, some approaches do not fuse features directly, but rather predict the features at multiple scales separately before integrating the predictions to obtain the final segmentation results. Rao *et al.* [18] built separate CNN models for each of the four modalities (T1, T1c, T2, and Flair), extracted features from the four networks, concatenated the features, and then used Random Forest to obtain segmentation results. Similarly, Li and Shen [19] extracted features from four different modalities and concatenated them as input to the subsequent discriminant model. Noori *et al.* [20] designed a low-parameter network based on 2D UNet, augmented with attention modules and adaptive weighting for each channel to prevent model clutter. Finally, multi-view fusion was employed to extract more detailed features from the 3D background information of the input image, thus further improving the accuracy of brain tumor segmentation. However, the independent layer-by-layer processing of 3D medical images through 2D fusion ignores the correlation between adjoining layers, leading to the loss of spatial contextual information of the volume data. To remedy this issue, in their paper [21], the researchers extracted features from the four modalities independently through four channel-independent encoder paths. These features were then fused using feature fusion blocks, and a decoder path was utilized to finally segment the tumor, all while taking into account the spatial correspondence of the labels. Additionally, He *et al.* [22] combined attention mechanisms with multi-scale feature extraction, introducing attention features by concatenating four different scales of feature maps, fully utilizing the complementary feature information at several sizes, optimizing features at each layer. This method can more comprehensively and effectively capture tumor boundary information and high-level semantic information.

However, these methodologies merely overlay MRI images or their low-level features onto the model input and lack refinement of multi-modal features. Different modalities of MR images capture varying pathological features and focus on different segmentation targets. In order to make full use of the rich multimodal information of MR images, Liu *et al.* [23] not only use pixel-level fusion but also feature-level fusion, which can effectively improve the segmentation accuracy of each tumor sub-region. For pixel-level fusion, a CNN named PIF-Net is proposed for 3D MRI image fusion, and the fused modality can strengthen the association between different types of pathological information captured by multiple source modalities. For feature-level fusion, a module is designed that adaptively adjusts the weights of each modal feature based on an attention mechanism that was devised to refine multi-modal features and address variations in segmentation targets across multiple modalities, allowing for a more efficient and refined use of multi-modal information.

Upon analyzing the aforementioned information, it becomes apparent that the feature fusion-based approach boasts the following merits:

1. By combining low-level features with specific information and high-level features with semantic information, the feature fusion technique is suitable for tasks requiring multi-scale target information. It gradually refines the segmentation results and addresses problems like high memory consumption and extensive computation.
2. The attention mechanism assigns various weights to various model components in order to retrieve significant and critical information from them. The model can easily obtain global and local connections, and computation can be parallelized. Comparing the model to CNN reveals that it is simpler and has less parameters.
3. The correlation between the lesion information obtained from various source modalities is improved by performing pixel-level and feature-level fusion on the pictures, enabling the fusion of multi-modal data for a more precise and effective utilization of multi-modal information.

However, the approach discussed also presents certain challenges that require addressing:

1. The use of simplistic fusion techniques to merge feature maps from different layers and regions may lead to the loss of low-level feature information, thus effective feature fusion strategies should be explored in the future.
2. While the attention mechanism can effectively extract critical information from the model, it requires a large amount of data and may not perform as well as CNN in capturing information in small datasets. Additionally, it cannot capture location information, which can be addressed by incorporating location

vectors. Due to the depth of the network and the abundance of parameters, distant feature capture may also be less effective, making it essential to avoid overfitting while designing the module.

3. FPN models can enhance semantic information of different scale features; however, their network architecture is manually designed and the fusion effect may not be optimal. Thus, a solution could be to search for an optimal FPN model based on network architecture.

2.3.1.3 Methods based on encoder–decoder

To enable pixel-level prediction with input images of various sizes, the encoder–decoder method uses a convolutional layer rather than the fully connected layer of CNN. Once the downsampled feature map is reduced to a certain level, upscaling methods such as transpose convolution are employed to effectively restore the spatial and temporal information of the original image, thus matching the ground truth. Currently, the prevalent frameworks are the encoder–decoder models, which include fully convolutional networks (FCN) [24] with contracting and expanding paths, 2D-Unet networks inspired by FCN designs [25], SegNet [26], and other such models.

The encoder–decoder architecture is an end-to-end semantic segmentation model that utilizes a skip-connection structure to fuse high- and low-level feature information, achieving accurate segmentation by combining shallow texture information with deep semantic information. To this end, Zhao *et al.* [27] employed 2D fully convolutional neural network (FCNN) models for feature extraction on each MRI cross-section individually and employed a voting-based fusion strategy to combine three 2D FCNNs and CRF to carry out brain tumor segmentation, successfully eliminating false positives and enhancing the accuracy of tumor boundary segmentation. Nevertheless, 2D FCNNs lack the capacity to fully exploit the 3D information of MRI data, including its background information.

Since most medical images are 3D, combining interlayer information in 3D networks can be more effective. However, 3D networks have greater performance requirements compared to 2D networks, leading to a decrease in segmentation accuracy. As a result, 2.5D networks have emerged, which can combine interlayer information. Hu *et al.* [28] implemented a multi-cascaded convolutional neural network and fully connected CRFs in a hierarchical manner for tumor segmentation. They also fused information from three different viewpoint segmentation models to enhance segmentation performance. Nevertheless, simply stitching 2D segmentation results into 3D segmentation results may lead to problems such as jaggedness and discontinuity. To address this issue, some researchers have started using 3D convolutions for brain tumor segmentation. Çiçek *et al.* [29] replaced all 2D operations with 3D operations, expanding the 3D convolution kernel into a 3D-Unet network. Isensee *et al.* [30] maximized brain tumor segmentation performance on top of 3D U-Net by using deep supervision in the upsampling phase and a pre-activation residual block in the downsampling phase. These approaches have shown good segmentation results, with real-time elastic deformation used during training for efficient data enhancement. Myronenko [31] used an innovative approach for image segmentation by incorporating a variational auto encoder into an asymmetric 3D U-Net encoder structure, which helped extract deep image features despite the limited data scale. This method achieved the first place in the BraTS 2018 challenge. In Isensee *et al.*'s [32] paper, the pixel point multi-classification problem was transformed into a three-channel binary classification using the nnU-Net model, and through post-processing, data enhancement, and region training, the segmentation performance was significantly improved, leading to winning the championship in the BraTS 2020 segmentation track. However, the limited receptive field of the convolutional kernel in the U-Net model hinders the learning of global context and long-range spatial dependencies, which are crucial for dense prediction tasks like segmentation. In order to overcome this challenge, taking cues from the triumphs of Transformer in long-distance sequence learning in natural language processing, several researchers have reframed the issue of 3D medical image segmentation as an inter-sequence prediction task. The vision transformer (ViT) [33] has been employed more effectively in image classification by partitioning the data into chunks and modeling the correlation among these chunks.

While the transformer has its own global self-attention mechanism, it is not as effective as CNN in acquiring local information. The location encoding in the transformer is artificially designed in the semantic space, rendering it non-transformable and inadequate in characterizing location information. The lack of shallow detail results in limited localization ability. In response, Chen *et al.* [34] proposed an approach by integrating CNN and transformer models. Specifically, the encoder part employed the transformer structure, while the decoder part used the upsampling structure in the UNet architecture. By combining the high-resolution spatial information of CNN with the global contextual information captured by the transformer model, the proposed approach compensates for the limitations of each individual model. While CNN compensates for the loss of feature resolution in the transformer model, the latter overcomes the limited field of sensation of CNN. This 2D network has the advantages of both transformers and U-Net, with a better way to perform self-attention and deeper fusion of shallow features for better segmentation accuracy. However, processing high-precision 3D images poses problems of high computational power and spatial complexity. To address this issue, Xie *et al.* [35] proposed combining CNN and transformer (CoTr) and introduced an efficient deformable transformer (DeTrans) as a long-range dependent model for the extracted feature maps. By focusing only on key locations, computational and spatial complexity is reduced, enabling the handling of multi-scale and high-resolution feature maps. Likewise, TransBTS [36] extracts local spatial features via 3D-CNN, which are then fed into a transformer to construct a global feature map. Subsequently, the encoder upsamples the embedded features acquired by the transformer to generate the ultimate segmentation map. This results in improved utilization of the continuous information between slices, as well as a reduction in the number of parameters and computational effort of the model, thus increasing operation speed. The UNET TRansformers, a combination of the UNet model and the Transformer model (UNETR) [37] leverages the transformer architecture as an encoder, similar to

TransUNet. The upsampling component efficiently captures multi-scale information globally by merging with a CNN-based decoder through skip connections at various resolutions. Presently, MedNeXt [38] structure yields optimal segmentation results in ViT series models. ConvNeXt [39] has integrated several pivotal design concepts from transformer, which focuses on limiting computational costs while simultaneously enhancing the perceptual field of the network to learn global features. The authors have presented a new network architecture, MedNeXt, that follows the design pattern of ConvNeXt and is similar to 3D-UNet. In this architecture, the researchers have proposed a ConvNeXt upsampling module with residuals to achieve semantic enrichment across scales, as well as a novel approach to up-sample small convolutional networks in a way that iteratively increases the convolutional kernel size through upsampling. This method helps to avoid performance saturation when dealing with limited medical data. The technique has achieved state-of-the-art performance in both CT and MRI modalities across four tasks with varying dataset sizes. Nonetheless, most of the MRI segmentation methods mentioned earlier combine MRI images at the early or mid-network stage, lacking complementary information within and between channels, and disregarding non-linear dependencies between modalities. However, Xing and colleagues [40] put forth a novel nested modality-aware transformer (NestedFormer) by proposing a new nested perceptual transformer. The two transformer modules are not serially or in parallel connected as in the past, but rather fused using a nested form. The proposed multimodal fusion method utilizes a tri-orientated spatial-attention transformer. Furthermore, a new gating strategy selectively transfers low-resolution features to the encoder for a more efficient jump connection, thus effectively extracting and merging different modal features hierarchically.

Most current studies enhance the nonlinear representation of the model by increasing network complexity and depth, but this also escalates the model's parameters and computation, thus slowing down its performance. In light of this, many lightweight approaches have emerged for brain tumor segmentation. To achieve high accuracy in volumetric medical image segmentation, 3D contextual information is the key. The most effective way to capture feature information is through 3D convolution, but its excessive usage can significantly increase computation and make model learning more complex, thus reducing the network's efficiency and effectiveness. In this regard, Chen [41] proposed the S3D-UNet model based on the separable 3D convolution and S3D models [42]. For the purpose of limiting the amount of parameters that can be learned, each 3D convolution is split into three parallel branches. Moreover, each 3D convolution can be changed to a 2D convolution for learning spatial features and a 1D convolution for learning temporal features, while a residual structure is added for the multi-segmentation task. To address the challenges of high parameters and computational intensity in 3D UNet, several researchers have optimized and adapted it. ESPNet [43] provides a faster and more efficient alternative to the U-Net model. The work of Nuechterlein and Mehta [44] extends the ESP blocks for brain tumor segmentation by replacing the spatial 2D convolution with volumetric 3D convolution. The model learned only 3.8 million parameters, thus reducing network parameters and redundant computations. Similarly, a dilated multi-fiber network was developed in the literature by Chen *et al.* [45]. The network utilizes a 3D multi-fiber unit to obtain different sizes of sensory fields while significantly reducing computational cost, with only 3.88 M parameters and approximately 27 G FLOPs. This model has been shown to ensure high accuracy in brain tumor segmentation while reducing computational cost. Brügger *et al.* [46] have designed a U-Net architecture that can employ partially reversible structures, inspired by the i-RevNet model. This system increases the depth of the network, improves the accuracy of segmentation, and reduces memory usage. Walsh *et al.* [47] used a lightweight UNet network to transform 3D MRI brain scans into 2D images through three view planes to simplify segmentation, reducing the number of model parameters and increasing the runtime while ensuring segmentation accuracy. Another paper proposes an upgraded version of the 3D UNet model [48]. These newly designed modules effectively gather information from multiple views and scales, minimize redundant feature information, and improve network performance. Although the performance of this method is slightly lower than other models on the same dataset, the number of parameters is only 0.35 M, resulting in significantly reduced FLOPs* and training time.

Based on the aforementioned analysis, it is apparent that the encoder–decoder-based approach offers the following benefits:

1. The encoder–decoder structure efficiently restores target spatial information, enhances feature image resolution, addresses the semantic prediction issue of generating pixel-level outputs for input images of varying sizes, and is applicable for graph-to-graph tasks that preserve image spatial information.
2. In contrast to 2D networks, 3D networks allow for the integration of data between picture layers and guarantee the consistency of modifications made to interlayer image masks.
3. The transformer model is equipped with a self-attentive mechanism, which effectively acquires global information. Multiple attention heads can be mapped to multiple spaces, enabling them to perform diverse tasks. This excellent modal fusion capability provides the model representation with significant power while simultaneously improving model segmentation speed.

However, the aforementioned method is also plagued by several pressing issues that require immediate attention:

1. Due to the intricate composition of brain tumor MRI images and the varied shapes of lesions, the segmentation efficacy of the network can be significantly compromised. To further enhance the performance of the U-Net model,

*"FLOPs" is "Floating Point Operations", a metric used to measure the complexity of an algorithm or model by calculating the number of multiplication/addition operations performed throughout the network model. A smaller FLOPs value implies a lower complexity of the model or algorithm, indicating fewer floating-point operations are needed. This typically means less computational resources are required for training and prediction, leading to faster execution. Therefore, when performance is comparable, we generally prefer models or algorithms with smaller FLOPs values.

processing modules such as attention module, residual connectivity, dense connectivity, and post-processing can be incorporated to acquire more effective, detailed information.

2. The 3D network must be divided into a number of 3D patches as input due to memory restrictions, which reduces the maximum perceptual field that the network can access. This causes a loss of global information as well as problems with multi-scale and imbalanced positive and negative samples. It may be difficult for the network to learn the target's general structural information if the segmentable target is significantly larger than the patch itself.
3. 3D networks are typically focused on local features and details. A 2D network can be employed for coarse segmentation initially, followed by a 3D network for fine optimization of details as a second-level network.
4. Transformer and 3D CNN architectures exhibit high network complexity, an excessive number of parameters, significant computational effort, long training times, and a predisposition towards overfitting, which could render these models ineffective in practical large-scale applications.

2.3.1.4 Methods based on network fusion

Despite the good performance of some of the above single-network-based model improvement methods in the field of segmentation, they extract a single type of features that cannot cover all the detailed information, resulting in blurred segmentation boundaries. Multi-network brain tumor segmentation methods are a class of deep learning segmentation methods with relatively high segmentation accuracy, which can be classified into dual-path CNN, cascade-path CNN, and ensembled networks. The method of combining different network architectures for segmenting brain tumors makes full use of the unique properties of each network, improves feature representation, stages the extraction of case features, assembles global and local feature information, and increases model segmentation accuracy.

The dual-path CNN simultaneously captures local and global features of MRI images by parallel processing, with one pathway extracting visual details of the region around the central pixel, while the other extracts global features, including brain tumor location information [12,49]. Kamnitsas *et al.* [50] proposed DeepMedic, a 3D brain tumor segmentation model that utilizes a dual-pathway architecture based on 3D local blocks and multi-scale dense CNNs, with inputs of image blocks of different sizes and a recombination of high and low resolution paths to obtain segmentation results. However, DeepMedic has less efficiency on local image blocks and limited flexibility in the fixed topology of the CRF used for false positive removal. In response, Razzak *et al.* [51] proposed a group CNN architecture with a cascade model combining two parallel CNNs, replacing the usual convolution layers with group equivariance convolution layers. The model uses a local CNN to obtain pixel details for accurate prediction of each pixel label. Simultaneously, the dual-pathway CNN is employed to acquire contextual information of pixels and preserve high-resolution details. By utilizing two paths to capture both local and global features, parameter sharing instability and overfitting are effectively reduced without increasing the number of parameters. While this approach has proven to enhance the precision of brain tumor segmentation, poor segmentation outcomes still occur due to the similarity in tumor location and brain tissue luminance levels. Ranjbarzadeh *et al.* [52] responded to this challenge by processing only key regions of the image, as opposed to the entire image. Their proposed model is designed to extract both local and global features through two distinct pathways. Moreover, they introduced a new distance-wise attention (DWA) mechanism that analyzes the spatial correlation among different modal images while considering the central location of the tumor in each modal slice, filters out irrelevant image features, and selects useful features to enhance the segmentation accuracy of brain tumors at the boundary while resolving the overfitting issue.

The fundamental concept of the cascade-path CNN is to utilize the outputs of the prior network as input to the subsequent one, thereby expanding the range of image features available. In order to address the challenge of high false positive rates encountered in image segmentation tasks, Wang *et al.* [53] transformed the multi-category image segmentation problem into three separate binary-segmentation problems, and fused the three networks together, such that the segmentation output of the preceding network serves as the Region of Interest (RoI) for the subsequent network, leading to segmentation of the WT, TC, and ET. This approach ensures that detailed information is preserved while effectively integrating residual connectivity and multi-scale feature information, thereby reducing false positive predictions. The networks utilize multi-view fusion alongside anisotropic and dilated convolution filters. Although the model may be effective, it heavily relies on the performance of the upstream network, leading to poor results in segmenting complete tumors. However, researchers such as Sobhaninia *et al.* [54] have introduced the cascaded dual-scale network, which allows the first network to specialize in learning features in specific brain tumor regions, while the second network focuses on boundary regions and detailed features. Additionally, Jiang *et al.* [55] proposed a novel two-stage cascaded U-Net model, in which a refined fine segmentation is performed in the second stage by leveraging the coarse segmentation results generated in the first stage. The proposed approach utilizes two distinct decoders, one using deconvolution and the other employing trilinear interpolation, to achieve superior segmentation performance. The model was successfully employed to secure victory in the BraTS2019 segmentation competition.

The concept of cascade networks involves linking multiple models in series to enable continuous refinement of lesion segmentation. However, the performance of these networks is often limited by the performance of the upstream network and they are computationally intensive, requiring substantial computational resources. The outcomes of the model are sensitive to initialization parameters and difficult to duplicate due to the high variance of neural networks. To address this issue, an effective solution is to use a network ensemble, which aggregates the segmented outputs of multiple networks to reduce the high variance of a neural network. Rivera *et al.* [56] proposed a model that utilizes three parallel contraction paths, each receiving inputs at different resolutions, and views each path as a subnetwork. The results are eventually integrated using three fully connected layers, effectively reducing the variance of the network ensemble. Zhou *et al.* [57]

developed a single-channel multitasking network that uses cross-task guided attention to segment brain tumors by combining various segmentation subtasks into a single depth model. This technique of weight sharing after feature extraction via sub-paths is known as an implicit ensemble. However, weight sharing can lead to increased sensitivity to training data details, training scheme choices, and the possibility of a single training run. To mitigate this problem, researchers often choose to allow different single models to use different training data. A common approach is to use k-fold cross-validation, which produces k sub-training sets from the overall training set, and uses each subset to train the models separately. Finally, the k models are combined into an ensemble. Silva *et al.* [58] split the training data into five folds to train the ensemble. The neural network was trained using the remaining four folds after each fold was divided into two pieces, half of which were allocated for validation and testing. An architecture made up of three deep aggregation neural network cascades was utilized to build the response; each step took the input of the MRI channel and feature maps and probabilities from the stage before it. This method, however, leads to a highly complicated system and ignores model interaction. To address these shortcomings of the cascade approach, the network structure has been modified by certain researchers in an effort to lessen computational load caused by memory usage. The paper [59] uses the one-pass multi-task network (OM-Net) model [60], incorporating subtasks into an end-to-end holistic network to exploit the potential correlation between each subtask, requiring only a single-pass computation for coarse-to-fine partitioning, and saving a significant number of parameters.

Each model possesses its own unique characteristics, which results in the prediction of different errors. Combining these models into an ensemble can help mitigate the high variance to some extent. In order to introduce high variance between models, Kamnitsas and Sun utilized three different 3D CNN architectures, which were configured and trained in different ways. Kamnitsas *et al.* [61] proposed the ensembled architecture, which includes the DeepMedic [50], 3D FCN, and 3D U-net networks. This model pair was first introduced in BraTS2017 and is insensitive to the unique errors of each component. Sun *et al.* [62] used cascaded anisotropic convolutional neural network (CA-CNN) [53], DFKZ Net [30], and 3D U-net. Evaluation results demonstrated that ensemble models outperformed individual models and that using the ensemble approach was effective in reducing model bias and improving performance. Similarly, Murugesan and colleagues [63] employed an ensemble of 2D multi-resolution networks (DensNET-169, SERESNEXT-101, and SENet-154) to produce robust segmentation maps, mitigate overfitting, and enhance model generalization. Kao and team [64] integrated multiple DeepMedics and block-based 3D U-Nets derived from the literature [29], with distinct parameters and training strategies. This enabled them to learn a more diverse set of mapping functions, and a lower correlation in terms of prediction and prediction error, resulting in more resilient tumor segmentation. Feng *et al.* [65] improved segmentation accuracy by integrating six 3D U-Net models trained with different hyperparameters. Meanwhile, Sundaresan *et al.* [66] proposed a three-plane integration architecture that consisted of 2D U-Nets in each MRI plane. Each plane's U-Net is trained individually and then integrated into a 3D volume, resulting in more accurate segmentation.

Based on the previous analysis, it is apparent that the network fusion method offers the following benefits:

1. Combining the advantages of different networks, building multi-category feature fusion, reducing the problem of bias arising from information loss due to a single model, and improving model robustness.
2. Improved model recognition rates and better scale invariance of the fusion model than a single model, and in less time than using a complex model.

However, the method also has the following problems that need to be addressed:

1. The ensemble network fuses MRI image information from the feature representation level, so choosing a reasonable feature extraction layer to reduce feature redundancy is a research direction that needs to be explored in the future.
2. There are problems such as difficult model design, large computation, long training time, and large memory consumption. Future research directions can be improved by combining multiple GPUs and exploring reasonable learning strategies.

2.3.2 Non-fully supervised brain tumor segmentation

Although deep learning-based methods that are fully supervised have proven effective for lesion region detection, they demand a great deal of semantically annotated sample data for pixel analysis. Unfortunately, a common challenge in the medical data arena is small data availability. Additionally, annotations for brain tumor image data necessitate input from multiple, experienced experts in the field and are time-consuming and costly, which results in very limited high-quality annotated samples for training purposes. Therefore, improving the segmentation accuracy of the model with limited labeled data has become a current research priority. To address this issue, non-fully supervised detection models, such as semi-supervised and unsupervised learning, have been proven effective. This section discusses these methods and classifies non-fully supervised detection models into **segmentation methods based on semi-supervised learning** and **segmentation methods based on unsupervised learning**.

2.3.2.1 Methods based on semi-supervised learning

Segmenting pixels or voxels of brain tumors can be a laborious and time-consuming task, as it requires a substantial amount of semantically annotated pixel samples. In view of this challenge, there is a growing interest in using unannotated data to enhance the model's segmentation performance. Semi-supervised learning (SSL) [67] combines supervised and unsupervised learning, using both labeled and unlabeled data for pattern recognition tasks. This learning method is particularly well suited for assessing medical images with a small number of high-quality annotated examples because it successfully uses unlabeled data to lessen the dependence on a large number of labeled samples during model training.

From this, a range of semi-supervised segmentation models have been proposed, incorporating innovative ideas such as self-training, co-training, auto-encoders, and knowledge distillation. For instance, Zhan *et al.* [68] utilized a multi-classifier collaborative training (co-training) approach, using SVM and SRC as base classifiers, along with superpixel maps to establish spatial and clinical constraints that leverage a priori knowledge that neighboring pixels in an image belong to similar classes, and clinical information. This further enhanced the accuracy of brain tumor segmentation. Similarly, Cui *et al.* [69] employed the mean teacher model [70] for brain tumor image segmentation, which enables the model to learn smoother outputs and improve the accuracy of the output labels via adding consistency loss weighting for regularization. More accurate labeling allows for shorter feedback loops between teacher-student models, thus increasing the accuracy of segmentation. Zhao and colleagues [71] employed a multi-space semi-supervised technique to assign labels to an unlabeled dataset by training various student models into a teacher model. The training process is iterated until the student model attains a certain degree of precision, utilizing a combination of human-labeled and machine-labeled samples as newly added training data. Hu *et al.* [72] employed a technique that combines generalized knowledge distillation [73] with encoder–decoder architectures. They transferred knowledge from a multimodal segmentation network (the teacher network) to a unimodal network (the student network), which were both constructed with encoder–decoder frameworks, to accomplish brain tumor segmentation. Inspired by classical joint segmentation and alignment techniques, Ito [74] *et al.* added pseudo-annotations to original images lacking labels through image alignment, thus constructing a dataset to train deep neural networks using images with a low number of annotations and original images with a relatively high number of pseudo-annotations. "Meissen *et al.* [75] performed abnormality segmentation in brain MRI through simple threshold segmentation of the input FLAIR images, exploiting the fact that lesions are high signal and tumors are usually bright in FLAIR images." The authors employed SAS models based on auto-encoder and GAN to learn the distribution of normal anatomical structures from healthy patient images, and to detect abnormalities from the residual images. Additionally, other scholars have approached the segmentation issue with a weakly supervised learning strategy [76], while Ji *et al.* [77] utilized a weakly supervised approach with Scribble tagging for segmentation. The whole tumor scribbling is used to train a WT segmentation network, which then attempts to approximate the training data's WT mask. After utilizing the global label as a guide to cluster the WT area, a second network is trained using the rough sub-structure segmentation that was obtained from the clustering as a weak label.

After analyzing the aforementioned research, it becomes apparent that the mean teacher and co-training-based network framework is among the most prevalent semi-supervised models in the domain of brain tumor segmentation. This approach mitigates the challenge of training the network with high-quality labeled data, by leveraging the benefits of learning pseudo-labeling and employing an iterative process of continuous updating.

However, the algorithm also has the following problems that need to be addressed:

1. Its performance depends on the quality of the generated pseudo-labels, the network will continuously amplify the learned mislabels and thus affect the final detection results. Due to the irregular shape and small inter-class variation of lesions in brain tumor images, it is not possible to generate stable and high-quality pseudo-labels by referring to the feature information provided by the network itself.
2. Future research directions can be improved by combining semi-supervised learning and active learning strategies [78]. For the semi-supervised learning method to obtain those sample data that are more difficult to classify, let the manual re-confirm and review. The manually annotated data is then trained again using supervised or semi-supervised learning models, incorporating the human experience into the machine learning model and gradually improving the model effect.

2.3.2.2 Methods based on unsupervised learning

Supervised methods necessitate the initial training of the machine and then testing, thus increasing computational expenses. Therefore, segmentation is more readily achievable with unsupervised learning methods [79]. Unsupervised learning enables data manipulation without any guidance, and it can be contended that the model will unearth concealed patterns and insights from the provided data. In other words, unsupervised learning aims to identify features of the image itself. This method of learning is akin to humans' ability to think independently based on their experiences, which makes it closer to true artificial intelligence.

Clustering-based segmentation is a powerful unsupervised technique used for region-based segmentation, which has been extensively studied in the field of medical image segmentation. Many researchers have proposed various unsupervised segmentation models based on clustering concepts [80].

The objective of clustering in MRI images is to partition image points into distinct categories based on the level of similarity between them within a category and the level of dissimilarity between image points belonging to different categories. This feature can assist researchers in rapidly classifying the regions of benign and malignant tumors for segmentation. K-means and fuzzy c-means (FCM) algorithms have been widely used in clustering due to their superior results and simpler methodologies compared to other clustering algorithms. Demirhan and colleagues [81] employed the stationary wavelet transform to extract features from MR images, without the need for an additional neural network. The extracted features were then clustered using a self-organizing map, with the supervised learning vector quantization algorithm used to determine the ideal network position and then change the output neurons of the network. Similarly, Khan *et al.* [82] utilized a k-means clustering approach to perform accurate feature extraction on brain tumors, with a focused ROI used for segmentation. The tumor classification (benign/malignant) was then obtained by fine-tuning VGG-19.

However, the k-means algorithm suffers from the incomplete division of tumor regions and is sensitive to outliers. In response, the researchers adopted FCM, an improved approach to k-means. Compared to the k-means algorithm in which samples belong to only one category, in the FCM algorithm samples belong to each category, except for different affiliation sizes. As each value contributes to each type of center point, iterations of the center point are more likely to be globally optimal and perform better on relatively noise-free images. Kaya and colleagues [83] added dimensionality reduction to the clustering approach for T1-weighted MRI image segmentation. By reducing errors caused by redundant information through dimensionality reduction, the essential structural features within the data can be identified, improving the accuracy of recognition. Prior to comparing the performance of k-means and FCM clustering algorithms, five common principal component analysis (PCA) algorithms were employed to reduce the image data complexity through dimensionality reduction, allowing for the identification of high-dimensional data through an appropriate low-dimensional representation. The EM-PCA and PPCA-assisted K-Means algorithms exhibited superior clustering performance in most cases. Both clustering algorithms provided better segmentation results for all sizes of T1-weighted MRI images. However, FCM performance is severely affected in medical images like brain MRI that are vulnerable to unknown noise. Numerous research endeavors have been undertaken to address FCM limitations. Kumar *et al.* [84] proposed a modified intuitionistic fuzzy c-means algorithm (MIFCM), involving adding hesitations, which comprise two new negation functions, to the FCM algorithm. The proposed method is designed to mitigate measurement inaccuracies and noise-induced uncertainties, while being insensitive to parameters. Simaiya and his colleagues [85] have proposed the HKMFSRT-Model, a hybrid model that combines hierarchical k-means clustering with fuzzy c-means and super rule tree. The model is capable of detecting early brain tumors in MR datasets, irrespective of the tumor's size, intensity variation, and location. The method boasts an impressive accuracy of 88.9%, representing a 3.5% increase over the accuracy of the k-means algorithm.

Self-supervision is a unique form of unsupervised learning that is driven by its supervised tasks instead of pre-defined prior knowledge. It can also be described as a method of replacing manual marker annotation by establishing pseudo-supervised tasks using certain properties of MRI image pictures. Self-supervised learning holds immense potential to replace fully supervised learning in representation learning. In the field of brain tumor segmentation, Fang *et al.* [86] utilized a self-supervised network to aid the learning of a supervised network in brain tumor segmentation. The self-supervised network was designed to extract regions related to brain tumors, thereby improving the overall learning capability and segmentation robustness of the network. The authors discovered that this approach was successful in mitigating overfitting, leading to a notable improvement in the accuracy of WT segmentation. However, due to their small size and complex edges, the segmentation performance of TC and ET regions was suboptimal.

From the above analysis, it is clear that the unsupervised learning-based segmentation method has the following advantages:

1. To some extent, it overcomes the difficulties of lack of data and data unavailability, and reduces the cost of annotation.

2. The use of clustering methods does not require a training set and the algorithms are simple and fast. It can be used either as a stand-alone segmentation algorithm or as a precursor process to other learning tasks such as classification to improve segmentation accuracy.

Self-supervised networks can be constructed because the data itself provides supervised information for the learning algorithm, allowing for some representation learning algorithms that do not focus on pixel details. It is easier to optimize by encoding higher-level features to differentiate between different objects, and also by not having to reconstruct at pixel level.

However, the algorithm also has the following problems that need to be addressed: unsupervised learning lowers computational costs but fully disregards prior knowledge that may result in weak performance because it does not require the labels of the training data for decision-making. The issue of greater sample sizes in supervised learning methods is solved by semi-supervised learning techniques, which often have a higher accuracy rate than unsupervised techniques. The unsupervised lesion detection framework alone is difficult to achieve detection results that meet expectations. Therefore it is an important research direction to try to extend the target region, to explore supervised information, to combine various approaches such as fully supervised learning, and to develop reasonable joint algorithms for collaborative learning.

2.3.3 Summary

In summary, the diversity and efficiency of deep learning methods provide technical support for the implementation of detection tasks, improving the detection accuracy of the models through methods such as detailed convolution, feature fusion, encoder–decoder structure, network fusion, semi-supervised learning, and unsupervised learning. The above segmentation methods are comprehensively analyzed and compared and summarized in terms of main ideas, advantages and disadvantages, and improvement measures, as shown in Table 2.3.

2.4 Small sample size problems

In current deep learning, a common issue is the inadequacy of available data, which presents economic and practical challenges in building large medical datasets. Furthermore, each model requires significant amounts of annotated data to prevent overfitting of the training set. The potential of data-driven medical imaging segmentation is hampered by the lack of well-labeled tumor masks. Several factors contribute to the scarcity of medical data, including the time-consuming and costly manual annotation of brain imaging diagnoses by experts, the variation in annotation by different experts due to clinical expertise and subjective judgement, and the limited utilization of spatial information in medical image data by 2D convolution, while 3D convolution demands expensive computational costs and memory requirements. We identified three primary difficulties in the task of brain tumor segmentation: class imbalance, lack of data, and missing modalities.

Table 2.3 Summary of deep learning-based methods for brain tumor segmentation

Method	Main methods	Advantage	Disadvantage	Improvement measures
Methods based on expanding receptive fields	Small convolution kernel Residual module Detailed convolution	Increase visual perception of the wild, maintain spatial dimensionality, extract dense features and refine segmentation structures	Convolution kernel too small, poor feature representation; detailed convolution kernel relatively fixed, loss of local information or acquisition of distant irrelevant information	Optimized convolutional structures; combined with deformable convolution
Methods based on feature fusion	Late fusion Attention module FPN	Fusing lesion features at different levels and in different regions to enrich semantic information	Loss of low-level features due to lack of effective fusion strategy; Attention mechanism is ineffective in small datasets, unable to capture lesion location information and inefficient in capturing more distant features; unable to perform intensive prediction	Development of effective fusion strategies and attention modules; search for optimal models based on network architecture
Methods based on encoder–decoder	Encoder–decoder structure	Restore feature map dimensions, support input information of any size, and recover target space information	Rough U-Net skip structure and fusion effect; limited receptive field, limiting the ability to learn global context and long-range spatial dependencies; 3D patch loses information; transformer has weak local information acquisition; many network parameters	Add other modules such as attention to explore new lesion features; combine 2D and 3D networks
Methods based on network fusion	Dual-path CNN Cascade-path CNN Ensemble network	Establish multiple types of feature fusion to solve the shortcomings of single feature extraction; improve model recognition rate and scale invariance	Difficult model design due to feature redundancy; high computational cost, high storage overhead, and time-consuming	Selecting a reasonable feature extraction layer; exploring optimal models and task allocation strategies; combining multiple GPUs
Methods based on semi-supervised learning	Mean teacher Co-training	Alleviates the difficulty of training networks due to the lack of high-quality labeled data	Less stable due to the quality of the generated pseudo-tags	Integrating multiple annotation methods and combining them with other methods to achieve complementary benefits
Methods based on unsupervised learning	Cluster Self-supervision	Can be used alone or as a precursor model; easy to optimize	Missing detail information, accuracy needs to be improved	Propose a network model that is more suitable for unsupervised training methods and optimize the structural model

2.4.1 Class imbalance

Class imbalance and inter-class interference are frequent and difficult issues in the segmentation of multi-label brain tumors. The BraTS dataset, for instance, contains a higher number of HGG images than LGG images, leading to a training model that is more biased towards HGG. Not simply the network's design but also the loss function used affects how well the segmentation model performs overall. To mitigate the category imbalance problem, some studies have employed a weighted loss function. This function assesses the degree to which the model predictions differ from the ground truth, and assigns weights to voxels or pixels belonging to different classes based on their distribution in the training data. This guarantees that LGG and HGG equally contribute to the model loss, and a better loss function generally leads to better model outcomes.

Various loss functions, including cross-entropy loss [49,52,87,88], modified cross-entropy loss such as weighted cross-entropy loss [89], binary cross-entropy loss [31,47,90], and categorical cross-entropy loss [9,15,30,91], can enhance the model's training performance. Among these, cross-entropy loss is the most commonly used loss function for image semantic segmentation tasks to evaluate the correctness of each pixel class. However, its effectiveness may decline after a few training sessions, and adjusting the weights of difficult samples artificially can increase the difficulty of fine-tuning. To address these issues, the dice loss function is a viable alternative. The dice loss function, [21,53,63] is useful for medical image segmentation tasks that suffer from extremely unbalanced proportions of foreground and background regions. Specifically, it measures the overlap between two samples, which can solve the problem of foreground regions with too small a proportion. Dice loss, although useful for image segmentation tasks, may pose challenges when optimizing with gradient descent algorithms due to its arithmetic problems. To address this, soft dice loss optimization techniques [36,55,92,93] can effectively alleviate these issues. Furthermore, both the generalized dice loss (GDL) function and multiclass dice loss [41,93] can effectively tackle the class imbalance problem. GDL [45] balances the focal regions and dice coefficients, while dice loss can be detrimental to smaller lesions because even one incorrect prediction can result in large changes in dice, leading to unstable training. Focal loss,[19,94] is a promising technique that enhances focus on positive samples, and is more effective than previous methods in addressing target imbalance. Moreover, researchers have proposed combining losses to improve pixel-level reconstruction accuracy [20,62,66,95–97]. Additionally, several studies have suggested the development of a tailored loss function to adjust the learning gradient. As a result, the model may better focus on smaller samples and lessen the influence of imbalanced datasets. Customized loss functions increase the weight of challenging samples while decreasing the weight of samples that are difficult to categorize in the loss function. In particular, one such study [98] proposes a loss function that incorporates tumor tissue and boundary information. Segmentation accuracy is enhanced by incorporating boundary information to the loss function.

The training dataset exhibits an uneven distribution of voxel classes, with a vast majority of healthy voxels and very few non-healthy ones. In the case of brain tumor segmentation, healthy voxels constitute 98% of the total voxels, leading to a potential bias towards the majority class and lower model accuracy. Hence, alternative research approaches have been proposed, which utilize cascade structures to convert the initial multi-label segmentation task into multiple binary segmentation subproblems. One such approach, proposed by Li *et al.* [94], leveraged 3D U-net as the foundation to build a multi-level cascade network model, proceeding from coarse to fine, which effectively addressed the evident imbalance between the tumor area and the background. Another cascade model, designed by Liu *et al.* [99], devised a cascade model that first segmented the whole tumor and then proceeded to segment its substructures, while implementing a loss weighted sampling scheme to tackle the issue of class imbalance during network training.

The sharing of comparable properties by various tumor tissues, on the other hand, results in inter-class interference, which hinders the ability to distinguish between classes and eventually affects prediction and segmentation. To address this issue, Chen and colleagues [100] utilized a method that disregards previously segmented tumor tissues before performing predictive segmentation of the next tumor label. Their proposed FSENet network model also includes a series of classifier branches that constrain the prediction to the tumor region of interest and eliminate the interference of non-tumors within the region. However, it lacks the exploration of the task-modality structure. To address the aforementioned issue, Zhang *et al.* [101] considered the inter-relatedness between multiple subtasks and formulated the prediction of each subtask as a structured prediction problem. In order to emulate the expertise of clinicians in discriminating brain tumors, the authors utilized a weighted combination of various modalities to identify different tumor regions. To this end, they proposed a task-structured brain tumor segmentation network that is capable of handling both task-modality relationships (i.e., segmenting out ET tumor regions) and task–task relationships (i.e., searching around ET and segmenting out TC and WT). Specifically, the authors employed a modality-aware feature embedding module to determine the relevance of each modality data for the segmented region.

2.4.2 Data lack

Improving the generalization performance of a model is a significant topic in contemporary research. To augment the training set, one approach is to generate synthetic data and include it in the training set. Data augmentation and transfer learning (TL) are effective techniques to address the challenge of data scarcity, and have become increasingly prevalent in various fields of deep learning.

2.4.2.1 Data augmentation

Therefore as result of data augmentation, it is possible to significantly increase the quantity and quality of training samples, lowering machine learning's generalization error and enabling the development of deeper neural networks. This technique involves artificially increasing the training dataset by generating equivalent data from a limited set of original data, which can be achieved through two approaches: raw data transformations and synthetic data generation.

Raw data transformations comprise affine transformations (e.g. rotation, scaling, clipping, flipping, and panning), elastic transformations, and pixel-level transformations [102], while some scholars have used affine transformations to transform the original data and overcome the issue of insufficient training data. Equally, there is some work that combines affine transformations with pixel-level spatial transformations. Most of the spatial transformations are all gamma transformations, elastic transformations, intensity shifts, and added noise are used to augment the data. Although this method is simple to use and frequently employed in studies, the performance improvement is not significant. Additionally, because the generated data differs from the real data, it frequently results in inaccurate images, which inevitably causes noise issues and impairs segmentation accuracy.

In addition, certain scholars have utilized the test time augmentation (TTA) methodology for refining images [30,35,36,103]. TTA creates multiple versions of an original image during the testing phase, by cropping and scaling various regions, which are then inputted into the model and averaged to produce the final output. Implementing TTA can remedy the issue of critical information being lost in specific parts of the original image, and enhance the overall performance of brain tumor segmentation, albeit at the cost of increased computational effort.

Conversely, a technique for generating artificial data samples involves the use of generative adversarial networks (GAN). GAN is a sophisticated deep network model that employs an adversarial learning approach and comprises a generator, a discriminator, and a real dataset. The generator creates brain tumor images, while the discriminator refines the generator's training. During training, the generator and discriminator operate concurrently, improving and producing optimal outcomes in the adversarial process. Two commonly used GAN-based methods for addressing the data scarcity problem include data-level and algorithm-level approaches.

The method based on data-level approach can be accomplished by resampling the data space [104]; however, this approach removes or introduces new samples, which can lead to a reduction in image quality. As a result, most existing techniques rely on the algorithm-level approach. Given that most brain tumor images belong to healthy regions while the proportion of tumor or non-healthy regions is small, Rezaei *et al.* [105] propose the voxel-GAN adversarial network. This module employed cGAN and new weighted adversarial loss function that are utilized to alleviate unbalanced training data with biased complementary labels in semantic segmentation tasks, and every form of brain imaging can be used with this framework, regardless of size.

In practice, researchers commonly employ data pre-processing methods such as rotating and flipping a limited dataset to perform data enhancement. However, these methods are insufficient in representing changes in shape, location, and pathology. In response, Mok and Chung [106] proposed a GAN architecture that incorporates a coarse-to-refine model and a boundary loss function to improve segmentation quality and enhance multi-modal MRI data generation. In addition to enhancing rotation and scaling invariance, this architecture also captures sophisticated data such as tumor structure and contextual information. Similarly, Sun *et al.* [107] used a parasitic GAN for brain tumor segmentation to make more efficient use of unlabeled data. The segmentor and generator produced a labeled graph, which was then used to generate supplementary labeled graphs, allowing the discriminator to learn more accurate truth bounds and improve the generalization ability of the segmentor. The discriminator is empowered to discern pseudo-labeled graphs from the genuine value distribution, thanks to the assistance of the generator. However, GAN is susceptible to network collapse and instability, where it maps diverse inputs to identical data or causes the same inputs to generate varied outputs, due to the gradient disappearance problem in the optimization process. Li *et al.* [108] proposed a novel approach to address the issue of limited data in brain tumor segmentation by introducing a mapping function between semantically labeled images and multimodality. They developed an image-to-image translation model, TumorGAN, which can generate a greater number of synthetic images from a smaller set of real data pairs. The framework merges brain tissue regions and tumor regions from different patients, while preserving the color and texture of the brain tissue through the use of attentional regions and region perception loss. This approach significantly improves the tumor segmentation performance for both unimodal and multimodal data.

2.4.2.2 Transfer learning

The rigidity of deep learning renders it unsuitable for tasks concerning the distribution and dimensionality of data, as well as changes in model outputs. In contrast, TL relaxes these assumptions and provides a viable means of training neural networks with limited data and labels. TL involves the transfer of knowledge from a source domain (existing knowledge) to a target domain (new knowledge to be learned). TL can be classified into three categories: directTL, semi-supervised transfer, and cross-domain adaptation. Direct TL methods pre-train neural networks with large-scale datasets and fine-tune the pre-trained networks for use in segmentation and classification tasks. Through the use of semi-supervised TL, the problem of small sample learning is resolved and training time is reduced. In cases where data labels are absent, this approach seeks to address the data imbalance between the source and target domains. The wealth of data and labels in the source domain helps to overcome the problem of poor performance in the target domain due to sparse data. Meanwhile, cross-domain adaptation endeavors to enhance model performance by acquiring relevant knowledge pertaining to disease diagnosis from source domain data replete with informative features.

Today, TL techniques are extensively employed in the medical domain [109], particularly in brain tumor segmentation tasks. The intricate nature of the U-Net model can result in a high computational and time overhead during its execution. To address this, Pravitasari *et al.* [110] conducted TL on the training data component of the UNet-VGG16 amalgamated model, which helped to streamline the U-Net architecture for brain tumor segmentation objectives. Although TL-based segmentation models have shown promise, they suffer from a surplus of parameters in the feature output, which can lead to overfitting and a reduction in the number of deep features acquired. Some researchers have opted for a fixed backbone layer structure by initializing the model with pre-trained weights [111] and later fine-tuning the network. Although this method reduces the number of parameters to some extent, the question of which layers to freeze for fine-tuning and how to fine-tune them to improve results without increasing parameters remains unresolved.

Meanwhile, several researchers have opted to pre-train on the ImageNet dataset. Stawiaski and Wacker designed models with pre-trained convolutional encoders. The convolutional layers, which were pre-trained on ImageNet, are utilized as the encoder part of the encoder–decoder architecture. This approach enables the model to be trained from scratch using only the decoder, thereby enhancing segmentation accuracy while accelerating convergence. Among them, Wacker *et al.* [112] applied FCN networks, and using three of the four MRI modalities (discarding T1) as RGB input channel for a pre-trained encoder architecture capable of interpreting spatial information. This approach stabilizes the training process and yields more robust segmentation outcomes. Arbane and Srinivas both conducted a performance comparison of three migration learning-based CNN pre-training models to determine which one is the superior model for automatically predicting brain tumor segmentation. Arbane *et al.* [113] utilized ResNet-50, MobileNet-v2, and Xception models to compare their performance for brain tumor segmentation. The experimental outcomes indicated that the MobileNet-v2 model exhibited exceptional precision, accuracy, and F1 values, surpassing the other two models. Srinivas *et al.* [114] examined VGG-16, ResNet-50, and Inception-v3 models and found that VGG-16 model attained competitive results. Subsequently, Ullah *et al.* [115] extended this study by applying TL to nine deep neural networks such as Inception-resnet-v2, Inception-v3, as well as seven other models. According to the experimental findings, the Inception-resnet-v2 model produced the most effective results for brain tumor classification. In all the aforementioned methods, pre-trained models from standard public datasets used for image segmentation are utilized as the backbone network for brain tumor segmentation. However, there is a significant distinction between the abstract characteristics of the source and target domain samples, which results in a reduced ability of the models to identify the tumor's detailed regions. Shen and Anderson [116] investigated the feasibility of transferring knowledge from the BraTS dataset to other neuroimaging datasets through the use of a pre-trained model for segmentation of images in the Rembrandt dataset. Although the transfer of pre-trained models from the BraTS dataset to other neuroimaging datasets shows potential, additional work is still required to fully validate the approach.

From the above analysis, it is clear that the approach based on TL has the following advantages:

1. Using pre-trained networks as the basis, it applies to all brain tumor segmentation tasks, is effective at obtaining multi-scale features and initialization parameters for the network, and lowers the training cost of the network.
2. It can effectively solve the problem of weak model generalization due to data scarcity, continuously improve model robustness, and obtain more accurate lesion detection results.

However, this method also has the following drawbacks:

1. Using ImageNet pre-trained models can speed up the operation in the early stage of training, but it cannot ensure the eventual improvement of brain tumor

segmentation accuracy or bring about regularization effect and is prone to negative transfer problems.
2. The model structure is more fixed, as well as the image input size is fixed and less flexible.
3. More study is required to choose appropriate fine-tuning procedures to utilize TL more successfully in the field of brain tumor segmentation due to the complexity of brain tumor images and the wide variation in lesion morphology.

2.4.3 Missing modalities

Data comprises of one or more modalities, each of which possesses its unique representation. It is a common scenario in the real world for some modalities to be absent, leading to reduced performance when carrying out multi-modal learning. In clinical practice, the absence of one or more MRI modalities has resulted in significant segmentation performance degradation due to differences in brain anatomy among patients, glioma size and shape variations, as well as MRI intensity range variability, which is prone to artifacts and blurred tumor contours. Hence, it is essential to design an automated brain tumor segmentation method that addresses the modality deficit problem. Zhou *et al.* [117] combined related representations from several modalities into a single shared representation by using an attention mechanism to acquire weight maps of the channels and spaces of various modalities. This method can produce stable results even in the absence of modalities. Unlike other modalities, each modality exhibits unique appearances, resulting in differing sensitivities to distinct tumor areas. In response, Ding *et al.* [118] proposed the region-aware fusion network (RFNet), which fuses modality-region relationships. The region-aware fusion module generates attention weights for distinct regions, thereby adaptively segmenting incomplete multi-modal image sets. As the demand for medical image segmentation of brain tumors continues to increase, novel research ideas are required. In addition to traditional changes to model structure or the addition of segmentation modules, the development of GAN [119] has given rise to a fresh line of inquiry into the segmentation of brain tumors. Another approach has been taken on the issue of insufficient finely annotated samples and modality loss in the brain tumor segmentation problem.

Several techniques have been suggested to address the issue of missing modalities in medical image segmentation [120]. One of the earliest efforts in learning with missing modalities was by Yu and colleagues [121], who employed 3D cGAN models and local adaptive fusion to generate FLAIR-like modal images from T1 uni-modal inputs, ensuring overall image and local block-level similarity. Meanwhile, Wang and colleagues [122] proposed a knowledge distillation method for extracting unimodal knowledge from multimodal models. The network uses two unsupervised learning modules to align the complete modality with the missing modality and learn couplingly, complementary domains and feature representations to achieve recovery of the missing modality. While it is possible to obtain some results by training a model for each missing modality, this approach is intricate and time-consuming. The above-mentioned methods involve segmentation by combining multiple modalities; however, they do not account for the differences in valuable information across various modalities. In the medical field, multiple modalities are commonly used to provide complementary information, which can lead to a combined effect that is superior to that of a single modality. However, effective fusion of multiple inputs is a major challenge in achieving multi-modal synthesis. To address this, some researchers have tackled the problem by first synthesizing the missing modalities and then using the full set of modalities as inputs. For instance, the hybrid-fusion network [123] leverages the interrelationships among various modalities to establish a mapping between multimodal source images (modalities already in existence) and target images (missing modalities). Following this, the network fuses the existing modal images with one another after independently modeling each modality, ultimately synthesizing multiple target (or missing) modal images. The individual modalities are trained using a modality-specific network to learn features that can be generalized, whereas a fusion network is utilized in a multimodal scenario to learn potential shared representations. MGM-GAN, as proposed by Zhan and colleagues [124], is a multi-scale generative adversarial network model based on the gate mergence mechanism. The network may automatically learn the relevance of various modalities at various points according to this method, thus enhancing important information and suppressing irrelevant details. By incorporating missing modalities from other modalities, the multi-scale fusion technique takes full advantage of the complementary information present in hierarchical features, thus leading to improved synthesis performance. However, these methods are computationally demanding, and the segmentation process requires additional networks for synthesis, which further increases the computational cost and time. Moreover, the segmentation performance is strongly impacted by the quality of the synthesis, and the segmentation and generation tasks have separate optimization targets.

The application of shared representation learning has been proven to be a highly successful method for utilizing correlations between multi-modal data [125]. Currently, numerous cutting-edge techniques are available for learning shared representations of tumor regions by fusing available modalities and features in the latent space. This makes it possible to map features to solve the problem of missing modalities. Compared to previous algorithms, this method is more efficient as it does not require learning multiple potential subsets of multi-modalities, and it is not impacted by the quality of the synthesized patterns. The fusion method is essential to achieving excellent segmentation accuracy in multi-modal medical image segmentation applications. Hence, identifying which potential features to learn and how to learn them are two critical questions. Zhou [126] has suggested an approach for conditionally generating the missing modality via multi-source correlation in order to replace it with a usable one. The network is made up of three sub-networks: a U-Net segmentation network, a correlation constraint block, and a feature enhancement generator for creating 3D feature-enhanced images of the missing modality. The incorporation of a correlation constraint network to extract potential multi-source correlations between modalities has proven to be instrumental in aiding the conditional generator to emphasize the correlation between tumor features of missing modalities. Although it is possible to create a single model to account for all missing states, doing so is biased and most models don't function well when more than one modality is missing. In this context, Chen *et al.* [97] used feature entanglement to encode the input images into two categories, one for each image's apparent encoding and the other for multimodal content encoding, which were then combined into a single common representation for segmentation using a gating method. The proposed fusion method reweighted the content codes spatially, but not channel-wise, leading to an average Dice performance over the whole tumor segmentation by mode-invariant features (removing mode-specific information) exceeding 16%. Vadacchino and colleagues [127] propose the hierarchical adversarial knowledge distillation network, HAD-Net, to improve the performance of the student network when faced with domain shifts that arise from the unavailability of segmentation-enhanced tumor images (missing T1ce). However, this approach is limited in its ability to reconstruct local details of the object, such as texture, through the missing modalities. To address this problem, Azad and colleagues [128] use the content-style matching mechanism to extract information from the full modal network and apply it to the missing models, resulting in the style matching U-Net (SMU-Net), which is capable of reconstructing the missing information. Similarly, Zhou and colleagues [129] propose a correlation model to capture potential multi-source correlation models that effectively fuse cross-modal correlation representations into a shared representation. This is achieved through an attention mechanism that emphasizes the most important segmentation features. Their proposed approach, which relies on shared knowledge extraction, is straightforward and capable of capturing correlations between multi-modal features, thereby enabling the network to gradually refine segmentation results. However, this approach is vulnerable to training instability and feature resolution entanglement. To address these issues, Yang *et al.* [130] introduced the dual disentanglement network (D2-Net) that explicitly separates modality-specific information based on frequency and space domains. The D2-Net consists of a modality disentanglement stage that separates information related to a modality and a tumor-region disentanglement stage that uses knowledge distillation to disentangle representations particular to a tumor and extract general features.

From the above analysis, it is clear that the generative adversarial network-based approach has the following advantages:

1. GAN uses adversarial learning to effectively model unknown distributions, thereby generating clear, realistic sample data and effectively solving the problem of data scarcity.
2. Due to the augmentation of limited labeled data or the filling of modal missing information, the deep network is adequately trained and eventually achieves good segmentation results.

However, the method also has the following problems that need to be addressed:

1. The GAN-based image generation algorithm relies on the non-linear fitting ability of the neural network, so the quality and diversity of the generated images are directly related to the neural network structure. How to design a suitable network architecture while guaranteeing a certain performance is a problem worth investigating.

Table 2.4 Solutions to problems based on small sample datasets

Problems to be solved	Main methods	Advantage	Disadvantage	Improvement measures
Class imbalance	Loss function	Balancing difficult sample weights to reduce performance loss due to class imbalance	The selection and integration of loss functions significantly impact segmentation performance, and the determination of an optimal function is not standardized.	Fine-tune, combine, or design own loss functions and optimize function formulas
	Change the model structure	By considering the relationship between multiple related subtasks and task pattern structures	Difficult model design; computationally expensive; difficult to meet real-time requirements	Optimization of task and model allocation strategies
Data lack	Data augment	Easy to access; meet massive medical data requirements; improve overall model performance	Chance of generating incorrect images; noise problems; insignificant performance improvements	Using a multi-sample data transformation
	Transfer learning	Obtain multi-scale features and initialize network parameters; improve model convergence speed and generalization ability	The properties of the source and target domains are different; insufficient flexibility of the model; choice of transfer layer and amount of transfer to be confirmed by experiment	Optimize model structure to reduce domain variation; develop effective transfer strategies and fine-tuning strategies
Missing modalities	GAN	Increase the number and diversity of training samples; enhance lesion segmentation	Poor stability; difficulties with model training and optimization of model and objective functions; problems with model collapse	Combining a priori knowledge; optimizing model structure and improving image quality

2. In real-world applications, GAN is unstable, hard to converge, and sensitive to network collapse, which frequently prevents the training process from continuing and even results in the network failing to converge because gradients vanish.

2.4.4 Summary

In summary, network models constructed from small sample datasets are poorly stable and cannot be extended to other sample sets for use, making it difficult to popularize deep learning models in the clinic. The three main problems in brain tumor segmentation are addressed by adapting and improving the segmentation methods, or introducing new methods that reduce the negative effects of small samples to a certain extent. The above segmentation methods were comprehensively analyzed and compared, and summarized in terms of main ideas, advantages and disadvantages, and improvement measures, as shown in Table 2.4.

2.5 Model interpretability

The segmentation model was enhanced by effectively fusing brain tumor images from various levels, regions, and modalities to boost its accuracy. Nonetheless, the model's neural network-based features rely on filter responses derived from extensive training data, and the model's interpretability remains poor, rendering it difficult to generalize in medical applications. Thus, there is a need to explore effective techniques to improve the model's interpretability.

Interpretability of neural models [131] pertains to the capacity of the network to explicate its workings in a comprehensible manner to humans. In medical image-based diagnosis, where doctors place more emphasis on the logical procedure of result derivation when scrutinizing model results, it is highly beneficial to visualize and mitigate the obscurity caused by black-box systems. An ideal brain tumor segmentation framework should not only provide dependable segmentation outputs, but also elucidate the reasoning process underpinning the outcomes. In 2016, Zhou *et al.* [132] proposed class activation maps (CAM) that generate feature maps consistent with the number of target classes by means of the final convolutional layer. After global average pooling, the results are attained via the softmax layer, providing vivid insights into each feature map for purposes of visualization. While CAM is a simple technique, it necessitates modifications to the original model's structure, which in turn demands retraining of the model, severely limiting its use cases. Selvaraju *et al.* [133] proposed gradient-weighted class activation maps (Grad-CAM) to address these issues. Grad-CAM calculates weights by average gradients globally, identifying the area of interest on the input image that has the greatest influence on class prediction. This simplifies the visualization of the network's focus on the input image. Compared to CAM, which can only extract the heat map of the last layer of the feature map, Grad-CAM can extract the heat map of any layer, allowing for the visualization of CNNs of any structure without the need for network modification or retraining.

Numerous studies have employed Grad-CAM for the purpose of visualizing neural networks that segment brain tumors. Due to its simpler visualization and analysis of interpretable correlations in 2D models, Natekar *et al.* [134] and Dasanayaka *et al.* [135] have opted to utilize this technique to create a more comprehensible deep learning model, generating a heat map that explains the contribution of each input region to the model's predicted output. This helps to clarify the functional organization of the model, providing insight into how neural networks are visualized. Esmaeili *et al.* [136] used 2D Grad-CAM to assess the effectiveness of three DL models for classifying brain tumors. For processing MRI images of brain tumors, Zhao *et al.* [137] proposed an interpretable deep learning image segmentation framework, where Grad-CAM was introduced to visualize and interpret the segmentation model based on the pyramid structure. The usage of CAM in visualizing the features of interest in each layer of the model after constructing global contextual information through features after multiple pooling layers has improved the image segmentation performance and achieved interpretation of PSPNet in the pyramid structure. However, the previous studies have some limitations. For instance, no models for 3D medical applications have been explored yet, and Grad-CAM relies on gradients which may not be able to correctly identify multiple occurrences of the same lesion in an image. To address these limitations, Saleem and the team [138] developed a 3D heat map based on a CAM approach to extract 3D interpretations for understanding the model's strategy in brain tumor segmentation. This approach uses gradient-free interpretability compared to gradient-based interpretability methodologies for more accurate coverage of class regions. Despite the high class discriminability and interpretability of the model, its accuracy still requires improvement. To tackle this issue, Zeineldin and colleagues [139] propose a NeuroXAI framework for interpretable artificial intelligence in deep learning networks, which implements seven gradient-based interpretation methods that generate visual maps to increase transparency in deep learning models. 2D and 3D interpretable sensitivity maps are derived to help clinicians comprehend and have faith in the performance of DL algorithms during clinical procedures. As mentioned in several studies in the field of brain tumor segmentation, the study of model interpretability is quite homogeneous and mainly focuses on heat maps to visualize CAM. However, this approach only provides a coarse-grained annotation of results, and there is still a need to explore methods that can meet the requirement for high-precision interpretation of results in the medical diagnosis field.

Methods such as CAM provide model interpretation by analyzing the extent to which features contribute to segmentation results, helping researchers to understand how models represent information about tumor regions through the model and how they change with layers. Meaningful visualization in the network can help medical professionals to evaluate the correctness of the predicted results and expand the scope of clinical applications. However, the method has the following problems that need to be addressed urgently:

1. The structural framework of the original model needs to be modified and retrained when using the CAM method, resulting in more cost and time spent in practical applications. Combining the gradient weighting method to increase the model's ability to identify lesions and improve the speed of the model is an important research direction to enhance the model's interpretation.

2. The model interpretation process is subject to interference from factors such as random perturbations in the sample and has certain limitations. Incorporating expert knowledge into the model design process, guiding model construction through expert feedback, and prompting clinical experts to control the model decision-making process are potential research directions for improving model interpretation.
3. Interpretability methods lack usable evaluation metrics to assess the quality of the generated interpretations. Therefore, it is also worthwhile to investigate how to quantitatively assess the interpretation of the results of integrating visual interpretability in segmentation models.
4. The inherent contradiction between model performance and interpretability prevents both from being optimal at the same time. Integrating multi-modal medical data for decision making and analyzing the contribution of each modality to the decision as a way to simulate a doctor's clinical diagnosis workflow can achieve a more comprehensive interpretation while ensuring network performance.

2.6 Conclusion and outlook

In recent times, the swift advancement of deep learning techniques has yielded commendable results in the domain of brain tumor segmentation. Starting from fully supervised detection techniques, where the receptive field is increased, and feature fusion is implemented, to U-net, Transformer, multi-model fusion, and cascade models, as well as numerous non-fully supervised learning algorithms, such as loss function adjustment, TL, and GAN to tackle the small sample problem, and model interpretability; the models are progressing in all technical aspects. However, they also encounter various challenges. Therefore, this chapter provides a suitable overview of lesion region detection methods based on deep learning frameworks, and the research hurdles are delineated as follows.

1. **Sample-related**: The issue of limited samples has consistently been a challenging factor for brain tumor segmentation. Despite having a substantial amount of available brain tumor images, most of these belong to small-scale datasets with inadequate annotation, and researchers often struggle with creating efficient and precise segmentation models due to label inconsistencies among different datasets.
2. **Small target lesion segmentation**: In the process of segmentation, the neural network tends to lose some of the smaller lesion information as a result of convolutional pooling operations, ultimately diminishing the accuracy of the model. Hence, finding effective means to mitigate or eliminate this loss of small lesion features during the segmentation process is a significant avenue of current research.
3. **Model interpretability**: Due to the "black box" property of deep learning, the inherent structure of the network model is not fully transparent and the interpretability is poor, which hinders the popularity of brain tumor screening systems in the medical field. The inherent tension between model performance and interpretability prevents both from being optimal.
4. **Clinical relevance**: Most researchers lack communication and exchange with hospitals when designing brain tumor auxiliary diagnostic systems, which

results in diagnostic models that are not applicable to clinical settings. Furthermore, although existing models such as Brainlab Elements AI software, CureMetrix cmAssist, and Aidoc, which are AI-driven medical imaging systems, have been put into use and are undergoing clinical trial evaluations, there is still a certain gap between their actual clinical application and they have not truly alleviated the burden of hospital doctors.

To tackle the challenges encountered in brain tumor segmentation, future research endeavors could concentrate on the following aspects:

1. **Model development for limited datasets**: The robustness of models trained on small sample datasets is poor, rendering the acquisition of stable detection outcomes difficult. In the absence of large-scale training sample sets, it is critical to combine TL, data augmentation, and GAN to develop a network architecture that is suitable for small-scale datasets. This constitutes an important link between practical applications and technology.
2. **Enhancing the diversity of label annotations**: The personal experiences and subjective differences in evaluating lesion features in brain tumor images can result in significant diagnostic discrepancies among different doctors. Therefore, obtaining label information from different experts or developing appropriate algorithms for automatic annotation to accommodate the variability in subjectivity is an important research direction in the future.
3. **Enhancing the utilization of multi-modal data**: To improve the accuracy of brain tumor segmentation, researchers could focus on incorporating missing modal data through the use of GANs or stage-wise learning strategies. Additionally, effective strategies for fusing multi-modal data can also be explored to enhance the performance of segmentation models.
4. **Enhancing image feature information**: To further improve segmentation accuracy, researchers could explore the use of multi-modal data, in combination with other relevant diseases, and employ multiple deep neural networks to extract richer features from brain tumor images. This approach would enable the models to achieve more accurate segmentation results.
5. **Improving model interpretability**: Using novel frameworks to learn various lesion features in brain tumor images, analyzing the contribution of each feature to decision-making, and simulating the clinical diagnosis work of doctors to achieve a more comprehensive explanation while ensuring model performance is also an important research direction in this field.
6. **Strengthen practical clinical need**: Integrating the brain tumor auxiliary segmentation system with hospital information systems such as image archiving, communication systems, and electronic medical records can facilitate its widespread application in clinical diagnosis.

In conclusion, the ongoing enhancement and advancement of deep learning techniques will generate more precise and efficient auxiliary diagnostic tools for brain tumor segmentation in the future, thereby offering effective support for clinical diagnosis and treatment.

References

[1] Oronsky, B., Reid, T.R., Oronsky, A., Sandhu, N. and Knox, S.J.: "A review of newly diagnosed glioblastoma," *Front. Oncol.*, 2021, **10**, p. 574012.

[2] Stupp, R. and Roila, F.: "Malignant glioma: ESMO clinical recommendations for diagnosis, treatment and follow-up," *Ann. Oncol.*, 2009, **20**, pp. 126–128.

[3] McFaline-Figueroa, J.R. and Lee, E.Q.: "Brain tumors," *Am. J. Med.*, 2018, **131**(8), pp. 874–882.

[4] Liu, Z., Tong, L., Jiang, Z., *et al.*: "Deep learning based brain tumor segmentation: a survey," *Complex Intell. Syst.*, 2023, *9*(1), pp. 1001–1026.

[5] Işın, A., Direkoğlu, C. and Şah, M.: "Review of MRI-based brain tumor image segmentation using deep learning methods," *Procedia Comput. Sci.*, 2016, **102**, pp. 317–324.

[6] Menze, B.H., Jakab, A., Bauer, S., *et al.*: "The multimodal brain tumor image segmentation benchmark (BRATS)," *IEEE Trans. Med. Imaging*, 2015, **34**(10), pp. 1993–2024.

[7] Baid, U., Ghodasara, S., Mohan, S., *et al.*: "The RSNA-ASNR-MICCAI BraTS 2021 benchmark on brain tumor segmentation and radiogenomic classification," arXiv preprint arXiv:2107.02314, 2021.

[8] Antonelli, M., Reinke, A., Bakas, S., *et al.*: "The medical segmentation decathlon," *Nat. Commun.*, 2022, **13**(1), p. 4128.

[9] Pereira, S., Pinto, A., Alves, V. and Silva, C.A.: "Brain tumor segmentation using convolutional neural networks in MRI images," *IEEE Trans. Med. Imaging*, 2016, **35**(5), pp. 1240–1251.

[10] Shaikh, M., Anand, G., Acharya, G., et al.: "Brain tumor segmentation using dense fully convolutional neural network," *Brainlesion: Glioma, Multiple Sclerosis, Stroke and TraumaticBrain Injuries: 3th International Workshop*, Quebec City, QC, Canada, September 2017, pp. 309–319.

[11] Huang, G., Liu, Z., van der Maaten, L. and Weinberger, K.Q.: "Densely connected convolutional networks," *2017 IEEE Conference on Computer Vision and Pattern Recognition (CVPR)*, Honolulu, HI, July 2017, pp. 2261–2269.

[12] Havaei, M., Davy, A., Warde-Farley, D., *et al.*: "Brain tumor segmentation with deep neural networks," *Med. Image Anal.*, 2017, **35**, pp. 18–31.

[13] Moreno Lopez, M., and Ventura, J.: "Dilated convolutions for brain tumor segmentation in MRI scans," *Brainlesion: Glioma, Multiple Sclerosis, Stroke and Traumatic Brain Injuries: 3th International Workshop*, Quebec City, QC, Canada, September 2017, pp. 309–319.

[14] Li, W., Wang, G., Fidon, L., et al.: "On the compactness, efficiency, and representation of 3D convolutional networks: Brain parcellation as a pretext task," *Information Processing in Medical Imaging: 25th International Conference*, Boone, NC, USA, June 2017, pp. 348–360.

[15] Sun, J., Peng, Y., Guo, Y. and Li, D.: "Segmentation of the multimodal brain tumor image used the multi-pathway architecture method based on 3D FCN," *Neurocomputing*, 2021, **423**, pp. 34–45.

[16] He, K., Zhang, X., Ren, S., and Sun, J.: "Deep residual learning for image recognition," *2016 IEEE Conference on Computer Vision and Pattern Recognition (CVPR)*, Las Vegas, NV, USA, July 2016, pp. 770–778.

[17] Zhou, Z., He, Z. and Jia, Y.: "AFPNet: a 3D fully convolutional neural network with atrous-convolution feature pyramid for brain tumor segmentation via MRI images," *Neurocomputing*, 2020, **402**, pp. 235–244.

[18] Rao, V., Sarabi, M.S. and Jaiswal, A.: "Brain tumor segmentation with deep learning," *MICCAI Multimodal Brain Tumor Segmentation Challenge (BraTS)*, 2015, **59**, pp. 1–4.

[19] Li, Y., and Shen, L.: "Deep learning based multimodal brain tumor diagnosis," *Brainlesion: Glioma, Multiple Sclerosis, Stroke and Traumatic Brain Injuries: 3th International Workshop*, QC, Canada, September 2017, pp. 149–158.

[20] Noori, M., Bahri, A., and Mohammadi, K.: "Attention-guided version of 2D UNet for automatic brain tumor segmentation," *2019 9th International Conference on Computer and Knowledge Engineering (ICCKE)*, Mashhad, Iran, October 2019, pp. 269–275.

[21] Zhou, T., Ruan, S., Guo, Y., and Canu, S.: "A Multi-Modality Fusion Network Based on Attention Mechanism for Brain Tumor Segmentation," *2020 IEEE 17th International Symposium on Biomedical Imaging (ISBI)*, IEEE, Iowa City, IA, USA, April 2020, pp. 377–380.

[22] He, X., Xu, W., Yang, J., Mao, J., Chen, S. and Wang, Z.: "Deep convolutional neural network with a multi-scale attention feature fusion module for segmentation of multimodal brain tumor," *Front. Neurosci.*, 2021, **15**, p. 782968.

[23] Liu, Y., Mu, F., Shi, Y., Cheng, J., Li, C. and Chen, X.: "Brain tumor segmentation in multimodal MRI via pixel-level and feature-level image fusion," *Front. Neurosci.*, 2022, **16**, p. 1000587.

[24] Long, J., Shelhamer, E. and Darrell, T.: "Fully convolutional networks for semantic segmentation," *Proceedings of the IEEE conference on computer vision and pattern recognition*, Boston, MA, USA, June 2015, pp. 3431–3440.

[25] Ronneberger, O., Fischer, P. and Brox, T.: "U-net: convolutional networks for biomedical image segmentation," *Medical Image Computing and Computer-Assisted Intervention–MICCAI 2015*, October 2015, pp. 234–241.

[26] Badrinarayanan, V., Kendall, A. and Cipolla, R.: "SegNet: a deep convolutional encoder–decoder architecture for image segmentation," *arXiv*, 2016.

[27] Zhao, X., Wu, Y., Song, G., Li, Z., Zhang, Y. and Fan, Y.: "A deep learning model integrating FCNNs and CRFs for brain tumor segmentation," *Med. Image Anal.*, 2018, **43**, pp. 98–111.

[28] Hu, K., Gan, Q., Zhang, Y., *et al.*: "Brain tumor segmentation using multi-cascaded convolutional neural networks and conditional random field," *IEEE Access*, 2019, **7**, pp. 92615–92629.

[29] Çiçek, Ö., Abdulkadir, A., Lienkamp, S.S., Brox, T. and Ronneberger, O.: "3D U-Net: learning dense volumetric segmentation from sparse annotation," *Medical Image Computing and Computer-Assisted Intervention–MICCAI 2016*, Athens, Greece, October 2016, pp. 424–432.

[30] Isensee, F., Kickingereder, P., Wick, W., Bendszus, M. and Maier-Hein, K. H.: "Brain tumor segmentation and radiomics survival prediction: Contribution to the BRATS 2017 challenge," *Brainlesion: Glioma, Multiple Sclerosis, Stroke and Traumatic Brain Injuries: 3th International Workshop*, Quebec City, QC, Canada, September 2017, pp. 287–297.

[31] Myronenko, A.: "3D MRI brain tumor segmentation using autoencoder regularization," *Brainlesion: Glioma, Multiple Sclerosis, Stroke and Traumatic Brain Injuries: 4th International Workshop*, Granada, Spain, September 2018, pp. 311–320.

[32] Isensee, F., Jaeger, P.F., Full, P.M., Vollmuth, P. and Maier-Hein, K.H.: "nnU-Net for brain tumor segmentation," *Brainlesion: Glioma, Multiple Sclerosis, Stroke and Traumatic Brain Injuries: 6th International Workshop*, Lima, Peru, October 2020, pp. 118–132.

[33] Dosovitskiy, A., Beyer, L., Kolesnikov, A., *et al.*: "An image is worth 16×16 words: transformers for image recognition at scale," arXiv preprint arXiv:2010.11929, 2020.

[34] Chen, J., Lu, Y., Yu, Q., *et al.*: "TransUNet: transformers make strong encoders for medical image segmentation," arXiv preprint arXiv:2102.04306, 2021.

[35] Xie, Y., Zhang, J., Shen, C. and Xia, Y.: "CoTr: efficiently bridging CNN and transformer for 3D medical image segmentation," *Medical Image Computing and Computer Assisted Intervention–MICCAI 2021: 24th International Conference*, Strasbourg, France, September 2021, pp. 171–180.

[36] Wang, W., Chen, C., Ding, M., Li, J., Yu, H. and Zha, S.: "TransBTS: multimodal brain tumor segmentation using transformer," *Medical Image Computing and Computer Assisted Intervention–MICCAI 2021: 24th International Conference*, Strasbourg, France, September 2021, pp. 109–119.

[37] Hatamizadeh, A., Tang, Y., Nath, V., *et al.*: "UNETR: transformers for 3D medical image segmentation," *Proceedings of the IEEE/CVF Winter Conference on Applications of Computer Vision*, Waikoloa, HI, USA, January 2022, pp. 574–584.

[38] Roy, S., Koehler, G., Ulrich, C., *et al.*: "MedNeXt: transformer-driven scaling of convNets for medical image segmentation," *International Conference on Medical Image Computing and Computer-Assisted Intervention*, Vancouver, BC, Canada, October 2023, pp. 405–415.

[39] Liu, Z., Mao, H., Wu, C.-Y., Feichtenhofer, C., Darrell, T. and Xie, S.: "A convNet for the 2020s," *Proceedings of the IEEE/CVF Conference on Computer Vision and Pattern Recognition*, New Orleans, Louisiana, USA, June 2022, pp. 11976–11986.

[40] Xing, Z., Yu, L., Wan, L., Han, T. and Zhu, L.: "NestedFormer: nested modality-aware transformer for brain tumor segmentation," *International Conference on Medical Image Computing and Computer-Assisted Intervention*, Singapore, September 2022, pp. 140–150.

[41] Chen, W., Liu, B., Peng, S., Sun, J., and Qiao, X.: "S3D-UNet: separable 3D U-Net for brain tumor segmentation," *Brainlesion: Glioma, Multiple*

Sclerosis, Stroke and Traumatic Brain Injuries: 4th International Workshop, Granada, Spain, September 2018, pp. 358–368.

[42] Xie, S., Sun, C., Huang, J., Tu, Z. and Murphy, K.: "Rethinking spatio-temporal feature learning: speed-accuracy trade-offs in video classification," *Proceedings of the European Conference on Computer Vision (ECCV)*, Munich, Germany, September 2018, pp. 305–321.

[43] Mehta, S., Rastegari, M., Caspi, A., Shapiro, L., and Hajishirzi, H.: "ESPNet: efficient spatial pyramid of dilated convolutions for semantic segmentation," *Proceedings of the European Conference on Computer Vision (ECCV)*, Munich, Germany, September 2018, pp. 552–568.

[44] Nuechterlein, N., and Mehta, S.: "3D-ESPNet with pyramidal refinement for volumetric brain tumor image segmentation," *Brainlesion: Glioma, Multiple Sclerosis, Stroke and Traumatic Brain Injuries: 4th International Workshop*, Granada, Spain, September 2018, pp. 245–253.

[45] Chen,C., Liu, X., Ding, M., Zheng, J. and Li, J.: "3D dilated multi-fiber network for real-time brain tumor segmentation in MRI," *Medical Image Computing and Computer Assisted Intervention–MICCAI 2019: 22nd International Conference*, Shenzhen, China, October 2019, pp. 184–192.

[46] Brügger, R., Baumgartner, C.F. and Konukoglu, E.: "A partially reversible U-net for memory-efficient volumetric image segmentation," *Medical Image Computing and Computer Assisted Intervention–MICCAI 2019: 22nd International Conference*, Shenzhen, China, October 2019, pp. 429–437.

[47] Walsh, J., Othmani, A., Jain, M. and Dev, S.: "Using u-net network for efficient brain tumor segmentation in MRI images," *Healthcare Anal.*, 2022, **2**, p. 100098.

[48] Xiao, Z., He, K., Liu, J. and Zhang, W.: "Multi-view hierarchical split network for brain tumor segmentation," *Biomed. Signal Process. Control*, 2021, **69**, p. 102897.

[49] Casamitjana, A., Puch, S., Aduriz, A. and Vilaplana, V.: "3D convolutional neural networks for brain tumor segmentation: a comparison of multi-resolution architectures," *International Workshop on Brainlesion: Glioma, Multiple Sclerosis, Stroke and Traumatic Brain Injuries*, Quebec City, QC, Canada, September 2017, pp. 150–161.

[50] Kamnitsas, K., Ledig, C., Newcombe, V.F.J., *et al.*: "Efficient multi-scale 3D CNN with fully connected CRF for accurate brain lesion segmentation," *Med. Image Anal.*, 2017, **36**, pp. 61–78.

[51] Razzak, M.I., Imran, M. and Xu, G.: "Efficient brain tumor segmentation with multiscale two-pathway-group conventional neural networks," *IEEE J. Biomed. Health Inform.*, 2019, **23**(5), pp. 1911–1919.

[52] Ranjbarzadeh, R., Bagherian Kasgari, A., Jafarzadeh Ghoushchi, S., Anari, S., Naseri, M. and Bendechache, M.: "Brain tumor segmentation based on deep learning and an attention mechanism using MRI multi-modalities brain images," *Sci. Rep.*, 2021, **11**(1), p. 10930.

[53] Wang, G., Li, W., Ourselin, S., and Vercauteren, T.: "Automatic brain tumor segmentation using cascaded anisotropic convolutional neural

networks," *Brainlesion: Glioma, Multiple Sclerosis, Stroke and Traumatic Brain Injuries: Third International Workshop*, Quebec City, QC, Canada, September 2017, pp. 178–190.

[54] Sobhaninia, Z., Rezaei, S., Karimi, N., Emami, A., and Samavi, S.: "Brain tumor segmentation by cascaded deep neural networks using multiple image scales," *2020 28th Iranian Conference on Electrical Engineering (ICEE),* Tabriz, Iran, August 2020, pp. 1–4.

[55] Jiang, Z., Ding, C., Liu, M., and Tao, D.: "Two-stage cascaded U-Net: 1st place solution to BraTS challenge 2019 segmentation task," *Brainlesion: Glioma, Multiple Sclerosis, Stroke and Traumatic Brain Injuries: 5th International Workshop*, Shenzhen, China, October 2019, pp. 231–241.

[56] Rivera, L.C., Castillo, L., Daza, L.A., and Arbelaez, P.A.: "Volumetric multimodality neural network for brain tumor segmentation," *13th International Conference on Medical Information Processing and Analysis*, San Andres Island, Colombia, October 2017, pp. 97–104.

[57] Zhou, C., Ding, C., Wang, X., Lu, Z. and Tao, D.: "One-pass multi-task networks with cross-task guided attention for brain tumor segmentation," *IEEE Trans. Image Process.*, 2020, **29**, pp. 4516–4529.

[58] Silva, C.A., Pinto, A., Pereira, S. and Lopes, A.: "Multi-stage deep layer aggregation for brain tumor segmentation," *Brainlesion: Glioma, Multiple Sclerosis, Stroke and Traumatic Brain Injuries: 6th International Workshop*, Lima, Peru, October 2020, pp. 179–188. *arXiv*, 2021.

[59] Zhou, C., Chen, S., Ding, C., and Tao, D.: "Learning contextual and attentive information for brain tumor segmentation," *Brainlesion: Glioma, Multiple Sclerosis, Stroke and Traumatic Brain Injuries: 4th International Workshop*, Granada, Spain, September 2018, pp. 497–507.

[60] Zhou, C., Ding, C., Lu, Z., Wang, X., and Tao, D.: "One-pass multi-task convolutional neural networks for efficient brain tumor segmentation," *Medical Image Computing and Computer Assisted Intervention–MICCAI 2018: 21st International Conference*, Granada, Spain, September 2018, pp. 637–645.

[61] Kamnitsas, K., Bai, W., Ferrante, E., *et al.*: "Ensembles of multiple models and architectures for robust brain tumour segmentation," *Brainlesion: Glioma, Multiple Sclerosis, Stroke and Traumatic Brain Injuries: Third International Workshop*, Quebec City, QC, Canada, September 14, 2017, pp. 450–462.

[62] Sun, L., Zhang, S., Chen, H. and Luo, L.: "Brain tumor segmentation and survival prediction using multimodal MRI scans with deep learning," *Front. Neurosci.*, 2019, **13**, p. 810.

[63] Murugesan, G. K., Nalawade, S., Ganesh, C., et al.: "Multidimensional and multiresolution ensemble networks for brain tumor segmentation," *Brainlesion: Glioma, Multiple Sclerosis, Stroke and Traumatic Brain Injuries: 5th International Workshop*, Shenzhen, China, October 2019, pp. 148–157.

[64] Kao, P.-Y., Ngo, T., Zhang, A., Chen, J. W., and Manjunath, B. S.: "Brain tumor segmentation and tractographic feature extraction from structural MR images for overall survival prediction," *Brainlesion: Glioma, Multiple*

Sclerosis, Stroke and Traumatic Brain Injuries: 4th International Workshop, Granada, Spain, September 2018, pp. 128–141.

[65] Feng, X., Tustison, N.J., Patel, S.H. and Meyer, C.H.: "Brain tumor segmentation using an ensemble of 3D U-nets and overall survival prediction using radio." *Front. Comput. Neurosci.*, 2020, 14, p. 25.

[66] Sharma, A.K., Nandal, A., Dhaka, A. and Dixit, R.: "A survey on machine learning based brain retrieval algorithms in medical image analysis," *Health Technol.*, 2020, **10**(6), pp. 1359–1373.

[67] Sharma, A.K., Nandal, A., Ganchev, T. and Dhaka, A.: "Breast cancer classification using CNN extracted features: a comprehensive review," In *Application of Deep Learning Methods in Healthcare and Medical Science (ADLMHMS-2020)*, Taylor and Francis, CRC Press, 2022, p. 147.

[68] Zhan, T., Shen, F., Hong, X., *et al.*: "A glioma segmentation method using cotraining and superpixel-based spatial and clinical constraints," *IEEE Access*, 2018, **6**, pp. 57113–57122.

[69] Cui, W., Liu, Y., Li, Y., *et al.*: "Semi-supervised brain lesion segmentation with an adapted mean teacher model," *Information Processing in Medical Imaging: 26th International Conference*, Hong Kong, China, June 2019, pp. 554–565.

[70] Tarvainen, A. and Valpola, H.: "Mean teachers are better role models: weight-averaged consistency targets improve semi-supervised deep learning results," *Adv. Neural Inf. Process. Syst.*, 2017, 30.

[71] Zhao, Y.-X., Zhang, Y.-M., and Liu, C.-L.: "Bag of tricks for 3D MRI brain tumor segmentation," *Brainlesion: Glioma, Multiple Sclerosis, Stroke and Traumatic Brain Injuries: 5th International Workshop,* Shenzhen, China, October 2019, pp. 210–220.

[72] Hu, M., Maillard, M., Zhang, Y., *et al.*: "Knowledge distillation from multi-modal to mono-modal segmentation networks," *Medical Image Computing and Computer Assisted Intervention–MICCAI 2020: 23rd International Conference*, Lima, Peru, October 2020, pp. 772–781.

[73] Lopez-Paz, D., Bottou, L., Schölkopf, B. and Vapnik, V.: "Unifying distillation and privileged information," arXiv preprint arXiv:1511.03643, 2015.

[74] Ito, R., Nakae, K., Hata, J., Okano, H. and Ishii, S.: "Semi-supervised deep learning of brain tissue segmentation," *Neural Netw.*, 2019, **116**, pp. 25–34.

[75] Meissen, F., Kaissis, G. and Rueckert, D.: "Challenging current semi-supervised anomaly segmentation methods for brain MRI," *International MICCAI Brainlesion Workshop, Virtual Event*, September 2021, pp. 63–74.

[76] Sharma, A.K., Nandal, A., Dhaka, A. and Dixit, R.: "Medical image classification techniques and analysis using deep learning networks: a review," In: Patgiri, R., Biswas, A., Roy, P. (eds.), "Health Informatics: A Computational Perspective in Healthcare." *Studies in Computational Intelligence*, vol. 932. Springer, Singapore, pp. 233–258, 2021.

[77] Ji, Z., Shen, Y., Ma, C. and Gao, M.: "Scribble-based hierarchical weekly," *Medical Image Computing and Computer Assisted Intervention–MICCAI 2019: 22nd International Conference*, Shenzhen, China, October 2019, pp. 175–183.

[78] Sharma, A.K., Nandal, A., and Dhaka, A.: "Brain tumor detection and classification using machine learning techniques," *RAMSACT* 2022.
[79] Radford, A., Metz, L., and Chintala, S.: "Unsupervised Representation Learning with Deep Convolutional Generative Adversarial Networks," arXiv preprint arXiv:1511.06434, 2015.
[80] Sharma, A.K., Nandal, A., Dhaka, A., *et al.*: "Enhanced watershed segmentation algorithm based modified ResNet50 model for brain tumor detection," *BioMed. Res. Int.*, 2022, 2022, pp. 1–14.
[81] Demirhan, A., Toru, M. and Guler, I.: "Segmentation of tumor and edema along with healthy tissues of brain using wavelets and neural networks," *IEEE J. Biomed. Health Inform.*, 2015, **19**(4), pp. 1451–1458.
[82] Khan, A.R., Khan, S., Harouni, M., Abbasi, R., Iqbal, S. and Mehmood, Z.: "Brain tumor segmentation using K-means clustering and deep learning with synthetic data augmentation for classification," *Microsc. Res. Technol.*, 2021, **84**(7), pp. 1389–1399.
[83] Kaya, I.E., Pehlivanlı, A.Ç., Sekizkardeş, E.G. and Ibrikci, T.: "PCA based clustering for brain tumor segmentation of T1w MRI images," *Comput. Methods Programs Biomed.*, 2017, **140**, pp. 19–28.
[84] Kumar, D., Verma, H., Mehra, A. and Agrawal, R.K.: "A modified intuitionistic fuzzy c-means clustering approach to segment human brain MRI image," *Multimed. Tools Appl.*, 2019, **78**(10), pp. 12663–12687.
[85] Simaiya, S., Lilhore, U.K., Prasad, D. and Verma, D.K.: "MRI brain tumour detection and image segmentation by hybrid hierarchical k-means clustering with FCM based machine learning model," *Ann. Romanian Soc. Cell Biol.*, 2021, **25**(1), pp. 88–94.
[86] Fang, F., Yao, Y., Zhou, T., Xie, G. and Lu, J.: "Self-supervised multimodal hybrid fusion network for brain tumor segmentation," *IEEE J. Biomed. Health Inform.*, 2022, **26**(11), pp. 5310–5320.
[87] Havaei, M., Guizard, N., Chapados, N. and Bengio, Y.: "HeMIS: heteromodal image segmentation," *Medical Image Computing and Computer-Assisted Intervention–MICCAI* 2016: *19th International Conference*, Athens, Greece, October 2016, pp. 469–477.
[88] Jungo, A., Balsiger, F. and Reyes, M.: "Analyzing the quality and challenges of uncertainty estimations for brain tumor segmentation," *Front. Neurosci.*, 2020, **14**, p. 282.
[89] Pereira, S., Oliveira, A., Alves, V. and Silva, C.A.: "On hierarchical brain tumor segmentation in MRI using fully convolutional neural networks: a preliminary study," *2017 IEEE 5th Portuguese Meeting on Bioengineering (ENBENG)*, Coimbra, Portugal, February 2017, pp. 1–4.
[90] Ottom, M.A., Rahman, H.A. and Dinov, I.D.: "Znet: deep learning approach for 2D MRI brain tumor segmentation," *IEEE J. Transl. Eng. Health Med.*, 2022, **10**, pp. 1–8.
[91] Zhang, W., Yang, G., Huang, H., *et al.*: "ME-net: multi-encoder net framework for brain tumor segmentation," *Int. J. Imaging Syst. Tech.*, 2021, **31**(4), pp. 1834–1848.

[92] Sharma, M.K., Dhiman, N., Mishra, V.N., Mishra, L.N., Dhaka, A. and Koundal, D.: "Post-symptomatic detection of COVID-2019 grade based mediative fuzzy projection," *Comput Electr Eng*, Elsevier, vol. 101, 2022.

[93] Dhaka, A., Nandal, A., Rosales, H.G., *et al.*: "Likelihood estimation and wavelet transformation based optimization for minimization of noisy pixels," *IEEE Access*, 9, pp. 132168–132190, 2021.

[94] Li, X., Luo, G. and Wang, K.: "Multi-step cascaded networks for brain tumor segmentation," *Brainlesion: Glioma, Multiple Sclerosis, Stroke and Traumatic Brain Injuries: 5th International Workshop*, Shenzhen, China, October 2019, pp. 163–173.

[95] Zhou, L., Dhaka, A., Malik, H., Nandal, A., Singh, S., and Wu, Y., "An optimal higher order likelihood distribution based approach for strong edge and high contrast restoration," *IEEE Access*, 9, pp. 109012 – 109024, 2021.

[96] Nandal, A., Bhaskar, V., and Dhaka, A.: "Contrast based image enhancement algorithm using grey scale and color space," *IET Signal Process.*, 12 (4), pp. 514–521, 2018.

[97] Nandal, A., Dhaka, A., Gamboa-Rosales, H., *et al.*, "Sensitivity and variability analysis for image denoising using maximum likelihood estimation of exponential distribution," *Circuits, Syst. Signal Process.*, Springer, vol. 37, no. 9, pp-3903–3926, 2018.

[98] Jia, H., Cai, W., Huang, H., and Xia, Y.: "H2NF-Net for brain tumor segmentation using multimodal MR Imaging: 2nd place solution to BraTS challenge 2020 segmentation task," *Brainlesion: Glioma, Multiple Sclerosis, Stroke and Traumatic Brain Injuries: 6th International Workshop*, Lima, Peru, October 4, 2020, pp. 58 A. 68.

[99] Liu, H., Shen, X., Shang, F. and Wang, F.: "CU-Net: cascaded U-Net with loss weighted sampling for brain tumor segmentation," *Multimodal Brain Image Analysis and Mathematical Foundations of Computational Anatomy: 4th International Workshop, MBIA 2019, and 7th International Workshop, MFCA 2019*, Shenzhen, China, October 2019, pp. 102 A. 111.

[100] Chen, C., Dou, Q., Jin, Y., *et al.*: "Robust multimodal brain tumor segmentation via feature disentanglement and gated fusion," *Medical Image Computing and Computer Assisted Intervention–MICCAI 2019: 22nd International Conference*, Shenzhen, China, October 2019, pp. 447–456.

[101] Zhang, D., Huang, G., Zhang, Q., *et al.*: "Exploring task structure for brain tumor segmentation from multi-modality MR Images," *IEEE Trans. Image Process.*, 2020, **29**, pp. 9032–9043.

[102] Nandal, A., Gamboa-Rosales, H., Dhaka, A., *et al.*: "Image edge detection using fractional calculus with feature and contrast enhancement," *Circuits, Syst. Signal Process.*, Springer, vol. 37, no. 9, pp. 3946–3972, 2018.

[103] Jia, H., Cai, W., Huang, H. and Xia, Y.: "H2NF-Net for brain tumor segmentation using multimodal MR imaging: 2nd place solution to BraTS challenge 2020 segmentation task," *Brainlesion: Glioma, Multiple Sclerosis, Stroke and Traumatic Brain Injuries: 6th International Workshop, BrainLes 2020,* Lima, Peru, October 4, 2020, pp. 58–68.

[104] Kohli, M.D., Summers, R.M. and Geis, J.R.: "Medical image data and datasets in the era of machine learning—whitepaper from the 2016 C-MIMI meeting dataset session," *J. Digit. Imaging*, 2017, **30**(4), pp. 392–399.

[105] Rezaei, M., Yang, H., and Meinel, C.: "Voxel-GAN: adversarial framework for learning imbalanced brain tumor segmentation," *Brainlesion: Glioma, Multiple Sclerosis, Stroke and Traumatic Brain Injuries: 4th International Workshop, BrainLes 2018*, Granada, Spain, September 16, 2018, pp. 321–333.

[106] Mok, T.C.W. and Chung, A.C.S.: "Learning data augmentation for brain tumor segmentation with coarse-to-fine generative adversarial networks," In *Brainlesion: Glioma, Multiple Sclerosis, Stroke and Traumatic Brain Injuries: 4th International Workshop, BrainLes 2018, Held in Conjunction with MICCAI 2018*, Granada, Spain, September 16, 2018, Revised Selected Papers, Part I 4, Springer, 2019, pp. 70–80.

[107] Sun, Y., Zhou, C., Fu, Y., and Xue, X.: "Parasitic GAN for semi-supervised brain tumor segmentation," *2019 IEEE International Conference on Image Processing (ICIP)*, Taipei, Taiwan, pp. 1535–1539.

[108] Li, Q., Yu, Z., Wang, Y. and Zheng, H.: "TumorGAN: a multi-modal data augmentation framework for brain tumor segmentation," *Sensors*, 2020, **20** (15), p. 4203.

[109] Sharma, A.K., Nandal, A., and Dhaka, A., "Brain tumor classification via UNET architecture of CNN technique," In *International Conference on Cyber Warfare, Security (SpacSec2021)*, Springer, 9–11 Dec 2021, Manipal University, Jaipur, pp. 18–33, Aug. 2022.

[110] Pravitasari, A.A., Iriawan, N., Almuhayar, M., *et al.*: "UNet-VGG16 with transfer learning for MRI-based brain tumor segmentation," *TELKOMNIKA*, 2020, **18**(3), p. 1310.

[111] Sharma, A.K., Nandal, A., Ganchev, T. and Dhaka, A., "Brain tumor classification using modified VGG model-based transfer learning approach," In *20th International Conference on Intelligent Software Methodologies, Tools, and Techniques (SOMET-2021)*, Accepted in June 2021, Conference date 21–23 Sept. 2021, Cancun, Mexico, pp. 538–550.

[112] Wacker, J., Ladeira, M. and Nascimento, J.E.V.: "Transfer learning for brain tumor segmentation," *Brainlesion: Glioma, Multiple Sclerosis, Stroke and Traumatic Brain Injuries: 6th International Workshop, BrainLes 2020*, Lima, Peru, October 2020, pp. 241–251.

[113] Arbane, M., Benlamri, R., Brik, Y., and Djerioui, M.: "Transfer learning for automatic brain tumor classification using MRI images," *2020 2nd International Workshop on Human-Centric Smart Environments for Health and Well-Being (IHSH)*, Boumerdes, Algeria, February 2021, pp. 210–214.

[114] Srinivas, C., Nandini Prasad, K.S., Zakariah, M., *et al.*: "Deep transfer learning approaches in performance analysis of brain tumor classification using MRI images," *J. Healthc. Eng.*, 2022, **2022**, pp. 1–17.

[115] Ullah, N., Khan, J.A., Khan, M.S., *et al.*: "An effective approach to detect and identify brain tumors using transfer learning," *Appl. Sci.*, 2022, **12**(11), p. 5645.

[116] Shen, L. and Anderson, T.: "Multimodal brain MRI tumor segmentation via convolutional neural networks," *Stanford Reports*, 2017, **18**, pp. 2014–2015.

[117] Zhou, T., Canu, S., Vera, P. and Ruan, S.: "Brain tumor segmentation with missing modalities via latent multi-source correlation representation," *Medical Image Computing and Computer Assisted Intervention–MICCAI 2020: 23rd International Conference*, Lima, Peru, October 2020, pp. 533–541.

[118] Ding, Y., Yu, X., and Yang, Y.: "RFNet: region-aware fusion network for incomplete multimodal brain tumor Segmentation," *Proceedings of the IEEE/CVF International Conference on Computer Vision*, Virtual Event, October 2021, pp. 3975–3984.

[119] Goodfellow, I.J., Pouget-Abadie, J., Mirza, M., *et al.*: "Generative adversarial networks," *Commun. ACM*, 2020, 63(11), pp. 139–144.

[120] Conte, G.M., Weston, A.D., Vogelsang, D.C., *et al.*: "Generative adversarial networks to synthesize missing T1 and FLAIR MRI sequences for use in a multisequence brain tumor segmentation model," *Radiology*, 2021, **299**(2), pp. 313–323.

[121] Yu, B., Zhou, L., Wang, L., Fripp, J., and Bourgeat, P.: "3D cGAN based cross-modality MR image synthesis for brain tumor segmentation," *2018 IEEE 15th International Symposium on Biomedical Imaging (ISBI 2018)*, Washington, DC, USA, April 2018, pp. 626–630.

[122] Wang, Y., Zhang, Y., Liu, Y., et al.: "ACN: adversarial co-training network for brain tumor segmentation with missing modalities," *Medical Image Computing and Computer Assisted Intervention–MICCAI 2021: 24th International Conference,* Strasbourg, France, September 2021, pp. 410–420.

[123] Zhou, T., Fu, H., Chen, G., Shen, J. and Shao, L.: "Hi-net: hybrid-fusion network for multi-modal MR image synthesis," *IEEE Trans. Med. Imaging*, 2020, 39(9), pp. 2772–2781..

[124] Zhan, B., Li, D., Wu, X., Zhou, J. and Wang, Y.: "Multi-modal MRI image synthesis via GAN with multi-scale gate mergence," *IEEE J. Biomed. Health Inform.*, 2022, **26**(1), pp. 17–26.

[125] Chartsias, A., Joyce, T., Giuffrida, M.V. and Tsaftaris, S.A.: "Multimodal MR synthesis via modality-invariant latent representation," *IEEE Trans. Med. Imaging*, 2018, **37**(3), pp. 803–814.

[126] Zhou, T., Canu, S., Vera, P. and Ruan, S.: "Feature-enhanced generation and multi-modality fusion based deep neural network for brain tumor segmentation with missing MR modalities," *Neurocomputing*, 2021, **466**, pp. 102–112.

[127] Vadacchino, S., Mehta, R., Sepahvand, N. M., et al.: "HAD-Net: a hierarchical adversarial knowledge distillation network for improved enhanced tumour segmentation without post-contrast images," *Proceedings of the Fourth Conference on Medical Imaging with Deep Learning*, Lubeck, Germany, July 2021, pp. 787–801.

[128] Azad, R., Khosravi, N. and Merhof, D.: "SMU-Net: style matching U-Net for brain tumor segmentation with missing modalities," *Proceedings of the*

5th International Conference on Medical Imaging with Deep Learning, Zurich, Switzerland, July 2022, pp. 48–62.
[129] Zhou, T., Canu, S., Vera, P. and Ruan, S.: "Latent correlation representation learning for brain tumor segmentation with missing MRI modalities," *IEEE Trans. Image Process.*, 2021, **30**, pp. 4263–4274.
[130] Yang, Q., Guo, X., Chen, Z., Woo, P.Y.M. and Yuan, Y.: "D2-Net: dual disentanglement network for brain tumor segmentation with missing modalities," *IEEE Trans. Med. Imaging*, 2022, **41**(10), pp. 2953–2964.
[131] Doshi-Velez, F. and Kim, B.: "Towards a rigorous science of interpretable machine learning," arXiv preprint arXiv:1702.08608, 2017.
[132] Zhou, B., Khosla, A., Lapedriza, A., Oliva, A., and Torralba, A.: "Learning deep features for discriminative localization," *2016 IEEE Conference on Computer Vision and Pattern Recognition (CVPR)*, Las Vegas, NV, USA, June 2016, pp. 2921–2929.
[133] Selvaraju, R.R., Cogswell, M., Das, A., Vedantam, R., Parikh, D. and Batra, D.: "Grad-CAM: visual explanations from deep networks via gradient-based localization," *Int. J. Comput. Vis.*, 2020, **128**(2), pp. 336–359.
[134] Natekar, P., Kori, A. and Krishnamurthi, G.: "Demystifying brain tumor segmentation networks: interpretability and uncertainty analysis," *Front. Comput. Neurosci.*, 2020, **14**, p. 6.
[135] Dasanayaka, S., Silva, S., Shantha, V., Meedeniya, D., and Ambegoda, T.: "Interpretable machine learning for brain tumor analysis using MRI," *2022 2nd International Conference on Advanced Research in Computing (ICARC)*, Belihuloya, Sri Lanka, February 2022, pp. 212–217.
[136] Esmaeili, M., Vettukattil, R., Banitalebi, H., Krogh, N.R. and Geitung, J.T.: "Explainable artificial intelligence for human-machine interaction in brain tumor localization," *JPM*, 2021, **11**(11), p. 1213.
[137] Zhao, M., Xin, J., Wang, Z., Wang, X. and Wang, Z.: "Interpretable model based on pyramid scene parsing features for brain tumor MRI image segmentation," *Comput. Math. Methods Med.*, 2022, **2022**, pp. 1–10.
[138] Saleem, H., Shahid, A.R. and Raza, B.: "Visual interpretability in 3D brain tumor segmentation network," *Comput. Biol. Med.*, 2021, **133**, p. 104410.
[139] Zeineldin, R.A., Karar, M.E., Elshaer, Z., *et al.*: "Explainability of deep neural networks for MRI analysis of brain tumors," *Int. J. CARS*, 2022, **17** (9), pp. 1673–1683.

Chapter 3

Optimization algorithms and regularization techniques using deep learning

M.K. Sharma[1], Tarun Kumar[1] and Sadhna Chaudhary[1]

Artificial intelligence is going to be the driving force in the near future, so it is necessary to know and spread this information. The two major components of artificial intelligence are machine learning and deep learning. In this book chapter, we studied how deep learning is used in optimization and regularization techniques. We have discussed the approaches of deep learning and also the types of deep neural networks in this chapter. Some basic optimization algorithms are discussed next. Regularization techniques and their types have also been explained. A comparative study of all the main deep learning approaches and all the types of deep neural networks has been explained in this chapter. At the end, the application of deep learning in today's era has also been discussed.

3.1 Introduction

The process of transferring human intelligence, knowledge, and information to machines is known as artificial intelligence (AI). The main goal of AI is to build autonomous machines capable of thinking and functioning like humans. These machines can perform tasks by studying problems and replicating human behavior. Most AI systems use natural intelligence as a model to address complex problems. Machine learning, a subfield of AI, enables a system to learn from experience and improve over time without being entirely programmed. Machine learning uses data to train and deliver exact results. Developing computer software that can access data and use it to educate itself is the aim of machine learning.

One of the most fascinating areas of research is deep learning, which has emerged in response to some of the limits of machine learning and the significant advancement in the theoretical and technological capabilities at our disposal today. It is used in a broad range of technologies, including text translation between languages, picture recognition on social media sites, and self-driving cars. Figure 3.1 illustrates the main distinctions between deep learning, machine learning, and AI.

[1]Department of Mathematics, Chaudhary Charan Singh University, India

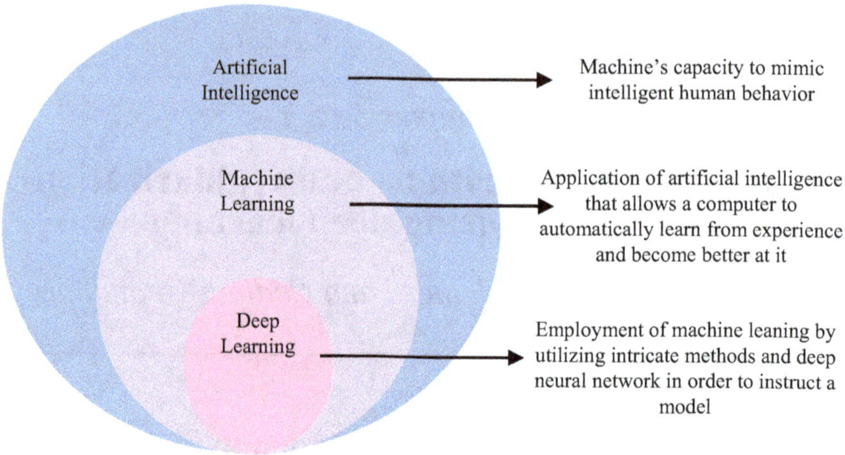

Figure 3.1 Artificial intelligence and its subfields

It is the only subset of machine learning that prioritizes teachingthe computer about human primal instincts. A computer algorithm learns to carry out classification tasks on complicated data in the form of images, text, or sound using deep learning. These algorithms are capable of achieving state-of-the-art accuracy and occasionally even outperforming human performance, and are developed by utilizing multi-layered neural network designs and a vast amount of labeled data. It is a key component of the technology that powers systems like driverless cars, virtual assistants, and facial recognition. It uses training data and experience-based learning to work. Since the neural networks quickly uncover new levels of data with each passing minute, the learning process is known as "Deep." Every time data is trained, the emphasis is on enhance efficiency. The training performance and deep learning capabilities have substantially improved with the rising depth of the data since it has been widely accepted by data professionals.

Deep learning models frequently take their influences from a variety of fields of study, including neuroscience and game theory, and many of them closely resemble the fundamental organization of the human nervous system. Many experts believe that as the area develops, a time will come when software won't need to be nearly as rigidly hard-coded as it does now, enabling more comprehensive, all-encompassing solutions to issues. Despite the fact that it started in a machine learning-related field, where the major goal was to satisfy constraints of varying complexity, deep learning now encompasses a broader definition of algorithms that can comprehend various levels of data representation that correlate to various complexity hierarchies. In other words, in addition to their capacity for prediction and classification, the algorithms are also capable of learning different levels of complexity. This is demonstrated in picture recognition, when a neural network progresses from detecting eyelashes to faces, people, and other objects. We can achieve the level of complexity required to develop intelligent software, which is the power in this. We may see this right now in functions like autocorrect, which simulates proposed adjustments to observed speech patterns that are unique to each person's vocabulary. The efficiency of conventional machine learning algorithms and deep learning techniques is shown in Figure 3.2. Traditional machine learning algorithms' performance stabilizes once they reach a certain amount of training data, whereas the performance of deep learning algorithms improves as the amount of data increases.

Deep learning models frequently feature layers of neurons, which are nonlinear units used to process data. Each layer in these models processes data at a different degree of abstraction. In Figure 3.3, the layers of deep neural networks (DNNs) are depicted.

DNNs are characterized by their abundance of hidden layers, because their inputs and outputs are not always clear to us beyond the fact that they are the result of the layer preceding them. An individual architecture can be distinguished from another by the inclusion of layers, and the functions found inside the neurons of these layers determine the many use cases for a particular model.

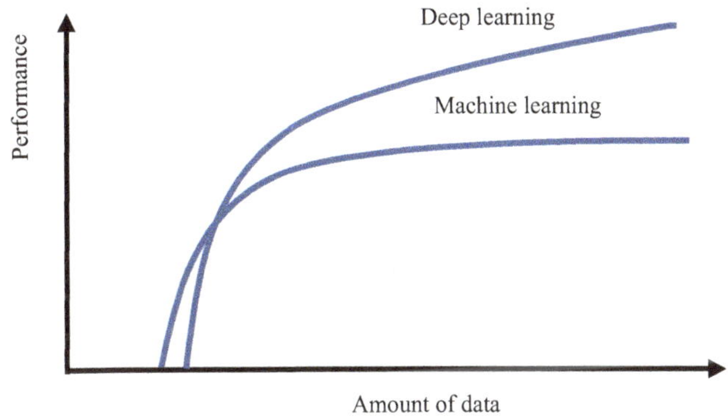

Figure 3.2 Performance of machine learning and deep learning

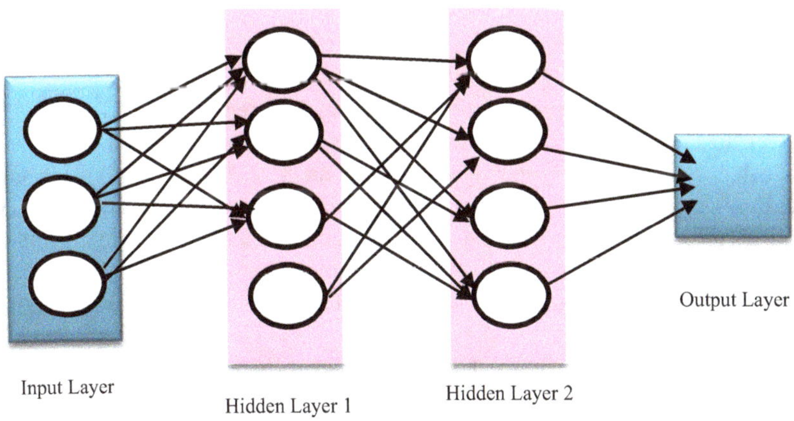

Figure 3.3 Deep neural network

3.2 Deep learning approaches

The three primary types of deep learning methods are unsupervised, partially supervised (semi-supervised), and supervised.

3.2.1 Deep supervised learning

In supervised learning, an algorithm is used to train the mapping function $Y = F(X)$, which converts variables represented as X in the input to variables represented as Y in the output. To predict the output Y for a new input X, the learning algorithm aims to roughly describe the mapping function. The learning algorithm's goal is to roughly represent the mapping function to forecast the output Y for a fresh input X. It deals with labeled data. Several supervised learning approaches are available for DL, including recurrent neural networks (RNNs), convolutional neural networks (CNNs), and DNNs. The main benefit of this method is the capacity to gather data or produce data output utilizing prior knowledge. when a class that belongs to the training set should have samples in it; this technique has the drawback that the decision boundary may get overstretched. Generally, this strategy is easier than others for learning with great performance.

3.2.2 Deep semi-supervised learning

This technique's learning is based on semi-labeled datasets. Furthermore, RNNs, which include long- and short-term memory (LSTM) and gated recurrent units (GRU) techniques, are used for partially supervised learning. One benefit of this approach is that it necessitates less tagged data. On the other hand, one downside of this technique is that irrelevant input features contained in training data may result in inaccurate judgments. Text document classifier is among the most well-known uses of semi-supervised learning. Due to the difficulties in obtaining a large number of labeled text documents, semi-supervised learning is suitable for text document categorization problems.

3.2.3 Deep unsupervised learning

The agent learns the essential traits or fundamental representations required to uncover the underlying structure or relationships in the input data during this phase. This category includes unsupervised learning methods including dimensionality reduction, grouping (clustering), and generative networks. The most recently created algorithms in the deep learning family, such as restricted Boltzmann machines, auto-encoders, and generative adversarial networks (GANs), have all demonstrated strong performance on dimensionality reduction and clustering in nonlinear fashion issues (GANs). RNNs, which include GRUs and LSTM techniques, have also been applied in a number of applications for unsupervised learning. Unsupervised learning's main shortcomings are its computing requirements and its inability to offer precise data-sorting information. Clustering is one of the most well-liked unsupervised learning methods.

3.2.4 Deep reinforcement learning

While supervised learning uses sample data that has already been provided, reinforcement learning works by engaging with the environment. It is far more difficult to achieve this learning than it is with conventional supervised methods since the reinforcement learning approach lacks a straightforward loss function. In reinforcement learning, complete access to the function is not possible because it requires optimization; therefore, it must be obtained through interaction Additionally, the environment that the state being interacted with is dependent on, and the input is generated from, prior actions.

For example, reinforcement learning strategies can be used to treat patients in the healthcare sector. Without having previously studied the mathematical representation of biological systems, using previous data, reinforcement learning can find the best policies. This justifies that reinforcement learning is more suitable than other control-based healthcare systems such as dynamic treatment regimens (DTRs), automated medical diagnostics, and other generic domains. DTRs are based on the clinical observations and assessments of a patient. The outcomes are the treatments that are available at each step. They are similar to Reinforcement Learning (RL) states. Because it can determine the best course of treatment for a patient at a specific time, the application of RL in DTRs is beneficial. The use of RL in healthcare also encourages the improvement of long-term results by taking into consideration the delayed effects of treatments. Reinforcement learning has also been used to find and develop the best DTRs for chronic illnesses. Comparative analysis of deep learning approaches is given in Table 3.1.

Table 3.1 Deep learning approaches

Deep learning approaches	**Merit**	**Demerits**	**Applications**
Deep supervised learning	Create a data output using the preceding knowledge.	Overstretching the decision boundary.	Hyperspectral data classification. Intrusion detection in the internet of things.
Deep semi-supervised learning	Reduce the quantity of labeled data required.	Training data with irrelevant input features could lead to inaccurate decisions.	Text document classification. Acoustic impedance inversion.
Deep unsupervised learning	It has the potential to unlock previously unsolvable problems.	Unable to accurately deliver data sorting and computational complexity information.	Clustering Ear recognition Face recognition.
Deep reinforcement learning	It permits us to determine the most effective strategy for achieving significant rewards. It gives the learning agent a reward function.	The speed of learning may be impacted by variables. When the workspace is large, it requires a lot of computing and takes time.	Robotics for industrial automation Communications and networking. Traffic signal control. Stock trading strategies and stock forecasting.

3.3 Deep neural network

The main types of DNNs are recursive neural networks (RvNNs), RNNs, and CNNs.

3.3.1 Recursive neural network

RvNN can make hierarchical predictions and classify outputs using compositional vectors. RvNN was inspired mostly by recursive auto-associative memory (RAAM). The RvNN architecture was designed to handle objects with randomly formed structures such as graphs or trees. RvNN creates a syntactic tree and computes a likely pair of scores for merging. Additionally, RvNN determines a score for each pair of units' merge plausibility. The pair that received the highest score is then combined into a composition vector. RvNN produces a larger area with many units after each merge, a compositional vector of the area, and a label for the class. The root of the RvNN tree structure represents the compositional vector for the entire region. In Figure 3.4, an example RvNN tree is exhibited.

3.3.2 Recurrent neural network

In the field of deep learning, RNNs are popular and frequently applied approach. Speech processing and natural language processing applications are where RNN is most frequently employed. Sequential data is used in the network by RNN, as opposed to traditional networks. This attribute is crucial for a number of applications because the inherent structure in the data sequence offers useful information. For instance, one needs to understand the statement's context for the purpose of interpreting a particular word in a sentence. The RNN, on the other hand, can be conceptualized as a unit of short-term memory, with x denoting the input layer, y the output layer, and s the state (hidden) layer.

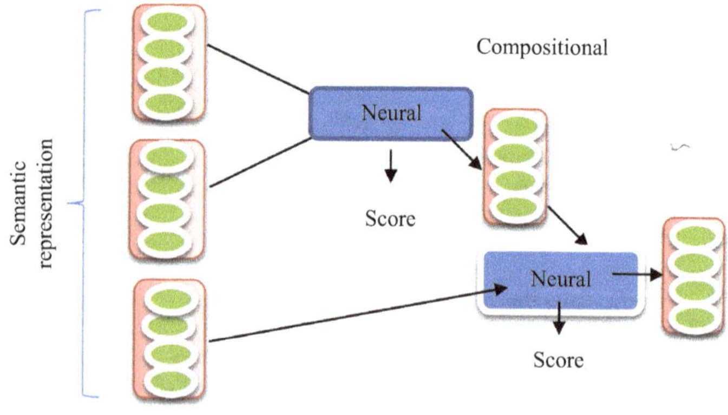

Figure 3.4 An example of RvNN tree

3.3.3 Convolutional neural network

An ANN family known as CNNs, or ConvNets is used most frequently in deep learning to interpret visual data. The primary characteristic of CNN over its predecessors is that it does it automatically, without the need for human participation, in order to identify important elements. Face recognition, audio processing, computer vision, and other areas have all seen extensive use of CNNs. Like conventional neural networks, CNNs were influenced by the neurons in animal and human brains. More specifically, the CNN mimics the complex network of cells that make up the visual cortex in a cat's brain.

A popular CNN form has several convolutional layers, followed by fully connected layers and subsampling (pooling) layers, similar to the multi-layer perceptron (MLP). Figure 3.5 displays a CNN architecture example for image classification.

3.3.3.1 NVIDIA self-driving car

The whole data center hardware, software, and workflows required to create autonomous driving technologies are covered by NVIDIA DRIVE® Infrastructure, from raw data gathering to validation. It offers the whole set of building pieces needed for the creation, validation, replay, and testing in the simulation of neural networks. For its self-driving car, Nvidia also uses a convolution neural network as the main algorithm. Parking lots and other inroads without lane markings can be operated by the network.

To recognize useful road features, it can also learn the features and the necessary representations. This end-to-end system simultaneously optimizes all processing steps in contrast to the problem's clear dissection, including path planning, lane marking detection, and control. Instead of focusing on human-selected intermediate criteria like lane detection, internal components self-optimize to maximize overall system performance. Such criteria are logically chosen for simplicity of human interpretation, but this does not inevitably result in the best system performance. Smaller networks are achievable because the system learns to solve the problem in the fewest number of processing steps possible. Three cameras are used, two on the sides and one in the front. A CNN framework of a NVIDIA self-driven car using CNN is depicted in Figure 3.6.

A comparison of several neural network types with their merits, demerits, and applications is given in Table 3.2.

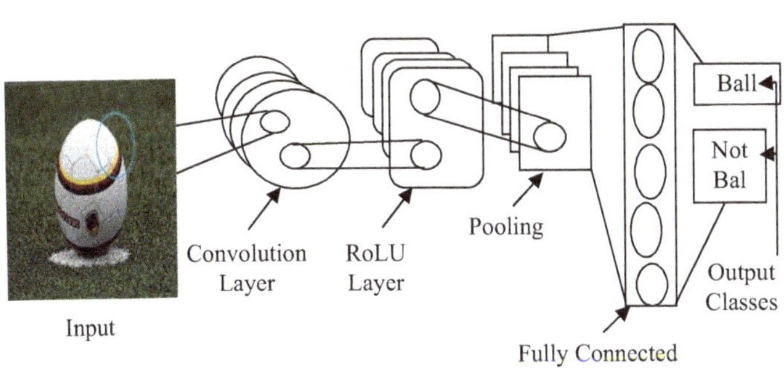

Figure 3.5 An illustration of the CNN system used for image classification

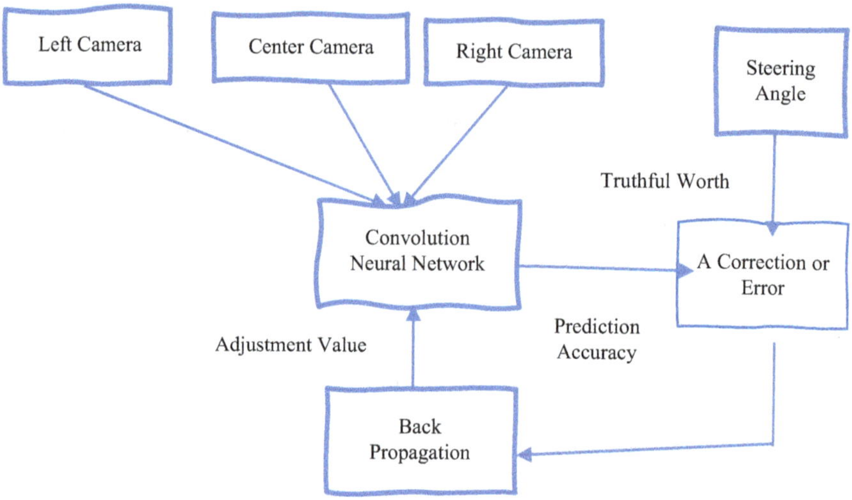

Figure 3.6 A CNN framework of a NVIDIA self-driven car

Table 3.2 A comparative study of deep neural networks

Deep neural network	Merits	Demerits	Applications
Recursive neural network	If it comes to learning hierarchical, tree-like structures, they can be quite effective	Every input sample's tree structure must be known at training time	Natural language processing
Recurrent neural network (RNN)	An RNN maintains all knowledge components throughout time It is only useful for time series prediction when it is able to remember previous inputs RNNs and convolutional layers are even coupled to improve the effective pixel neighborhood	Problems with gradient vanishing and explosions An RNN is extremely challenging to train The activation function cannot process very long sequences if tanh or relu is utilized	Text-to-speech conversions Speech recognition Machine translation Time-series forecasting Language modeling and generating text
Convolutional neural network (CNN)	Reduce computation in comparison to a conventional neural network Convolution grossly oversimplifies computing without sacrificing the significance of the data They are excellent at image classification. They apply the same information to all image places	CNN does not encode the object's position or orientation The inability of the incoming data to be spatial Lots of practice data are needed for invariant	Natural language processing, text digitization, and facial recognition

3.4 Optimization algorithms

Optimization algorithms are in charge of minimizing losses and producing the most accurate outcomes possible. Optimization algorithms serve as the foundation for a machine's ability to learn via experience. They use gradients to try to minimize the loss function. Learning can be applied in a variety of ways using various optimization techniques. The best results are obtained by employing some optimization strategies or algorithms known as optimizer. The algorithms investigated in this chapter are described below.

3.4.1 Gradient descent

An iterative technique known as gradient descent (GD) starts at a random position on the function and descends gradually until it reaches the function's lowest level. When it is difficult to locate optimal locations by equating the function's slope to 0, this method can be helpful. To get the function to the minimal value, the weights must be adjusted. Back propagation is used to transfer loss from one layer to the next, and to lessen loss, the "weights" parameter is also modified.

In the case of the GD algorithm, the complete dataset is loaded at once. As a result, it is computationally demanding. Another disadvantage is that the iteration values may become trapped at local minima or saddle points and never converge to minima. To find the best solution, global minima must be reached.

3.4.2 Stochastic gradient descent

Stochastic gradient descent (SGD) is an expansion of GD that addresses some of the algorithm's shortcomings. It also takes into consideration the data elements on the go and train them. By computing the derivative of one point at a time, SGD attempts to address the problem of being computationally costly. As a result, SGD requires more iterations than GD to attain the minimum and contains some noise as compared to GD.

3.4.3 Mini-batch-stochastic gradient descent

Mini-batch SGD spans the gap between the two prior approaches, GD and SGD. It randomly selects training samples from the complete dataset (the so-called mini-batch) and computes gradients solely from them. It seeks to approximate batch GD by sampling only a part of the data. This method is more efficient and trustworthy than earlier GD variants. The technique is more effective since the method uses batching, which prevents all of the training data from being held in memory. In addition, while the cost function in SGD is smoother than the cost function in mini-batch GD, the latter is noisier. Thus, mini-batch GD is the best alternative since it offers a reasonable balance between speed and accuracy.

3.4.4 Momentum

This is an adaptive optimization technique that accelerates optimization and stabilizes convergence by using weighted average gradients from prior iterations. This is accomplished by multiplying the previous iteration values by a fraction (gamma). When the gradient points are all pointing in the same direction, the momentum term increases and decreases. As a result, the loss function value converges faster than planned.

$$w_{t+1} = w_t - \alpha m_t$$

where $m_t = \beta m_{t-1} + (1-\beta)\left[\frac{\delta L}{\delta w_t}\right]$

m_t is the aggregate of gradients at time t, w_t at time t, α is learning rate at time t, δL is derivative of loss function, δw_t is derivative of weights at t, and β is the moving average parameter.

3.4.5 Nesterov momentum

Nesterov momentum, an alternative form of momentum, computes the update direction in a slightly different manner. We compute the gradient vector based on what it would have been if we had simply moved at our build-up velocity. This anticipatory update keeps us from moving too quickly and makes us more responsive. The Nesterov momentum update can be visualized in Figure 3.7.

The Nesterov accelerated gradient (NAG) technique is the most well-known method for using Nesterov momentum.

3.4.6 Adapted gradient (AdaGrad)

Adaptive gradient embraces based on its position throughout each cycle; the learning rate of parameters is increased, i.e., by using faster learning rates when features occur infrequently and slower learning rates when features do presently. It operates by dividing the learning rate parameter by gamma, which is the sum of all gradients squared. AdaGrad modifies each step's global learning rate and N is in the update rule depending on previous computations. The accumulation of squared gradients in the denominator is one of the most serious drawbacks. Because each new term is positive, the accumulative sum grows during the training. As a result, the learning rate decreases and eventually becomes negligible. This approach converges faster and is less sensitive to master step size.

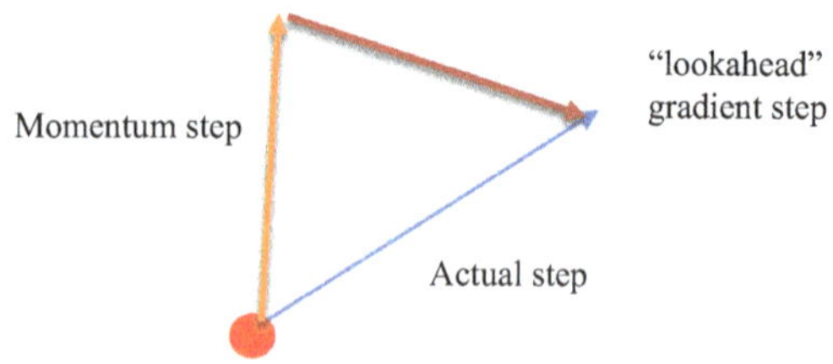

Figure 3.7 Nesterov momentum update

3.4.7 Adapted delta (AdaDelta)

It is simply an AdaGrad enhancement which slows down the learning rates monotonically. AdaDelta sets a limit on the number of summation values rather than summing all earlier gradients (w). The "Decaying Average of all Past Squared Gradients" is how AdaDelta defines the sum of past gradients (w). The prior average and the current gradient are the only factors that affect the current average at each iteration.

3.4.8 Root mean square propagation

Another adaptive learning rate technique that seeks to improve AdaGrad is called root mean square propagation. It uses the exponential moving average rather than AdaGrad's cumulative sum of squared gradients. The first step is the same for RMSProp and AdaGrad. By using RMSProp, the learning rate is simply divided by an average with exponential decay.

$$w_{t+1} = w_t - \frac{\alpha_t}{(v_t + \varepsilon)^{1/2}} * \left[\frac{\delta L}{\delta w_t}\right]$$

where $v_t = \beta v_{t-1} + (1 - \beta)\left[\frac{\delta L}{\delta w_t}\right]^2$

w_t is weight at time t, v_t is the sum of past gradients, α is learning rate at time t, δL is derivative of loss function, δw_t is derivative of weights at t, β is the moving average parameter, and ε is a small positive constant.

3.4.9 Adaptive moment estimation (Adam)

A learning technique called Adam was developed specifically for DNN training. Adam's advantages include better memory efficiency and lower computing requirements. Adam's method is to compute adaptive LR for each model parameter. It combines Momentum's and RMSprop's benefits. It uses the squared gradients to scale the learning rate as RMSprop and uses the moving average of the gradient to approximate momentum. Adam's equation is depicted as follows:

$$w_{t+1} = w_t - \widehat{m}_t \frac{\alpha_t}{\sqrt{\widehat{v}_t} + \varepsilon}$$

3.4.10 Nesterov-accelerated adaptive moment (Nadam)

The Adam algorithm has been somewhat modified to include Nesterov momentum in the Nadam (Nesterov-accelerated adaptive moment estimation) algorithm. Nadam often solves problems with very noisy gradients or slopes with significant curvature. It also normally provides a little bit less training time.

3.4.11 AdaBelief

AdaBelief is a new optimization algorithm. The only additional parameters in the optimizer AdaBelief, which is a descendant of Adam, are changes to one of the existing parameters. A decent generalization rate and quick convergence time are both provided by this optimizer. In accordance with its "belief" in the present gradient direction, it changes the step size.

3.5 Regularizations techniques

In a machine learning model, when training samples along with corresponding labels are provided, the model will begin to identify patterns in the data and adjust the model parameters accordingly; this process is known as training. The labels or outputs on a different set of samples that have not yet been observed by the model or the testing dataset are then predicted using these parameters or weights and this procedure is known as inference.

If the model is able to perform well on the testing dataset, the model can be said to have generalized well, i.e., correctly understood the patterns provided in the training dataset. This type of model is called a correct fit model. However, if the model performs really well on the training data and does not perform well on the testing data, it can be concluded that the model has memorized the patterns of training data but is not able to generalize well on unseen data. This model is called an overfit model.

When a model is able to execute well on the testing dataset, then this type of model is known as correct fit model since it accurately comprehended the patterns supplied in the training dataset. But if the model performs exceptionally well on the training data but poorly on the testing data, it might be inferred that the model has memorized the patterns of the training data but is unable to generalize effectively on unknown data; this type of model is known as types an overfit model.

The phenomenon known as overfitting occurs when a machine learning model learns patterns and performs well on data that it has been trained on but inadequately on data that has not yet been seen. Figure 3.8 demonstrates that the training error decreases when the model is trained for a longer period of time. However, at a certain point, the testing error starts to rise. This shows that overfitting has already begun with the model.

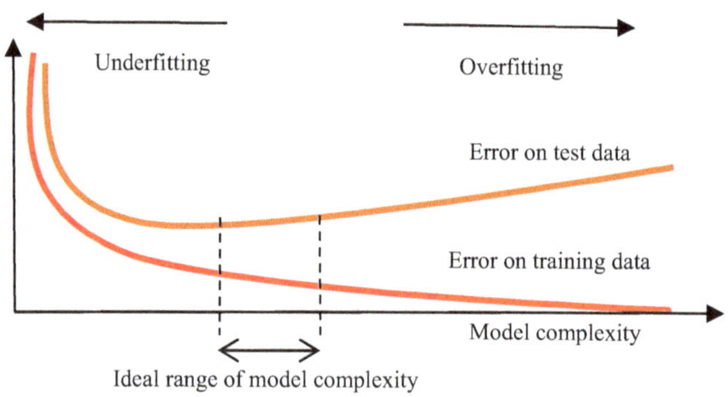

Figure 3.8 Relation between predictive error and complexity of model

Overfitting occurs when a machine learning model learns patterns and performs well on data that it has been trained on but inadequately on data that has not yet been seen. It can be handled through the following techniques:

1. Training on more training data to better identify the patterns.
2. Data augmentation for better model generalization.
3. Early stopping, i.e., stop the model training when the validation metrics start decreasing or loss starts increasing.
4. Regularization techniques.

These methods strive to enhance performance by changing the input data rather than changing the model architecture through data augmentation and more training data. Instead of explicitly addressing the issue of overfitting, early stopping is used to stop the model's training at the proper point, or before the model overfits. To prevent overfitting, regularization is a more reliable method that can be utilized.

A well-known and crucial machine learning technique called regularization reduces the complexities of trained models while enabling them to make accurate predictions over a collection of test data. Due to their highly nonlinear characteristics, which are increased by the addition of only one more layer, this strategy is even more important within the framework of deep learning and causes models to perform poorly when generalizing to new data. Five broad categories of deep learning regularization strategies might be identified.

The first category is data-based regularization; it seeks to either use data augmentation techniques to generate a huge number of data points or to simplify the representation of input data by carrying out specific changes. The focus of the second category is network modification architecture, including choosing an appropriate activation function and reducing the number of nodes and layers. The third type of this taxonomy is regularization via the error function, which attempts to add properties to the resilience of the error function to imbalanced data. The fourth category is based on updating the network learning optimization technique. This category includes methods such as termination, dropout, momentum, and weight initialization. The inclusion of a regularization term in the network loss function determines the final category. In this category, it is believed that the regularization term and the targets are independent of one another and do not depend on one another. This group frequently appears in the forms of weight decay and l_2 norm.

3.5.1 l_2 Regularization

l_2 Regularization is sometimes called weight decay. This method, sometimes referred to as ridge regression or Tikhonov regularization, works by including a l_2 term in the function that must be minimized, in this case, $\Omega(W) = \frac{1}{2}\|W\|_2^2$. The weights are forced to exist in a sphere whose radius is inversely proportional to the regularization parameter λ by this extra term in the l_2 norm. The updating rule utilizing the GD approach changes in this scenario:

$$w_{ij}^l \rightarrow \left(1 - \frac{\alpha\lambda}{n}\right) w_{ij}^l - \frac{\alpha}{n} \sum_{i=1}^{n} \frac{\partial C_{x_i}}{\partial w_{ij}^l}$$

This indicates that the weights are multiplied by a number somewhat less than 1 after each iteration. The model is frequently compelled to favor light weights. l_2 regularization when required, try to penalize the large weights by reducing them after each updating step so that their reduction is proportionate to w_{ij}^l. As seen in Figure 3.9, the l_2 a circle defined by regulation with a radius that is inversely proportional to its regularization value.

3.5.2 l_1 Regularization

The loss function is changed in l_1 regularization approaches by the addition of l_1 a term, i.e., $\Omega(W) = \sum_{w \in W} |w|$. The fundamental idea behind this strategy is to regularize the loss function by removing the incorrect characteristics from the training model. The modified version of rule is written as:

$$w_{ij}^l \rightarrow w_{ij}^l - \frac{\alpha\lambda}{n}\mathrm{sgn}(w_{ij}^l) - \frac{\alpha}{n}\sum_{i=1}^{n}\frac{\partial C_{x_i}}{\partial w_{ij}^l}$$

where $\mathrm{sgn}\left(w_{ij}^l\right)$ is the sign of w_{ij}^l. When necessary, it also attempts to penalize heavy weights by reducing them after each update step such that the weights gradually go in the direction of zero. As seen in Figure 3.10, the l_1 norm creates a

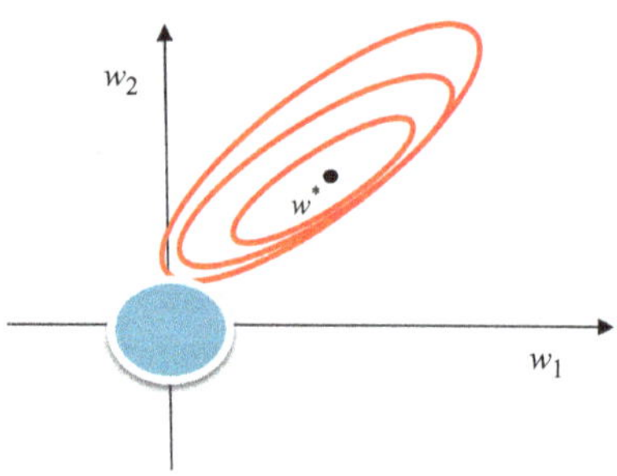

Figure 3.9 l_2 *Norm.*

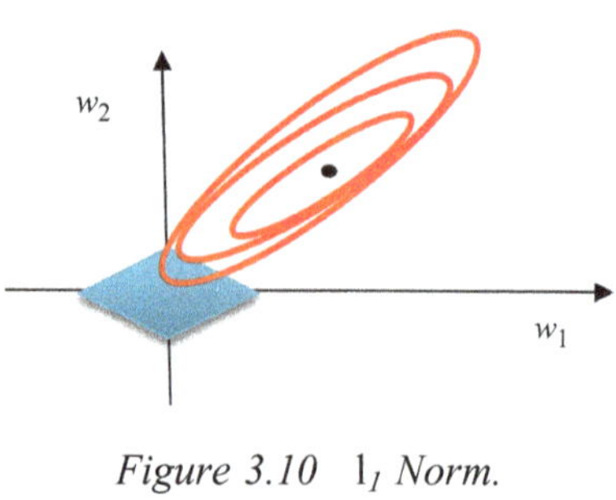

Figure 3.10 l_1 *Norm.*

parameter space in two dimensions that is constrained by a parallelogram at the origin.

In this scenario, the parallelogram's vertices are more likely to be hit by the loss function than its edges. The l_1 regularization removes some of the parameters, making it possible to employ the l_1 methodology as a feature selection method.

3.5.3 Entropy regularization

Entropy regularization is the method of the norm penalty that functions with probabilistic models. Additionally, it has been used into a number of reinforcement learning techniques such as A3C and policy optimization techniques. Similar to prior strategies, the loss function in this method is altered by the addition of a penalty term. When $p(x)$ is the probability distribution of output, the penalty term is

$$\Omega(X) = -\sum p(x)\log(p(x))$$

3.5.4 Dropout technique

Dropout technique, used to train neural networks, regularizes learning by eliminating some hidden units with a predetermined frequency. The neural network would be changed in the same way if certain hidden activation functions were set to zero. We may properly define the neural network using Dropout as follows:

$$\tilde{a}_j^l = \delta_j^l a_j^l,$$

where δ_j^l represent Bernouli random variable of parameter p^l. The output activation a_j^l at each neuron j in a hidden layer l is multiplied by a sampled variable δ_j^l to obtain thinning output activations. The same procedure is followed at each layer, with these thinned functions serving as inputs to the following layer. This application is comparable to taking a sample from a smaller network of neural networks.

3.6 Review of literature

AI includes a field called machine learning, where algorithms and methods are used to train computers keeping a purpose in mind. In 1959, **Samuel** invented the term "Machine Learning" [1]. Machine learning was used in electrocardiogram and speech patterns during the 1960s. In these, the reinforcement learning part of machine learning was applied. Deep learning is an integral part of machine learning, in which layers of algorithms are used to process data. The data from the first layer goes to the second layer, and after being optimized, it goes on from the second to the third layer; this sequence continues until we get the optimum solution to a problem. This whole body of deep learning is called neural network. Neural network systems and algorithms for deep learning have been discovered and refined over time.

In 1965, **Lvakhnenko and Lapa** developed deep learning algorithms [2]. Within these models, the polynomial equations were used for the activation function. **Kabrisky** in 1966 gave a model for processing visual information in the human brain. In a RNN, Grossberg investigated contour enhancement, short-term memory, and constancy in 1973 [3]. He examined a model of the nonlinear dynamics of networks of nerve cells or cell populations that reverberate on-center and off-surround. Fukushima used convolution neural network in 1975 [4]. They built neural networks using multiple pooling and convolution layers. In 1979, Fukushima created its own artificial neural network [5]. Hierarchical and multi-layer design was used to develop it and it was named "Neocognitron." The advantage of making this design was that the computer learned to recognize visual patterns. In 1979, Vapnik studied support vector machines under the theory of pattern recognition [6]. The action done under some conditions for making the best use of any situation or resource is called optimization. In deep learning too, we reach the best optimal solution by finding layer by layer the best optimal values. Over time, we have come across different optimization algorithms for deep learning optimization.

In 1991, Bottou studied the SGD algorithm [7]. In this study, he discussed certain cost and differentiability at certain points. Also, he developed two convergence theorems and proved these theorems. Gradient-based learning was utilized for document identification by Lecun *et al.* in 1998 [8]. Document analysis researchers can employ a list of practical recommended practices that Simard *et al.* published in 2003 to get successful outcomes when using neural networks [9]. They add a new type of distorted data to the training set to increase its size. In 2006, Hinton *et al.* studied fast learning algorithms for deep belief nets [10]. In highly linked belief nets with numerous hidden layers, they demonstrated how to employ "complementary priors" to remove the effects of explanation that lead to inference challenging. In 2008, a theoretical framework was developed by Bottou and Bousquet to approximate the optimization on learning algorithms [11]. By this analysis, different tradeoffs of small- and large-scale learning problems were observed. In 2009, many supervised learning models were trained using innovative online and batch techniques that were developed by Do *et al.* [12].

The conditions under consideration for developed models are the ideal regularization choice parameters resulting in functions with low curvature. In 2010, Zinkevich *et al.* presented the first parallel SGD algorithm [13]. This method does not have too strict latency requirements, which would only be possible in a multi-core environment, and it is guaranteed to accelerate in parallel. Snoek *et al.* presented a study on machine learning methods with realistic Bayesian optimization in 2012 [14]. In this problem, learning algorithm's generalization performance is based on the Gaussian process. They described new algorithms that consider utilizing multiple cores for concurrent testing and the variable cost of learning algorithm experiments. In 2013, to enhance the effectiveness of the artificial neural network, Roy *et al.* used particle swarm optimization and studied neural networks in IRIS plant classification [15].

In 2014, using meta-optimization methods, Camilleri *et al.* created an algorithm for parameter selection in machine learning [16]. In this work, to solve a classification problem, they compared grid search and SGA method, considering them as meta-optimizing systems, and with the help of this, they found the optimal parameter sets for ID3 learner. Young *et al.* optimized hyperparameters of deep learning by an evolutionary algorithm in 2015 [17]. In this, he gave a method called multi-node evolutionary neural network for deep learning for automatic network selection in computational clusters. In 2016, Shen developed a fast optimization method for general binary code learning [18]. They also dubbed the discrete proximal linearized minimization method to handle discrete constraints in the learning process. In 2017, Shin *et al.* improved the fixed-point optimization algorithm for the DNN [19]. The development of this algorithm helped to dynamically estimate the quantization step size during training.

In 2017, the internal parameters of processing layers are evolved using a modified version of particle swarm optimization by Khalifa *et al.* [20]. They used the seven-layer convolution neural network for handwriting digit classification. In 2019, Hoseini *et al.* developed an adapt-ahead optimization algorithm for deep learning convolution neural network and applied it to MRI segmentation [21]. They used multi-modality MR images from the BRATS 2015 and BRATS 2016 datasets were used to test the suggested optimization technique and resilient design for handling the enormous volume of data. In 2020, an optimization technique for machine learning based on ADAM (adaptive moment estimation) was presented by Yi *et al.* [22]. He constructed an optimization method for the non-convex cost function. Also, they proved the suggested approach's numerical advantage over GD and the proposed method's suggested sequences' convergence (GD, ADAM, and AdaMax). A dynamic butterfly optimization technique for feature selection was created by Tubishat *et al.* in 2020 [23].

Hussain *et al.* created an object recognition system in 2021 employing intelligent deep learning and a better whale optimization technique [24]. Convolution neural network (DenseNet201) was taken into consideration and adapted for the chosen dataset (Caltech101). The whale optimization algorithm was applied to the DNNs by Brodzicki *et al.* in 2021 [25]. Also, a comparison of the suggested approach to other widely used algorithms, such as grid and random search techniques, has been given. In 2021, Yao *et al.* presented an adaptive second-order optimizer for machine learning named "ADAHESSIAN" [26]. It is a new stochastic optimization technique with various significant performance-improving features that directly incorporates roughly curvature information from the loss function. Mohammed *et al.* applied the best deep-learning model for diagnosing COVID-19 in 2022 using the crow swarm optimization technique and selection method [27]. They used two datasets: the first contains total of 746 CT images, 349 of which are instances of COVID-19 with proof and the remaining 397 are persons in good health; the second is made up of lung computed tomography scans that have not been enhanced from 632 COVID-19 positive cases, and it is analyzed using 15 deep learning models that have already been trained, while incorporating nine evaluation metrics.

3.7 Deep learning-based neuro fuzzy system and its applicability in self-driven cars in hill stations

Self-driving cars are autonomous decision-making systems. Cameras, LiDAR, RADAR, GPS, and inertia sensors are just a few of the sensors that they can handle.

After that, these data are modeled using deep learning techniques, and as a result, decisions are made that are appropriate for the driving situation. One of the key technologies that made self-driving possible is deep learning. We are applying our approach over self-driving car in the hill station areas in order to develop a successful and commercial automated car. For this purpose of study, we develop a deep learning technique-based neuro fuzzy system. In this segment, we are adopting a framework of five input and one output (which describe the speed limit as low (0–0.3), moderate (0.28–0.6) and high (0.55–1)) based fuzzy inference system. Various layers of proposed deep neuro fuzzy system are as follows:

Layer 1:

Let us consider five input factors, namely, altitude of the road, road bend, traffic position, weather condition, and road condition.

Layer 2:

The ranges of triangular membership function of each input factors are as follows:

Input factors	Linguistic ranges		
Altitude	Low [0 – 20°]	Moderate [18° – 40°]	High (above 40°)
Road bend	Low [0 – 50°]	Moderate [50° – 90°]	High (above 88°)
Traffic condition	Low [0 – 0.3]	Moderate [0.25 – 0.6]	High [0.58 – 1]
Weather condition	Good [0 – 0.3]	Moderate [0.25 – 0.6]	Bad [0.58 – 1]
Road condition	Good [0 – 0.3]	Moderate [0.25 – 0.6]	Bad [0.58 – 1]

Layer 3:

Out of the various rules, ten fuzzy rules for the proposed system are given as follows:

Fuzzy rules	Antecedent part					Consequent part
	Altitude I_1	Road bend I_2	Traffic I_3	Weather I_4	Road condition I_5	
R_1	Low	Low	Low	Good	Good	y_1
R_2	Low	Moderate	Moderate	Moderate	Good	y_2
R_3	High	High	High	Moderate	Good	y_3
R_4	Low	Low	Moderate	Good	Moderate	y_4
R_5	Low	Moderate	Low	Moderate	Good	y_5
R_6	Low	Low	Low	Good	Good	y_6
R_7	Low	Low	Low	Good	Good	y_7
R_8	High	Low	High	Bad	Bad	y_8
R_9	Low	Moderate	Low	Bad	Good	y_9
R_{10}	Moderate	High	Moderate	Bad	Moderate	y_{10}

where $y_i = \sum_{i=1}^{5} w_i I_i$ and w_i represents the weight associated with layer 3. For the optimization of the weights, we may apply the genetic algorithm, teaching learning-based optimization, etc.

Layer 4:
Now we normalize the value obtained (y_i) in layer 3 by the following formula:

$$\bar{y}_i = \frac{y_i}{\sum_{i=1}^{5} y_i}$$

Layer 5:
The aggregated output of the proposed system is given by

$$y = \frac{\sum_{i=1}^{5} \bar{y}_i}{5}$$

Now, let us take some input values of the applicability of proposed model as altitude of the road is 15°, road bend is 45°, traffic position is 0.22, weather condition is 0.2, and road condition is 0.1. The corresponding membership values are 0.5, 0.2, 0.32, 0.67, and 0.67, respectively. Corresponding to these input values, the output value for the fuzzy fired rule is given by

$$y = \sum_{i=1}^{5} w_i I_i = 0.2(0.5 + 0.2 + 0.32 + 0.67 + 0.67) = 0.472$$

where 0.472 represents the moderate output category for the speed limit of vehicle (due to the complexity of computation, we are taking each weight with value 0.2).

3.8 Conclusion

The subclass of machine learning called "deep learning" is based on improving and learning on its own by analyzing computer algorithms. This research focuses on artificial neural networks and related machine learning algorithms. The deep learning model includes an activation function, input, output, hidden layer, loss function, and other elements. Every deep learning model employs a formula to try to generalize the data and make predictions based on never-before-seen information.

The regularized minimization of the loss function is referred to as optimization in neural networks. When mapping inputs to outputs, an optimization technique determines the value of the parameters (weights) that minimize error. The efficiency of the deep learning model is significantly impacted by these optimization methods or optimizers. The weights and learning rates of neural networks are modified using a function or algorithm known as an optimizer, among other properties. Due to this, it contributes in lowering total loss and enhancing precision. Choosing the appropriate model weights is a difficult challenge because deep learning models typically have millions of parameters. It emphasizes the necessity to select an effective optimization algorithm for respected optimization model. This chapter provides a thorough analysis of various optimization algorithms present in the literature. Several optimization algorithms such as GD, SGD, mini-batch SGD, momentum, Nesterov momentum, adapted gradient (AdaGrad), adapted delta (AdaDelta), root mean square propagation (RMSProp), adaptive moment estimation (Adam), Nadam, and AdaBelief are discussed in this chapter.

Deep learning uses a collection of techniques called regularization to reduce the generalization error. Regularization techniques try to minimize overfitting while maintaining the minimum amount of training error. This chapter provides a briefly described various regularization techniques such as l_2 regularization, l_1 regularization, and dropout technique.

References

[1] Samuel, A. L. (1959). Some studies in machine learning using the game of checkers. *IBM Journal of Research and Development*, 3(3), 210–229. DOI:10.1147/rd.33.0210.

[2] Ivakhnenko, A. G. and Lapa, V. G. (1965). *Cybernetic Predicting Devices*. CCM Information Corporation. Available from https://www.gwern.net/docs/ai/1966-ivakhnenko.pdf.

[3] Grossberg, S. (1973). Contour enhancement, short-term memory, and constancies in reverberating neural networks. *Studies in Applied Mathematics*, 52(3), 213–257. doi:10.1002/sapm1973523213.

[4] Fukushima, K. (1975). Cognitron: a self-organizing multilayered neural network. 20(3–4), 121–136. *Biological Cybernetics*, ACM: New York, NY, USA, doi:10.1007/bf00342633.

[5] Fukushima, K. (1979). Self-organization of a neural network which gives position-invariant response. In *Proceedings of the 6th International Joint Conference on Artificial Intelligence*, vol. 1, pp. 291–293. ACM: New York, NY, USA. https://dl.acm.org/doi/abs/10.5555/1624861.1624928

[6] Vapnik, V. and Chervonenkis, A. (1979). *Theory of Pattern Recognition*. Akademie-Verlag, Berlin.

[7] Bottou, L. (1991). Stochastic gradient learning in neural networks. *Proceedings of Neuro-Nımes*, 91(8), 12. https://leon.bottou.org/publications/pdf/nimes-1991.pdf

[8] LeCun, Y., Bottou, L., Bengio, Y. and Haffner, P. (1998). Gradient-based learning applied to document recognition. *Proceedings of the IEEE*, 86(11), 2278–2324. DOI: 10.1109/5.726791.

[9] Simard, P. Y., Steinkraus, D. and Platt, J. C. (2003). Best practices for convolutional neural networks applied to visual document analysis. In *ICDAR*, vol. 3, no. 2003, IEEE Xplore: New York, NY, USA. https://cognitivemedium.com/assets/rmnist/Simard.pdf

[10] Hinton, G. E., Osindero, S. and Teh, Y. W. (2006). A fast learning algorithm for deep belief nets. *Neural Computation*, 18(7), 1527–1554. https://doi.org/10.1162/neco.2006.18.7.1527

[11] Bottou, L. and Bousquet, O. (2008). Learning using large datasets. In *Mining Massive Data Sets for Security*, Françoise Fogelman-Soulié, Domenico Perrotta, Jakub Piskorski, Ralf Steinberger (eds.) pp. 15–26, IOS Press: Amsterdam, The Netherlands. https://leon.bottou.org/publications/pdf/mmdss-2008.pdf

[12] Do, C. B., Le, Q. V. and Foo, C. S. (2009). Proximal regularization for online and batch learning. In *Proceedings of the 26th Annual International Conference on Machine Learning*, pp. 257–264, ACM: New York, NY, USA. https://doi.org/10.1145/1553374.1553407

[13] Zinkevich, M., Weimer, M., Li, L. and Smola, A. (2010). Parallelized stochastic gradient descent. *Advances in Neural Information Processing Systems*, 23, pp. 1–9. https://proceedings.neurips.cc/paper/2010/hash/abea47ba24142ed16b7d8fbf2c740e0d-Abstract.html

[14] Snoek, J., Larochelle, H. and Adams, R. P. (2012). Practical bayesian optimization of machine learning algorithms. *Advances in Neural Information Processing Systems*, 25, pp. 1–9. https://proceedings.neurips.cc/paper/2012/hash/05311655a15b75fab86956663e1819cd-Abstract.html

[15] Roy, A., Dutta, D. and Choudhury, K. (2013). Training artificial neural network using particle swarm optimization algorithm. *International Journal of Advanced Research in Computer Science and Software Engineering*, 3(3), 430–434. https://www.academia.edu/5098268/Training_Artificial_Neural_Network_using_Particle_Swarm_Optimization_Algorithm.

[16] Camilleri, M., Neri, F. and Papoutsidakis, M. (2014). An algorithmic approach to parameter selection in machine learning using meta-optimization techniques. *WSEAS Transactions on Systems*, 13(1), 203–212. https://www.wseas.org/multimedia/journals/systems/2014/a165702-311.pdf.

[17] Young, S. R., Rose, D. C., Karnowski, T. P., Lim, S. H. and Patton, R. M. (2015). Optimizing deep learning hyper-parameters through an evolutionary algorithm. In *Proceedings of the Workshop on Machine Learning in High-Performance Computing Environments*, pp. 1–5, ACM: New York, NY, USA. https://doi.org/10.1145/2834892.2834896

[18] Shen, F., Zhou, X., Yang, Y., Song, J., Shen, H. T. and Tao, D. (2016). A fast optimization method for general binary code learning. *IEEE Transactions on Image Processing*, 25(12), 5610–5621. DOI: 10.1109/TIP.2016.2612883.

[19] Shin, S., Boo, Y. and Sung, W. (2017). Fixed-point optimization of deep neural networks with adaptive step size retraining. In *2017 IEEE International Conference on Acoustics, Speech and Signal Processing (ICASSP)*, IEEE, pp. 1203–1207, IEEE Xplore: New York, NY, USA. DOI: 10.1109/ICASSP.2017.7952347.

[20] Khalifa, M. H., Ammar, M., Ouarda, W. and Alimi, A. M. (2017). Particle swarm optimization for deep learning of convolution neural network. In *2017 Sudan Conference on Computer Science and Information Technology (SCCSIT)*, IEEE, pp. 1–5, IEEE Xplore: New York, NY, USA. DOI: 10.1109/SCCSIT.2017.8293059

[21] Hoseini, F., Shahbahrami, A. and Bayat, P. (2019). AdaptAhead optimization algorithm for learning deep CNN applied to MRI segmentation. *Journal of Digital Imaging*, 32(1), 105–115. https://doi.org/10.1007/s10278-018-0107-6.

[22] Yi, D., Ahn, J. and Ji, S. (2020). An effective optimization method for machine learning based on ADAM. *Applied Sciences*, 10(3), 1073. https://doi.org/10.3390/app10031073.

[23] Tubishat, M., Alswaitti, M., Mirjalili, S., Al-Garadi, M. A. and Rana, T. A. (2020). Dynamic butterfly optimization algorithm for feature selection. *IEEE Access*, 8, 194303–194314. DOI: 10.1109/ACCESS.2020.3033757.

[24] Hussain, N., Khan, M. A., Kadry, S., *et al.* (2021). Intelligent deep learning and improved whale optimization algorithm based framework for object recognition. *Human Centric Computing and Information Sciences*, 11, 34. http://hcisj.com/data/file/article/2021083002/11-34.pdf.

[25] Brodzicki, A., Piekarski, M. and Jaworek-Korjakowska, J. (2021). The whale optimization algorithm approach for deep neural networks. *Sensors*, 21(23), 8003. https://doi.org/10.3390/s21238003.

[26] Yao, Z., Gholami, A., Shen, S., Mustafa, M., Keutzer, K. and Mahoney, M. (2021). Adahessian: an adaptive second order optimizer for machine learning. In *Proceedings of the AAAI Conference on Artificial Intelligence*, vol. 35, no. 12, pp. 10665–10673, AAAI Press: Palo Alto, CA, USA. https://doi.org/10.1609/aaai.v35i12.17275.

[27] Mohammed, M. A., Al-Khateeb, B., Yousif, M., *et al.* (2022). Novel crow swarm optimization algorithm and selection approach for optimal deep learning COVID-19 diagnostic model. *Computational Intelligence and Neuroscience*, pp. 1–22. https://doi.org/10.1155/2022/1307944

Chapter 4

Computer-aided diagnosis in maritime healthcare: review of spinal hernia

İsmail Burak Parlak[1], Hasan Bora Usluer[2] and Ömer Melih Gül[3]

Computer-aided diagnosis (CAD) systems have become major support systems in current medical decision-making. Even though automatic decision precision is still debated in medicine, CAD offers accurate remote access, medical archival, and patient follow-up. In this study, we focused on maritime healthcare and major CAD approaches in hernia. Maritime healthcare is considered as a specific topic in medicine where the availability of accurate medical services is crucial for seafarers. Unfortunately, medical facilities have limited CAD capacity for onboard seafarers. The shipside doctors prefer remote survey and telemedicine for less emergent medical cases. Hernia is considered as a major medical case among maritime diseases. Seafarers have a broad range of orthopedic illnesses due to limited space in a vessel. In medical literature, the severity and progress of hernia are less studied due to the long service periods and limited repatriation. We reviewed major CAD tools for herniated discs and new approaches in medical imaging through deep learning mechanisms. We conclude that CAD offers reliable benefits for onboard seafarers where hospitalization procedures are limited.

4.1 Introduction

Over the last two decades, the vessel traffic density has increased all around the world leading to transport accidents in major waterways. The vessel density and the accidents have negative effects on seafarer life safety which figures out in medical cases. Due to limited vessel space conditions, medical diagnosis is generally achieved through remote access and telemedicine [1]. International regulation becomes a challenge in maritime healthcare in case of emergency and seafarer diseases [2]. The evolution of expert systems and artificial intelligence (AI)

[1]Department of Computer Engineering, Galatasaray University, Turkey
[2]Maritime Vocational School, Galatasaray University, Turkey
[3]Department of Computer Engineering, Bahceshir University, Turkey

promote accurate standards in healthcare. Big data analytics is a multidisciplinary field where social scientists and data engineers meet to analyze the insights of a task. Even if medical maritime big data problems have been less studied [3], new maritime safety regulations lead to maritime computer-aided diagnosis (CAD) systems in international trade routes. Therefore, CAD benefits from AI-based technology to decrease health issues in maritime transport.

AI algorithms are integrated into CAD to analyze patient status in several stages. Maritime healthcare involves new machine learning methods coupled with telemedicine and mobile applications where patient data is evaluated with other cases as a dataset. Therefore, AI provides new insights for imaging and tabular datasets through patient records and offers associated models to retrieve disease related knowledge. The prediction results are generally evaluated with previous records to keep an updated performance in CAD. Deep learning strategies offer new outcomes by generating image features in learning-training datasets. Since the CAD is regularly maintained and validated, the prediction results allow quick response in emergency cases and hospitalization procedures. Moreover, it would render a pre-classification of disease before the interpretation of physician and radiologist. Therefore, human-biased errors would be relatively minimized. In maritime healthcare, CAD systems are available in any medical decision-making process such as disease detection, recognition, staging, follow-up, and prognosis planning [4].

Maritime healthcare is a complex task where health issues are investigated in several constraints. The new era of CAD technologies such as the Internet of Medical Things (IoMT) emerged as a built-in biomedical tool that integrated different devices in a single network through medical software architecture. Therefore, IoMT becomes a medical support system by tracking the sensors attached to the patient in a vessel for cardiovascular follow-ups, cancer tracking, and diabetes mellitus management [5]. AI-integrated IoMT devices enable new features to improve the functionality and CAD accuracy in order to assess medical risks with true critical warnings in medical support systems.

Maritime is one of the most convenient and oldest modes of transportation in the world. People first started their maritime business to find food, and then used them for exploration and trade. Seventy percent of the inhabited earth is covered with water. Coastal areas are the places where the people on earth mainly live, and its ratio to the whole is 80%. Ninety percent of the transportation of endless human needs is done through sea transportation. There are three important elements in maritime transport: ships, ports, and shipping networks or lines. But the most important and still the most important factor that still causes controversy is human influence.

Seafarers are the basic elements of the maritime transportation sector, which tries to respond to the endless human needs due to the developing and changing technology. Seafarers are among the most difficult professions in the world due to factors such as increasing ship traffic, intense working hours, and irregular living conditions [6]. There are many standards for all occupational groups in the world. In the maritime industry, standards and contracts are indispensable like SOLAS and

STCW. In the recommendations sent by the International Maritime Organization-IMO in accordance with the member states and their domestic laws, the constant reference is to the safety of navigation and the safe work of seafarers on board and in the harbor [7]. Occupational health and safety are all of the systematic efforts to protect against the dangers arising from production and the conditions harmful to health in working life [8,9].

Maritime healthcare is coupled with telemedicine, disembarkation and repatriation procedures. In most medical cases, maritime follow-up care involves remote access of physicians. Regular medical checkups including physical exam, blood tests, and imaging tests are generally done at the end of service. Seafarer diseases could be detected in the pre-employment medical examination such as hypertension, diabetes mellitus, urinary tract problems, and cardiovascular problems. However, orthopedic diseases are generally discarded during this examination if the disease is not found severe or chronic. Also, accurate medical imaging techniques are required to detect early stages of orthopedic problems. Moreover, the physical condition and activities might lower the back pain in seafarers for the early stages of hernia. Chauhan and Sagaro *et al.* [10,11] have presented the incidence of diseases among seafarers. The spinal hernia and musculoskeletal problems become the second most encountered disease. In similar studies, the back/neck pain, injury, and hernia was denoted as the top disease in maritime healthcare [12,13]. Hernia is a significant cause of disability which would become recurrent in maritime healthcare. It is also a potential risk for unexpected disembarkation and unfitness. Back pain and hernia are correlated with the onboard occupation involving lifting and bending. Lefkowitz *et al.* have performed an analysis about the risk factors on repatriation [14]. It figures out that telemedicine and remote healthcare suggest opportunities for risk factors to prevent unpredicted repatriation and disembarkation.

In this study, we have started with a broad review of maritime healthcare and focused on new AI tools in maritime CAD systems. We have focused on the classification of techniques during the last two decades in order to limit the scope of our survey with modern approaches. Even if some techniques are relatively experimental in maritime healthcare, medical AI shows up better CAD by limiting repatriation and increasing service quality. Our survey is organized as follows. Section 4.2 presents similar studies for the common diseases of seafarers. Section 4.3 consists of the background of CAD. Section 4.4 reviews the spinal hernia and AI-based techniques for medical imaging and analysis. Consequently, we have concluded our survey by highlighting the CAD insights in maritime healthcare and future trends.

4.2 Similar studies and common diseases of the seafarers

Seafarers face serious risks while working on ships and in ports. With the occurrence of health problems, it is possible to experience serious injuries and life-threatening hazards. Working on board also causes some habits to end, some to change, and some to form new ones. While some of these habits are very beneficial, some of them are in a situation that will directly harm the health of the person.

Hand-arm vibration syndrome (HAVS) is also important and harmful. The large and powerful machine system of the ship and the tools used for repairing any work on the ships generally transmit vibrations and disturbances on a high scale. Most electrical devices like chipping machines and needle guns are mainly dangerous onboard. The tingling of fingers, numbness, blanching, and even pain in the arm and wrist are also common symptoms of HAVS. Cardiovascular disease (CVD) is also another important disease onboard. A population that is subject to high levels of stress, like seafarers, is more likely to have cases of this type of sickness. Apart from bad habits, old age, smoking, drinking alcohol, a particular combination of genes, and other factors influenced by conditions aboard such as stress, food, and lack of exercise, CVD can be caused by other factors [15]. Unfortunately, one of the most important diseases of today is hypertension. Although it has spread to all occupational groups with stress, it is known that there are other known symptoms as well. Especially on ships, loneliness, fatigue, consumption of alcohol and tobacco products, and lack of physical exercise are among the important known causes. A healthy diet, not using alcohol and smoking, and being able to control stress can significantly improve hypertension (www.marineinsight.com).

While a study listed multiple ailments faced by ship workers, it drew attention to orthopedic and bone diseases in particular. Musculoskeletal disorder (MSD) is the worldwide seafarer disease. As a result of the research conducted on Norwegian and Danish seafarers, they reported that they complained of musculoskeletal diseases for seafarers. As a result of the research, working cycles of up to six-hour shifts as a result of the expectation of doing a lot of work with less personnel on ships with advanced technology cause them to be exposed to a more difficult situation physically. However, on the ship, while it does not allow to do the exercises that the human body really needs, it is necessary to create an opportunity for physical movements like healthy individuals after they move to land. The most appropriate treatment, albeit limited, is exercise and stretching. In the developing ships, as a result of the guidance of the World Health Organization (WHO) and the international workers' organization, there are cabins with the possibility of sports on the ships. However, the working period on the ships cannot be used in real terms due to work stress and lack of motivation (osha.europa.eu).

Even though the ship is large, it becomes small for its employees after a while since the working hours on the ship are long, and the vessel size is big enough. One of the major problems encountered is actually mental complaints. Working in a ship with an isolated area and environments could present major challenge toward the seafarers' psychological state [16]. Seafarers work on ships with a shift work system for 24 hours. Each shift has different statuses and difficulties. Also, the ship's hierarchical commanding structure makes unclear working schemes a distinction between job and leisure timetables [17,18].

The mistakes that can be made in the maritime profession are not very similar to the examples in other occupational groups. At the end of the mistake, seafarer suffers many accidents that leave him/her dead and disabled. Therefore, the work discipline is provided at a high level and hierarchically inside the ship, at the ports thanks to the external control levels and port states control. Work discipline and controls create not only anxiety and pressure disease but also many other diseases on seafarers and affect their current mood. Seafarers, whose working hours continue during the contracts of 2, 4, or 6 months with the shipowner company, and who practice their profession with a working principle that requires returning to land at the end of the period, return to their homes and social lives at the end of the period. Seafarer, who is mentally tense at the end of the period, has difficulty in mixing social life with this state.

According to the experts of the International Labor Organization (ILO) and the WHO, "Occupational Health" is keeping, maintaining, and developing employees' physical, mental, and social well-being in all professions at the most outstanding level. The ILO and the WHO define occupational health [19]. Protecting the employees' mental and physical health from the harmful impacts of the workplace, taking safeguards against work accidents and occupational diseases, and making sure they work in a setting that is both comfortable and safe are all important aspects of occupational health and safety and constitute the main purpose of occupational health and safety studies. Besides psychological pressures, seafarers who perform 4 hours of continuous ship management duties in different shifts have to deal with many duties such as first aid, firefighting, and countering piracy, which are their duty not only during the shift but also in any emergency situation whole voyage. Therefore, they are also very worn out in terms of physical strength. At this stage, mainly skeletal systems and orthopedic disorders are encountered.

The Dreadnought Seamens Hospital's study in 1984 showed that orthopedical diseases had high importance for seafarers. The study covered 1409 seafarers' orthopedic operations over 4 years. The example seafarer group consisted of both retired and still on active duty [20]. Table 4.1 consists of many articles that are an essential study of the working period and explain that orthopedic problems are significant for seafarers and cause many complaints with old age during retirement.

4.3 Background

It has been tentatively estimated that there are between 110 and 120 million tons of goods shipped globally, and that about 750,000 sailors are employed on ships that travel abroad. It is crucial that every effort be made to protect the health of those working in a field with the size and level of specialization of this one. There is a dearth of precise or trustworthy data, making it challenging to draw a firm conclusion on the scope of the health issues faced by seafarers [33].

Motion sickness is a physiological condition brought on by oscillatory movements or whole-body vibrations, such as thóse on sea vessels like ships and boats, and it can have a significant impact on how well soldiers perform their duties at sea. There are a variety of psychological, social, and environmental elements that can affect an individual's vulnerability to motion sickness, and the profiles of these risk factors may range from person to person as a result of different neural mismatch composition. Most of the data pertains to populations in the West. According to

Table 4.1 Similar studies

	Related works	Authors
1	Incidence of occupational injuries and diseases among seafarers: a descriptive epidemiological study based on contacts from onboard ships to the Italian Telemedical Maritime Assistance Service in Rome	Sagaro GG, Dicanio M, Battineni G, Samad AM, and Amenta, F [11]
2	Injury, illness, and disability risk in American seafarers	Lefkowitz RY, Slade MD, and Redlich CA [21]
3	Comparative analysis of medical assistance to seafarers in the world and the Republic of Croatia. *Int Conf Transp*	Mulić R and Vidan P [22]
4	Fatal accidents and injuries among merchant seafarers worldwide. *Occup Med.*	Roberts SE, Nielsen D, Kotłowski A, *et al.* [23]
5	"Injury, illness, and work restriction in merchant seafarers." *Am J Ind Med*	Lefkowitz RY, Slade MD, and Redlich CA. [14]
6	Hospital contacts for injuries and musculoskeletal diseases among seamen and fishermen: a population-based cohort study. *BMC Musculoskeletal Disorders*	Kaerlev L, Jensen A, Nielsen PS, Olsen J, Hannerz H, and Tüchsen F [24]
7	Workload and musculoskeletal problems: a comparison between welders and office clerks (with reference also to fishermen). *Ergonomics*	Torner M, Zetterberg C, Anden U, Hansson T, and Lindell V [25]
8	Musculoskeletal symptoms among commercial fishers in North Carolina. *Appl Ergon*	Lipscomb HJ, Loomis D, McDonald MA, Kucera K, Marshall S, and Li L [26]
9	Work related shoulder disorders: quantitative exposure–response relations with reference to arm posture. *Occup Environ Med*	Svendsen SW, Bonde JP, Mathiassen SE, Stengaard-Pedersen K, and Frich LH [27]
10	The effects of anthropometrics, lifting strength, and physical activities in disc degeneration. *Spine*	Videman T, Levälahti E, and Battié MC [28]
11	Musculo-skeletal symptoms as related to working conditions among Swedish professional fishermen. *Appl Ergon*	Torner M, Blide G, Eriksson H, Kadefors R, and Karlsson R, Petersen I [29]
12	Musculo-skeletal symptoms and signs and isometric strength among fishermen. *Ergonomics.*	Torner M, Zetterberg C, Hansson T, Lindell V, and Kadefors R [30]
13	The influence of musculo-skeletal load, and other factors, on staff turn-over in fishery: a post employment questionnaire study. *Bull Inst Marit Trop Med Gdynia.* 1990, 41: 97–108	Torner MI, Nilsson E, and Kadefors R [31]
14	Knee pathology among seamen: a review of 299 patients. *Occup Med*	Pearce MS, Buttery YE, Brueton R.N. [32]

epidemiological population surveys conducted around the globe, on calm seas, the Caucasian populations of the USA and UK, respectively, experienced motion sickness rate of 25–30% and Indian populations of 28%. When seas were calm, 10–30% of the British naval personnel reported feeling queasy; when waves were rough, that number rose to 50–90%. The majority of people—90%—have suffered from motion sickness at some point in their lives [4].

When soldiers are at sea, motion sickness can significantly impact their ability to fulfill their duties. To provide information on the prevalence of motion sickness among Singaporean sailors (seafarers) and attached army troops (nonseafarers) aboard naval platforms, as well as an understanding of the risk factors for the condition. A total of 503 employees participated in a self-administered survey for a cross-sectional study during the 2001 monsoon season, which lasted from January to April. Throughout various sea situations, motion sickness was noticeably more significant among army soldiers (59.2%) than among navy personnel (38.3%). Headache, nauseousness, and vertigo were the most typical signs. For the purpose of determining susceptibility, the Motion Sickness Susceptibility Questionnaire was utilized, and it was found that non-seafarers appeared to correlate more favorably than seafarers did. It appeared that smoking provided protection against motion sickness, and it was considered that discomfort in one's environment contributed to the development of the condition. Frequent sailing seems to be a key element in reducing motion sickness. Despite the fact that we now understand that motion sickness is a continuum of physiological reactions to whole-body vibration, some people still get nauseous when they travel; it is nevertheless very common among people who have never traveled by sea. With repeated sailing, seafarers will grow less vulnerable on their own, and they are also more aware of the methods that are available to relieve symptoms.

To examine trends in work-related CVD mortality among seafarers working in British merchant shipping from 1919 to 2005; to compare work-related CVD mortality among British seafarers working in British shipping and ashore in Britain with that of the general British population; and to examine work-related CVD mortality in British shipping over the past few years based on variables such as rank, nationality, location, and type of vessel a long-term research project. The primary result indicators are standardized mortality ratios and population-based mortality rates. In most of the years from 1919 to 1962, the death rate from work-related CVD increased, but it started declining until 2005. Compared to the corresponding general population, seafarers employed by British ships had lower rates of work-related CVD and ischemic heart disease mortality (standardised mortality ratio SMRs = 0.35–0.46), although CVD mortality among British seafarers ashore in Britain was frequently higher. The crews of offshore ships in the North Sea were shown to have a higher than average risk of CVD death due to employment. This study demonstrates a protective effect of a healthy work environment against CVD mortality among seafarers employed in British shipping. However, it also shows greater risks for British seafarers who are onshore in Britain. This would include individuals who have been dismissed due to the morbidity of CVD as well as other ailments. It is possible that the particular hazards that are present in the workplaces of companies that operate supply ships in the North Sea are to blame for the high rates of CVD mortality among seafarers [34].

Medical facilities are typically not available on commerce ships. When a seafarer is unwell or has an accident, the ship's captain or the officers in charge will help them, but these people lack the medical expertise necessary to do so. We created the seafarer health expert system (SHES) as a solution to this problem, which enables telemedical assistance in an emergency. An in-depth examination of the medical problems that affect seafarers was done using the medical files of those who received assistance from the International Radio Medical Center (CIRM), Italy, on board ships. Epidemiological data analysis is handled in a two-phase setup with the use of data mining tools. In the initial stage, the most frequent pathologies that occurred on board were examined. Subsequently, a thorough questionnaire was created for each medical issue in order to give the onshore physician accurate symptomatic data. The SHES framework, design flow, and functionality were the key points of emphasis in this article. Aside from that, three players and nine creating policies with distinct functioning panels were explicitly outlined. The suggested method is simple to use even for those without prior computer knowledge, and it generates medical requests for the quick transfer of symptomatic information to an on-site physician [35,36]

A "seafarer" is anyone working in any position on a ship offshore, whether it be for hire or charter, and ships of war are not included in this definition. An environment that requires awkward motions and limits a person's usual mobility is something that seafarers must deal with every day as they live in a world that is constantly changing. A study was done on a sample of sailors for confirming high incidence of herniated discs among these specific employees in view of the recent implementation of INAIL-tabulated diseases. The results of the data analysis revealed that the sample of seafarers had a herniated lumbar disc in 48.3% of cases, 34.5% of these cases were related to deck work, and 65.5% were related to machinery. The study of the sample, which varied in age and the task being performed, provides support for the claim that the individual risk factors, particularly age and obesity, are not strongly implicated in the genesis of disk herniation suffered by seafarers, but that the work factors (vibrations) play a more significant role in the onset of this disease. The study of the sample, which provided support for the claim, was conducted using a sample that varied in age and the task being performed. This topic must be considered in the framework of legal medical evaluation, particularly the causal relationship, which at the moment appears to be fairly underrepresented in literature and scientific output [37].

Studying morbidity among merchant mariners who are currently at sea will help identify any potential occupational diseases as well as lifestyle- and work-related diseases, which will in turn help identify areas where preventative measures may be put in place. It was possible to identify a group of Danish commercial mariners who had been actively employed at sea in 1995 by consulting a registry kept by the Danish Maritime Administration. The ship, charge aboard, and all sea service durations for each seafarer were all known. Denmark's National In-patient Registry was connected to the cohort. Standardized hospitalization ratios (SHRs) for each major diagnostic category were computed using all gainfully employed individuals as the reference group. The inhomogeneity of seafarers was demonstrated by the considerable disparities in SHRs for the same illness groups among several groups of seafarers according to charge and ship type. SHRs for diseases associated with a sedentary lifestyle were high, while rates for acute problems, such as acute myocardial infarction, were low. This is likely due to referral bias since acute conditions are more likely to necessitate hospitalization abroad and are therefore excluded from the study. SHRs for poisoning and injuries were high, particularly for ratings and officers on tiny ships. Despite being pre-selected, a sizable part of sailors make up a group of employees that exhibit signs of ill health that are undoubtedly the result of lifestyle choices. The subgroups with a greater risk of hospitalization for diseases linked to lifestyle also had a higher risk of hospitalization because of poisoning and injury. Despite the underreporting of hospitalization due to work accidents overseas, the hospitalization burden among sailors is substantial due to diseases as well as trauma-related occurrences. The same factors that contribute to lifestyle-related disorders may also contribute to accidents. Depending on whether they are officers or ratings, distinct groups of seafarers within the group exhibit significant variances. Ratings had greater hospitalization ratios than officers, with a few outliers. This distinction can be apparent for external causes such as trauma, intoxication, and other factors in addition to hospitalization due to illness. Higher social classes on land generally have far better health compared to lesser social classes [38].

Text-based patient data, physician notes, and prescriptions are all widely available in digital health systems. A higher standard of healthcare will be provided, possibly with fewer medical errors, at lower costs thanks to this knowledge consolidated over the electronic clinical information. The work culture, climate variations, and personal habits of sailors also make them more likely to experience accidents and be exposed to health risks. Hence, the use of text mining to analyze the medical records of seafarers can produce greater insights into the medical problems that frequently arise at sea. CIRM, an Italian telemedical maritime support system, collects medical records from its digital health systems Telemedical Maritime Assistance Service (TMAS). Patient information from the three-year period (2018–2020) has been analyzed. The Naive Bayes technique and the Lexicon were both adopted for sentimental analysis, and trials were carried out using the R statistical tool. Using the use of word clouds, the symptomatic data was shown, and a 96% connection between the severity of the medical issues and the final diagnosis was attained. Around 80% accuracy and precision are used to confirm the sentiment analysis [39].

Before beginning employment on board a ship and on a regular basis later, every seafarer must pass a fitness test. Hence, unless an incidental disease should arise or they sustain an injury, it would be realistic to anticipate that they will stay in good health while aboard. However, this is not often the case, which puts seafarers' health and welfare at risk and increases costs for companies. This study aims to complement and deepen the investigation that we began in 2007 into the factors that lead sailors of various nationalities to disembark and be referred to specialists ashore. It was posed as follows: Is there a reason to broaden the scope of the pre-boarding medical examination, and if so, should this differ for any particular nationality group? The P & O Princess Cruises International fleet's officers and crew were considered in the study. Throughout the course of six months, data on diagnosis, age, and nationality were collected and analyzed. Calculations were made for rates and 95% confidence intervals. Accidents and injuries were reported to be the most frequent reasons for disembarkation (202 cases; 39.8% of all causes), followed by abdominal surgery issues (81 instances; 16% of all causes). Psychiatric cases were the second most prominent category (42 instances, or 8.3% of all reasons). Filipino seafarers were disembarked less frequently than other nationalities overall when the six most frequent groups were broken down by nationality, with the difference being most pronounced in orthopedic (0.58 CI 95%: 0.40–0.83) and psychiatric (0.08 CI 95%: 0.01–0.59) instances. It is significant that Filipino sailors have a low disembarkation rate for medical reasons, and it is important to look into the root causes. If there are no additional factors influencing the health of the crew from the Philippines, such as psychosocial support and the thoroughness of the medical examination, these findings may provide evidence that psychometric testing may be of some value, as it is required in the Philippines (as opposed to elsewhere) for pre-embarkation of medical examination [12].

The regulation of shipping and the issues it raises are typically viewed through the prism of traditional maritime nations with abundant wealth, as is the case with other globalized industries. Discourse on the globalization of the shipping sector has received a lot of attention from the perspectives of powerful and governing shipowners, as well as in relation to the ownership of ships, their registration, and the chain of employment for maritime personnel. Given the transformation and shift that has taken place in the maritime market over the past couple of decades, from traditional to emerging maritime crew supplying states that are mostly from developing countries, a study that focuses on how the seafaring industry is organized and regulated in these recently developed markets and how local and global economic forces impact these labor supplying states is the only one that can be considered legitimate in this context. Knowing the regulatory environment, labor supply arrangements, as well as the obstacles and problems they face may be helpful in determining how future policy changes can be implemented on both national and international levels.

This essay intends to discuss how the Philippines, a significant labor supplier, conducts and regulates its business dealings with the world shipping industry on the one hand, and its seafarer nationals who work on foreign-flagged ships on the other. A summary of the Philippines' historical seafaring heritage and its current condition as a maritime nation are given. Following that, there will be a study of the country's national labor laws and regulations to show how a determined effort is being made to secure the employment of its marine personnel abroad through a globalized development policy strategy. An analysis of regulatory approaches and strategies to maritime overseas employment, in general, and the occupational health and safety of seafarers serving foreign flags, in particular, is then undertaken in order to shed light on the competing tendencies of maintaining its global share of the seafaring market and ensuring the protection of its nationals while working outside its territorial jurisdiction. This is done in order to shed light on the competing tendencies of maintaining its global share of the seafaring market and ensuring the protection of its nationals while working outside its territorial jurisdiction. Lastly, the scope and constraints of the Filipino state's sovereign arm are examined to shed light on the numerous problems plaguing the maritime industry, particularly those that have an impact on seafarers' health and safety. The research consisted of a review of relevant laws, regulations, government documents, policy papers, and relevant periodicals, as well as interviews with a number of high-level government officials, labor union officials, and leaders of non-government organizations in the Philippines, which served as the main foundation for the data and conclusions generated in this paper.

Throughout the previous 150 years, a number of recurring elements in the body of knowledge for maritime health have emerged. They can be used to organize a study of the present body of knowledge and to pinpoint areas where there is solid evidence about the nature and scope of hazards as well as the efficacy of intervention to decrease harm. Additionally, it can highlight areas of knowledge gaps and indicate how to fill them. The dynamics of political, economic, and social interactions are essential to advancing knowledge and using it to enhance the health of seafarers, and past events, as outlined in the first article, also provide this information.

A single case report of a rare illness to lengthy studies of the incidence of prevalent chronic diseases is only a few of the sources of relevant information about seafarers' health. Clinically obvious illness, injury, or cause of death are the easiest events to document, but ongoing investigative research may also examine environmental dangers, individual risk factors, or early stages of disease. For in-depth analysis of health hazards or the efficacy of interventions, comparisons between subsets of a population are required. The ideal approach to accomplish this is if data on the population at risk can be utilized as the foundation for estimating the incidence or prevalence of sickness and if the populations being compared are as comparable as feasible in all respects, save for the one being researched. It is occasionally possible to gather data from sizable studies of land based populations that cannot be collected from sailors. It is possible to gather data on the health of sailors in a variety of locations, including at sea, upon arrival in port, during leave, and after retirement. For acute disease and injury, a single environment can serve as the basis for calculating risks, but for chronic disorders, cases occurring in several settings must be considered and the at-risk population must be determined in order to allow for the study of incidence. By paying attention to the conditions of living and working at sea and by choosing sailors who are deemed "fit for work," knowledge about the health of seafarers can be used to improve prevention. Determining the requirements for emergency care at sea and in ports is similarly critical. Regulational choices about the steps to be taken to reduce harm from illness and injury should consider the overall patterns of illness and injury among sailors and how they relate to those of other workers. The success of actions conducted with this objective in mind can be verified by indicators of improved seafarer health. Understanding the consequences of these impairments on performance and safety in normal and emergency duties of a seafarer is necessary to reduce the contribution of health-related impairment to accidents and other dangers at sea. With this information, it will be possible to decide if it is safe for someone with a disability to work at sea.

4.4 Computer-aided diagnosis of spinal hernia

The human spine is composed of vertebrae and discs stacked on top of each other which is divided into five segments; cervical spine (C1–C7), thoracic spine (T1–T12), lumbar spine (L1–L5), sacral spine (S1–S5), and the tailbone. The main task of our backbone is to provide protection to the cord during our movements and to ensure the body's load while preserving internal organs. The most widespread complaints of seafarers are encountered as lower back pain related to lumbar disc diseases. These orthopedic pains are encompassed in several types of herniation diseases. In a nutshell, herniation is defined as a disc substance compression problem within the spinal cord and roots through lumbar discogenic pain. The grade of a lesion is measured through the nucleus pulposus state. The progression of herniation is characterized by four stages, namely bulging, protrusion, extrusion, and sequestration. The bulging or pre-bulging phase is graded through the annulus fibrosus which is almost intact and the herniated disc has little compression on the spinal cord and ligaments. The protrusion is determined by the rupture of the herniated disc around the ligaments and the increase of compression on the spinal cord and roots. The extrusion is defined as the rupture of all ligament layers except the outer one where the compression is considerably increased through the spinal cord and roots. The sequestration, which is called a free disc fragment, corresponds to an extruded disc that has no continuity with the main disc. The medical imaging techniques are the main approaches in the diagnosis of herniated discs. The CAD generates meaningful results for the interpretation of disc grades. On the other hand, deep learning techniques provide quick responses to increase the precision of diagnosis and prevent human errors depending on the resolution and noise levels of medical images. The image noise level is coupled with the shape, angle, size, and imaging modality of the spinal cord. Low noise levels increase the precision of CAD and the quality of medical treatment. Moreover, herniated disc diagnosis might be confused in vertebral fusions, disc inflammations, or other auto-immune diseases due to noisy images. Figure 4.1 shows different pathology for hernia forms in T1 and T2 MRI views. The sagittal scans show lumbar discs. We have preprocessed the dataset provided by [40,41] in order to extract the spinal part and the location of herniated disc in a region of interest (ROI). The new era AI models perform automatic hernia classification through machine learning algorithms where the ROIs are transferred based on image features.

The primary clinical approach is magnetic resonance imaging (MRI) in the diagnosis of herniated discs. MRI modalities provide high-resolution medical images, which becomes a golden standard in the diagnosis of lumbar disc

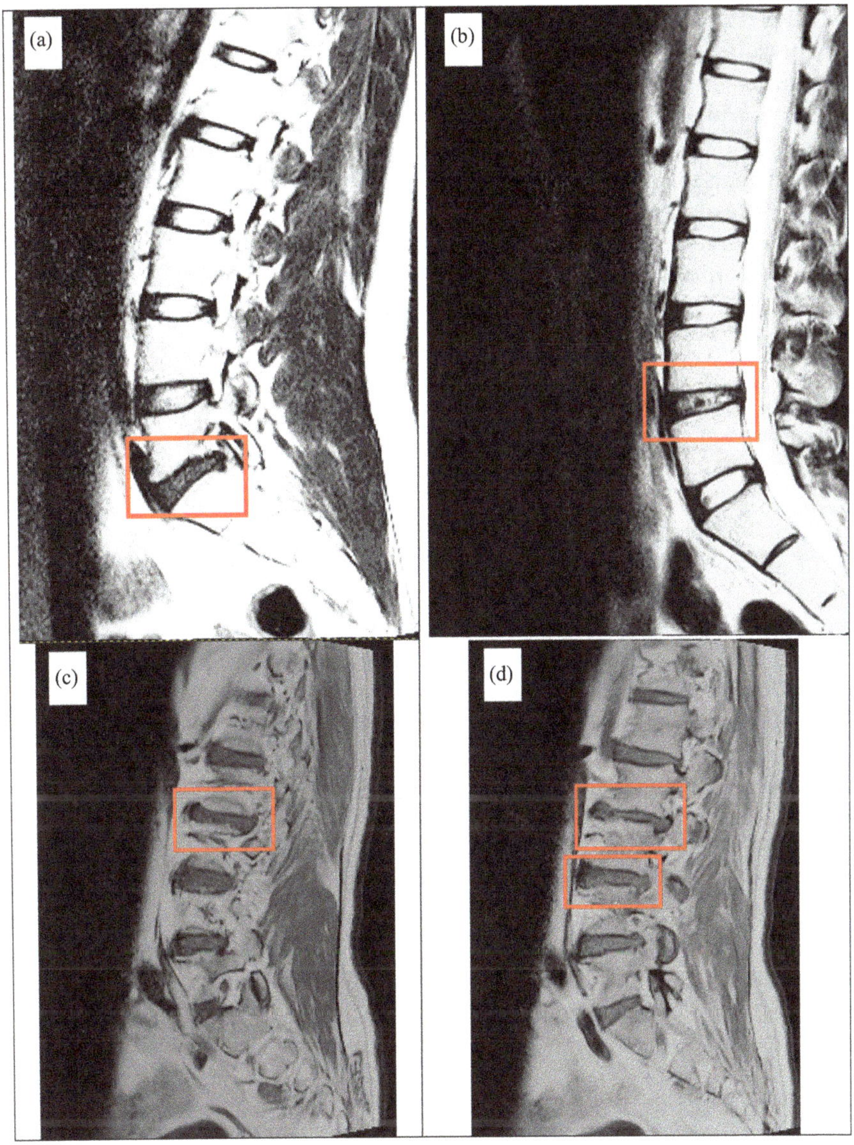

Figure 4.1 *T1–T2 MR images showing (a) Grade I spondylolisthesis of L5/S1 level, (b) mild compressing thecal sac at L4/L5 level, (c) spinal canal stenosis, and (d) ligamentum flavum hypertrophy at L3/L4 and L4/L5 level preprocessed and adapted from Sudirman* et al. *[40,41]. Region of interest (ROI) denotes herniated discs aka image features in machine learning.*

herniation. It allows to locate hernia morphology and the number of hernia floors. Some radiologists might prefer computed tomography (CT) or X-ray imaging for the detection of back pain. CT imaging combined with contrast root capsule would be an alternative imaging technique in hernia diagnosis where patients cannot undergo MRI. On the other hand, MRI–CT comparison would be required in some conditions due to patient status or noisy images. Even if CT is a technique with radiative emission, it would be suitable in hernia diagnosis where the disease is located in the inner bone structures and roots. X-ray imaging is a conventional modality in orthopedic radiology. Scoliosis, intervertebral stenosis, and spondylolisthesis would be monitored through X-ray imaging where herniated discs might be localized. Moreover, lumbar instability, waist defects, and herniated injuries would be analyzed via X-ray imaging. Lumbar back pain in seafarers is mainly caused by posture, body mass, and heavy lifting problems during the work time. Acute herniated disc or protrusion stage starts with the incorrect posture which is followed by soft tissue damage and degeneration on the joints and discs. Back pain for more than one month is considered as chronic hernia. The gradually increase in the duration of pain causes damages on vertebrae roots and discs.

In all levels of the herniated disc, the physicians might offer patients both surgical and drug treatments. Herniated disc surgery or discectomy is the most common technique for surgical intervention. On the other hand, drug therapy is preferred for the first stages of hernia. In order to increase maritime service and prevent emergency cases and repatriation, hernia treatment requires follow-ups and detailed medical imaging procedures. Rak and Tönnies [42] have reviewed the computerized methods for spinal imaging in MR. They have divided the state of the art into two parts: the localization and the segmentation problems for the vertebrae, the intervertebral disc, the spinal canal and cord. They have noticed that cervicothoracic and midsagittal imaging are in the majority of spinal MR studies. Then, they have emphasized the linear relation between landmark localization in hernia shape representation during CAD. This structural feature becomes a keypoint in statistical inference of hernia treatment. Moreover, Galbusera *et al.* [43] have described the AI applications of spine in different categories such as the CAD, the localization of vertebrae and discs, the biomechanics, the content-based spinal image retrieval of the spine segmentation, the medical decision systems, and the prediction analysis of spine treatment. Chiu *et al.* [44] have studied the regression of herniated discs for non-surgical cases. They have noted that the conservative treatment might resolve hernia according to reviewed medical cases. Also, they have concluded that sequestration would have a higher regression rate than extrusion. In conventional pain treatment, opioids are usually preferred to reduce back pain. However, surgical treatments might present a risk factor for prolonged postoperative opioids. Karhade *et al.* [45] have studied 5413 patients to generate a big dataset and to perform machine learning analysis to see the prolonged effects of opioids after the surgical operation. Yoon and Koch [46] have focused on the surgical bottlenecks to set an accurate decision for surgical intervention. They have divided the surgery lifetime according to spinal anatomy. In cervical disc herniation, they have proposed six months of increasing back pain with low response to

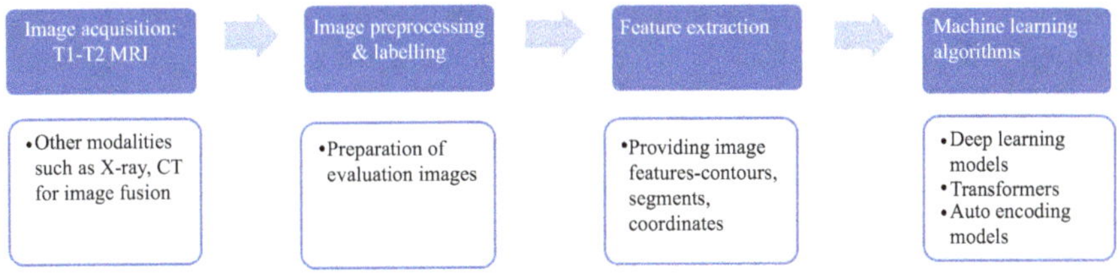

Figure 4.2 A general overview of herniated disc classification using machine learning techniques

conservative and drug therapy. In thoracic herniation, giant calcification on discs and roots, myelopathy, and neurological symptoms would be considered in surgical intervention. In lumbar disc herniation, six weeks of conservative and drug therapy would be considered before the surgery.

In a nutshell, CAD systems are considered as medical classification techniques where disease or not disease labels are assigned on images. Medical image classification is a fundamental task of AI-based decision-making. Spinal hernia detection and diagnosis is achieved by analyzing image features and assigning normal or herniated labels on medical images. MRI modality offers a broad range of image acquisition techniques for spinal monitoring. A radiologist can easily understand the hernia mechanism by considering sagittal and axial view of vertebrae discs. In CAD, image features related to herniated discs must be revealed and transferred to learning algorithms. Figure 4.2 represents the general flowchart of the new AI approach in spinal radiology. The herniated disc is evaluated in a dataset which is divided into training and test parts. The image acquisition step offers the physical property of the anatomical structure. Therefore, an appropriate image modality such as T1-T2 MRI would become a bottleneck in maritime healthcare where radiology facilities are not available in a vessel. In the patient follow-ups, recent image acquisitions must be considered in telemedicine and be adapted in machine learning approach.

Alsmirat *et al.* [47] have built a deep-learning architecture using convolutional neural networks. They have applied region extraction models to remove image noise in the learning phase of neural architecture. They have noticed that there are few studies in hernia detection and recognition using axial scans. They have called the recognition phase as the localization pathology on the related disc such as left, right, central, and diffuse. In a small MRI dataset, they have achieved more than 90% accuracy rate for hernia detection and recognition. In a similar study, Mbarki *et al.* [48] have focused on axial lumbar scans in MRI. They have used VGG16 deep learning architecture in spinal image classification. The study is composed of two steps. The first U-net deep learning architecture localizes the herniated discs. The second part finds the location of normal and pathological herniated discs. The model showed 94% accuracy in 450 patients. Alomari *et al.* [49] have studied lumbar disc localization by proposing a probabilistic model through T2-weighted MRI sagittal views. They have achieved more than 90% accuracy using three main features: location, appearance, and context of lower spine. Ghosh *et al.* [50] have analyzed MRI in lumbar discs to detect hernia abnormalities. They have performed this study using T2 sagittal scans and obtained 91% accuracy in detection evaluation. Koh *et al.* [51] have proposed a CAD system for lumbar spinal stenosis (LSS) using MRI. They have preprocessed the thecal sac to generate context features. They have applied a Gaussian blurring filter and Canny edge detection algorithm. The lumbar features have been manually coregistered on the coordinated lower spine. Ninety one percent accuracy has been obtained for 55 patients. Bhole *et al.* [52] have investigated hernia using ensemble classifier approach using K-means, least mean square, support vector machines, and perceptron classifiers. They have collected lumbar MR images and created the image features for the vertebrae, the disc, and the spinal cord in sagittal view and obtained 99% accuracy in 70 patients.

In future, we can apply data augmentation for improving classification performance. It is vitally crucial for essential infrastructures in medical systems such as those that hold personal health records in a safe setting, to protect themselves from malicious cyberattacks. In recent years, a number of researchers in the academic world have presented a method that makes use of deep learning referred to as RF fingerprinting as a means of resisting hostile attacks, in particular spoofing attacks [53–55]. The study that is described in [55–57] recommends a fine-grained data augmentation strategy as a means of boosting the performance of a deep learning-based RF fingerprinting method. This can be accomplished by adding additional data at a smaller scale.

4.5 Conclusion

Hernia is considered a major medical case among maritime diseases. Seafarers have a broad range of orthopedic illnesses due to limited space in a vessel. In medical literature, the severity and progress of hernia are less studied due to the long service periods and limited repatriation. We reviewed major CAD tools for herniated discs and new approaches in medical imaging through deep learning mechanisms.

Although recent AI techniques of CAD encompass a broad spectrum of applications in medical imaging, there are still bottlenecks in the patient care procedures such as invasive treatments, disease staging, prognosis planning, and asymptomatic and/or recurrent cases. In modern healthcare, CAD focuses on medical imaging and classification algorithms where disease and non-disease classes are identified. However, disease staging and lesion/inflammation localization aka recognition require new guidelines for training procedures. In herniated disc diagnosis, the correct localization of pathology would increase the accuracy rate in CAD.

Telemedicine has a critical tool in maritime healthcare by connecting the staff on board with onshore physicians. Even if telemedicine becomes crucial in CAD, correct communication links, fault-tolerant data transfer rate, and disease transcription are indispensable for onshore physicians to set appropriate decision rules. New AI-based

CAD interprets hernia with already trained image models. Analysis starts with training datasets, and engineers and radiologists collect disease and non-disease conditions during image acquisition. The labeling stage requires a sophisticated experience in spinal medicine in order to extract vertebrae and root features in ROIs. Image features are transferred into the AI model. Deep learning methods are also available in CAD where general spinal features would be transferred in the training step. The success rate is evaluated with previous acquisitions collected as a test dataset. If the accuracy score is above the target rate, physicians use the model as a decision-making tool in hernia detection. We conclude that new AI models offer reliable benefits in CAD for onboard seafarers where hospitalization procedures are limited.

References

[1] Feng Z, Li Y, Liu Z, and Liu RW (2021). Spatiotemporal Big Data-Driven Vessel Traffic Risk Estimation for Promoting Maritime Healthcare: Lessons Learnt from Another Domain than Healthcare. In *Artificial Intelligence and Big Data Analytics for Smart Healthcare*, (pp. 145–160). Academic Press.

[2] Gionis TA. (2000). Paradox on the high seas: evasive standards of medical care-duty without standards of care: a call for the international regulation of maritime healthcare aboard ships. *John Marshall Law Review*, *34*, 751.

[3] Lytras, MD, Visvizi A, Sarirete A and Chui KT. (2021). Preface: artificial intelligence and big data analytics for smart healthcare: a digital transformation of healthcare primer. *Artificial Intelligence and Big Data Analytics for Smart Healthcare*, Academic Press, xvii–xxvii.

[4] Chan G, Moochhala SM, Zhao B, Wl Y and Wong J. (2006). A comparison of motion sickness prevalence between seafarers and non-seafarers onboard naval platforms. *International Maritime Health*, *57*(1–4), 56–65. PMID: 17312694.

[5] Manickam P, Mariappan SA, Murugesan SM, *et al.* (2022). Artificial intelligence (AI) and internet of medical things (IoMT) assisted biomedical systems for intelligent healthcare. *Biosensors*, *12*(8), 562.

[6] Kuleyin B, Köseoğlu B and Töz AC. (2014). Gemiadamlarının sağlık ve emniyet koşullarının değerlendirilmesi: DEÜ denizcilik fakültesi Örneği. *Journal of ETA Maritime Science*, *2*(1), 47–60.

[7] Arslan O, Solmaz MS and Usluer HB. (2022). Determination of the perception of ship management towards environmental pollion caused by routine operations of ships. *Aquatic Research*, *5*(1), 39–52.

[8] Sabanci, A. İş sağlığı-İş Güvenliği ve Ergonomi. İş Sağlığı İş Güvenliği Kongresi Bildiriler Kitabı, MMO Yayın No: E/2001/263, *Adana*, pp. 279–298, 2001.

[9] IMO, London. SOLAS. International Convention for the Safety of Life at Sea, 1974, and 1998 Protocol relating thereto. (2001).

[10] Chauhan RS. (2022). Importance of fitness for marine cadets or seafarers. *Open Access Repository*, *8*(8), 79–90.

[11] Sagaro GG, Dicanio M, Battineni G, Samad MA and Amenta F. (2021). Incidence of occupational injuries and diseases among seafarers: a descriptive epidemiological study based on contacts from onboard ships to the Italian telemedical maritime assistance service in Rome, Italy. *BMJ Open*, *11*(3), e044633.

[12] Bell SSJ and Jensen OC. (2009). An analysis of the diagnoses resulting in repatriation of seafarers of different nationalities working on board cruise ships, to inform pre-embarkation medical examination. *Medicina Maritima*, *9*(1), 32–43.

[13] Abaya AR, Rivera JJL, Roldan S and Sarmiento R. (2018). Does long-term length of stay on board affect the repatriation rates of seafarers? *International Maritime Health*, *69*(3), 157–162.

[14] Lefkowitz RY, Slade MD and Redlich CA. (2015). Injury, illness, and work restriction in merchant seafarers. *American Journal of Industrial Medicine*, *58*, 688–696.

[15] Oldenburg M. (2014). Risk of cardiovascular diseases in seafarers. *International Maritime Health*, *65*(2), 53–57. doi:10.5603/IMH.2014.0012. PMID: 25231325.

[16] Palinkas, LA. (2003). The psychology of isolated and confined environments: understanding human behavior in Antarctica. *American Psychologist*, *58*(5), 353–363.

[17] Oldenburg, M, Baur, X, Schlaich, C. (2010). Occupational risks and challenges of seafaring. *Journal of Occupational Health*, *52*(5), 249–256.

[18] Zhao Z, Jepsen JR, Chen Z and Wang H. (2016). Experiences of fatigue at sea—acomparative study in European and Chinese shipping industry. *Journal of Biosciences and Medicine*, *4*(3), 65–68.

[19] Bilir N. (1997). *İş Sağlığı, Halk Sağlığı Temel Bilgiler* (Ed. Bertan M and Güler Ç.), Güneş Kitabevi, Ankara.

[20] Jamall, OA. (1984). Musculoskeletal diseases. In: *Handbook of Nautical Medicine*. Berlin: Springer, pp. 203–206.

[21] Lefkowitz RY, Slade MD and Redlich CA. (2018). Injury, illness, and disability risk in American seafarers. *American Journal of Industrial Medicine*, *61*, 120–129.

[22] Mulić R, Vidan P, & Bošnjak R. (2012). Comparative analysis of medical assistance to seafarers in the world and the Republic of Croatia. In *International Conference on Transport Sciences*, (pp. 1–8).

[23] Roberts SE, Nielsen D, Kotłowski A, *et al.* (2014). Fatal accidents and injuries among merchant seafarers worldwide. *Occupational Medicine*, *64*, 259–266.

[24] Kaerlev L, Jensen A, Nielsen PS, Olsen J, Hannerz H and Tüchsen F. (2008). Hospital contacts for injuries and musculoskeletal diseases among seamen and fishermen: a population-based cohort study. *BMC Musculoskeletal Disorders*, *9*, 1–9.

[25] Torner M, Zetterberg C, Anden U, Hansson T and Lindell V. (1991). Workload and musculoskeletal problems: a comparison between welders

and office clerks (with reference also to fishermen). *Ergonomics*, *34*(9), 1179–1196.

[26] Lipscomb HJ, Loomis D, McDonald MA, Kucera K, Marshall S and Li L. (2004). Musculoskeletal symptoms among commercial fishers in North Carolina. *Applied Ergonomics*, *35*, 417–426.

[27] Svendsen SW, Bonde JP, Mathiassen SE, Stengaard-Pedersen K and Frich LH. (2004). Work related shoulder disorders: quantitative exposure-response relations with reference to arm posture. *Occupational and Environmental Medicine*, *61*, 844–853.

[28] Videman, T, Levälahti, E and Battié, MC. (2007). The effects of anthropometrics, lifting strength, and physical activities in disc degeneration. *Spine*, *32*(13), 1406–1413.

[29] Torner M, Blide G, Eriksson H, Kadefors R, Karlsson R and Petersen I. (1988). Musculo-skeletal symptoms as related to working conditions among Swedish professional fishermen. *Applied Ergonomics*, *19*, 191–201.

[30] Torner M, Zetterberg C, Hansson T, Lindell V and Kadefors R. (1990). Musculo-skeletal symptoms and signs and isometric strength among fishermen. *Ergonomics*, *33*. 1155–1170.

[31] Torner MI, Nilsson E and Kadefors R. (1990). The influence of musculoskeletal load, and other factors, on staff turn-over in fishery: a post employment questionnaire study. *Bulletin of the Institute of Maritime and Tropical Medicine in Gdynia*, *41*, 97–108.

[32] Pearce MS, Buttery YE and Brueton RN. (1996). Knee pathology among seamen: a review of 299 patients. *Occupational Medicine (London)*, *46*, 137–140.

[33] Hutchison A. (1969). Health problems of seafarers. *Royal Society of Health Journal*, *89*(3), 117–121. doi:10.1177/146642406908900303.

[34] Roberts SE and Jaremin B. (2010). Cardiovascular disease mortality in British merchant shipping and among British seafarers ashore in Britain. *International Maritime Health*, *62*(3), 107–116.

[35] Battineni, G, Chintalapudi, N and Amenta, F. (2022). Maritime telemedicine: design and development of an advanced healthcare system called marine doctor. *Journal of Personalized Medicine*, *12*(5), 832.

[36] Battineni G and Amenta F. (2020). Designing of an expert system for the management of seafarer's health. *Digital Health*, 6, 2055207620976244, 2020. doi:10.1177/2055207620976244.

[37] Onofri E, Salesi M, Massoni F, Rosati MV and Ricci S. (2012). Medical legal issues associated with the evaluation of herniated discs in seafarers to merchant ships. *La Clinica Terapeutica*, *163*(5), e365–e371. PMID: 23099988.

[38] Hansen, HL, Tüchsen, F and Hannerz, H. (2005). Hospitalisations among seafarers on merchant ships. *Occupational and Environmental Medicine*, *62*(3), 145–150.

[39] Chintalapudi N, Battineni G, Di Canio M, Sagaro GG and Amenta F. (2021). Text mining with sentiment analysis on seafarers' medical documents.

International Journal of Information Management Data Insights, *1*(1), 100005, ISSN 2667-0968. https://doi.org/10.1016/j.jjimei.2020.100005.

[40] Sharma AK, Nandal A, Dhaka A, and Dixit R. A survey on machine learning based brain retrieval algorithms in medical image analysis, *Health and Technology*, Springer, vol. 10, pp. 1359–1373, 2020.

[41] Sharma AK, Nandal A, Dhaka A, *et al.* HOG transformation based feature extraction framework in modified Resnet50 model for brain tumor detection biomedical signal processing and control. *Biomedical Signal Processing & Control*, Elsevier, vol. 84, 2023.

[42] Rak M and Tönnies KD. (2016). On computerized methods for spine analysis in MRI: a systematic review. *International Journal of Computer Assisted Radiology and Surgery*, *11*, 1445–1465.

[43] Galbusera F, Casaroli G and Bassani T. (2019). Artificial intelligence and machine learning in spine research. *JOR Spine*, *2*(1), e1044.

[44] Chiu CC, Chuang TY, Chang KH, Wu CH, Lin PW and Hsu WY. (2015). The probability of spontaneous regression of lumbar herniated disc: a systematic review. *Clinical Rehabilitation*, *29*(2), 184–195.

[45] Karhade AV, Ogink PT, Thio QC, *et al.* (2019). Development of machine learning algorithms for prediction of prolonged opioid prescription after surgery for lumbar disc herniation. *The Spine Journal*, *19*(11), 1764–1771.

[46] Yoon WWand Koch J. (2021). Herniated discs: when is surgery necessary? *EFORT Open Reviews*, *6*(6), 526.

[47] Alsmirat M, Al-Mnayyis N, Al-Ayyoub M and Asma'A AM. (2022). Deep learning-based disk herniation computer aided diagnosis system from mri axial scans. *IEEE Access*, *10*, 32315–32323.

[48] Mbarki W, Bouchouicha M, Frizzi S, Tshibasu F, Farhat LB and Sayadi M. (2020). Lumbar spine discs classification based on deep convolutional neural networks using axial view MRI. *Interdisciplinary Neurosurgery*, *22*, 100837.

[49] Alomari RS, Corso JJ, Chaudhary V and Dhillon G. (2010). Computer-aided diagnosis of lumbar disc pathology from clinical lower spine MRI. *International Journal of Computer Assisted Radiology and Surgery*, *5*, 287–293.

[50] Ghosh S, Raja'S A, Chaudhary V and Dhillon G. (2011). Computer-aided diagnosis for lumbar MRI using heterogeneous classifiers. In *2011 IEEE International Symposium on Biomedical Imaging: From Nano to Macro*, IEEE, Chicago, IL, USA, 2011, pp. 1179–1182.

[51] Koh J, Alomari RS, Chaudhary V and Dhillon G. (2011). Lumbar spinal stenosis CAD from clinical MRM and MRI based on inter-and intra-context features with a two-level classifier. In Ronald M., Summers M.D., Bram van Ginneken (eds.), *Medical Imaging 2011: Computer-Aided Diagnosis*, vol. 7963, pp. 30–37, SPIE, Lake Buena Vista (Orlando), FL, USA.

[52] Bhole C, Kompalli S and Chaudhary V. (2009). Context sensitive labeling of spinal structure in MR images. In Nico Karssemeijer, Maryellen L. Giger (eds.), *Medical Imaging 2009: Computer-Aided Diagnosis*, vol. 7260, pp. 1064–1072, SPIE, Lake Buena Vista (Orlando Area), FL, USA.

[53] Comert C *et al.* (2023). Secure design of cyber-physical systems at the radio frequency level: machine and deep learning-driven approaches, challenges and opportunities. In: Traore, I, Woungang, I and Saad, S (eds.) *Artificial Intelligence for Cyber-Physical Systems Hardening. Engineering Cyber-Physical Systems and Critical Infrastructures*, vol 2. Springer, Cham.

[54] Comert C, Kulhandjian M, Gul OM, *et al.* (2022). Analysis of Augmentation Methods for RF fingerprinting under impaired channels. In *Proceedings of the 2022 ACM Workshop on Wireless Security and Machine Learning* (WiseML '22), Association for Computing Machinery, New York, NY, USA, pp. 3–8.

[55] Gul OM, Kulhandijan M, Kantarci B, Touazi A, Ellement C and D'Amours C. On the impact of CDL and TDL augmentation for RF fingerprinting under impaired channels. In *48th Wireless World ResearchForum (WWRF 2022)*, 07–09 November 2022, UAE, pp. 1–6. 2022 IEEE 27th International Workshop on Computer Aided Modeling and Design of Communication Links and Networks (CAMAD), Paris, France, 2022, pp. 115-120

[56] Gul OM, Kulhandjian M, Kantarci B, Touazi A, Ellement C and D'Amours C. (2022). Fine-grained augmentation for RF fingerprinting under impaired channels. In *2022 IEEE 27th International Workshop on Computer Aided Modeling and Design of Communication Links and Networks (CAMAD)*, Paris, France, pp. 115–120, IEEE.

[57] Gul OM, Kulhandijan M, Kantarci B, Touazi A, Ellement C and D'Amours C. (2023). Secure industrial IoT systems via RF fingerprinting under impaired channels with interference and noise. *IEEE Access*, *11*, 26289–26307.

Chapter 5
Diabetic retinopathy detection using AI

Anas Bilal[1]

One of the most prevalent causes of eyesight loss in the modern world is diabetic retinopathy (DR). Reduced risk of blindness may result from early detection of DR. In this chapter, we used several models to improve the robustness or error-proneness of the DR detection process while classifying the data using the majority voting approach in the early stages of the thesis. The system incorporates DR detection and assessment, feature extraction, categorisation, and picture pre-processing. This research's classification rate is raised by merging an enhanced SVM-RBF with DT and KNN. A total of 516 photos from the IDRiD dataset that were categorised into the following five categories: normal, mild, severe, moderate, and proliferative were used to evaluate the suggested Hybrid model. For DR detection and assessment tests, the suggested hybrid model achieved an accuracy of 98.89%, sensitivity of 85.76%, and specificity of 100%. The results of the experiments demonstrate that the suggested strategy works better than the standard methods.

5.1 Introduction

The eye is an essential part of the human body with nearly 40 interconnected modules, like the retina, iris, pupil, optic nerve, and lens. Many ophthalmic disorders are correlated with the eye such as diabetic retinopathy (DR), cataracts, glaucoma, macular degeneration, keratitis, ocular hypertension, uveitis, and trachoma. DR is a visual manifestation of diabetes and leads to blindness. It is the second leading cause of blindness globally, most frequent in working-age patients in developed nations because of diabetes. According to the World Health Organization (WHO) report on vision, around one billion people have suffered from various ophthalmic disorders such as DR, glaucoma, cataracts, and corneal opacity. DR is the second most occurring retinal disease that can cause blurred vision and blindness in later stages. Generally, DR patients are rising exponentially, projected to reach about 439 million by 2030 [1]. Therefore, the prior detection and diagnosis of DR tend to prevent vision loss in the early stages and have been shown

[1]College of Information Science and Technology, Hainan Normal University, China

to decrease blindness in working age along with collaborative care. To do this, people with diabetes should have frequent eye checkups, which use many resources. This is an expensive exercise for patients, physicians, and the healthcare system, with most clinical and imaging procedures involving mydriasis (a dilated ocular fundus examination). The availability of cost-effective periodic eye tests for DR has become a crucial challenge with growing diabetes prevalence in the first world, which can be addressed through routine screening programs. In such scenarios, the automatic detection of DR in retinal images could provide a better solution [2].

Over the last several years, researchers have developed various machine learning-based prediction models to aid ophthalmologists in detecting and classifying DR. The early detection of DR, including multi-stage DR classification utilising handcrafted feature extraction, has received much attention. The diagnosis of microaneurysms is crucial for the early identification of DR. With advancements in computer-aided diagnosis, numerous automated approaches for automatic DR identification and categorisation have been presented in recent years. Table 5.1 summarises a review of the literature on major approaches for detecting DR.

In this study, we categorise retinal pictures into five categories – normal, mild, moderate, severe, and proliferative – using a combination of enhanced SVM-RBF and unique models based upon that majority voting approach. The remainder of the essay is structured as follows. A comprehensive explanation of the process is provided in Section 5.2. Section 5.3 contains the findings and comments, while Section 5.4 summarises the proposed work.

5.2 Methodology

The suggested process for identifying and categorising DR is described in this section. We begin by summarising the basic design of the suggested technique, which includes the whole training and assessment stages. Then, in the following subsections, descriptions of the various modules are provided. As shown in Figure 5.1, the technique for the proposed approach consists of preprocessing, extracting features, and finally DR classification.

5.2.1 Preprocessing

Adaptive histogram equalisation, contrast stretching, and a median filtering method are all carried out in this stage with the goal of improving the provided fundus picture. The fundus image's brightness, chroma blue CB, as well as chroma red CR colour spaces are used for the preprocessing stage. In order to create the YCbCr picture (transformation of RGB image to YCbCr image), it adds two of the RGB's chromatic colour components to the intensity (Y). To use the median filtering technique, the Y component of the YCbCr picture is separated. The contrast is stretched after that, at which point the intensity normalisation technique is used. Later, the picture is converted from YCbCr to RGB.

Table 5.1 AI-Powered Approaches to VTDR Identification

Reference	Findings	Proposed methodology	Shortcomings
[3]	DR lesion detection	VGG-16 and Resnet-50	Inaccurate micro-aneurysm detection may result from fluorescein's ubiquitous presence in the bloodstream.
[4]	DR classification based on lesions severity	Dense-121, Dense-169, Inception-V3, Resnet-50, and Xception.	The method is quite computationally expensive.
[5]	Achieve more than 90% spec and sens using DL	DL- pixel, highlighting the significance of each input sequence. The assigned score served as the basis for the final categorisation decision.	It may improve the algorithm's evaluation performance by taking appropriate measures.
[6]	DR detection with a kappa value of 0.829	Inception V3	If there are no matching fundus photographs in the sample datasets, this could not work properly.
[7]	The proposed model achieved a spec and sens value of 97.7% and 97.5%, respectively.	DeepDR	In order to fully evaluate the model described, it has to be applied to more complex and larger datasets.
[8]	The proposed framework achieved sens, acc, and spec of 83.67%, 98.06%, and 100%.	SVM mixed models	Future studies may use deep learning methods in order to evaluate their efficacy in light of previous studies.
[9]	State-of-the-art performance on three publically available datasets.	A U-Net and transfer learning	Combining ensemble ML and DL techniques may enhance our model's categorisation of retinal illnesses like cataracts and glaucoma.
[10]	DR classification with 92.11% accuracy	The creation of this technique took into account morphological, geometrical, as well as orientational factors. SVM was used in the classification process.	There is still room for improvement in the detection accuracy
[11,12]	Achieve acc of 96.91% and 98.33%, respectively.	SVM with the optimisation algorithm.	Need to integrate methods into upcoming work using high-performance technology by offering proposed procedures for various real-world situations.

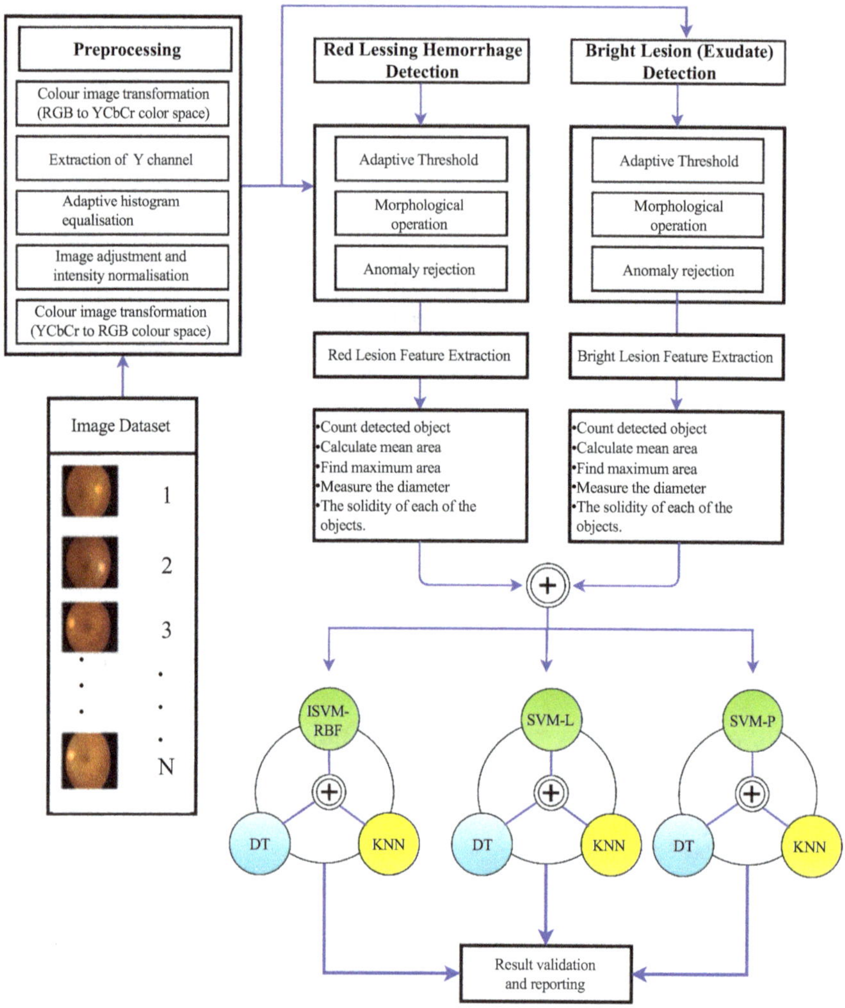

Figure 5.1 Block diagram of the proposed method

5.2.2 Feature extraction

Bright as well as red lesion identification algorithms get the preprocessed Fis and then extract features from the places they have identified.

5.2.3 Classification

The characteristics are extracted, and then three classifiers, ISVM, KNN, and DT, begin the classification process. With a sensitivity level of 0.85, each technique enhances pictures using an adaptive threshold algorithm. After that, the morphological procedure removes the noise from the image's output.

5.2.4 Proposed method algorithm

In Figure 5.1, we can see an example of the suggested work algorithm. The following segmentation stages are:

Step 1: Preprocessing

This module defines the necessary preprocessing for the better detection of the MAs and to remove the inherent and external noise induced in the fundus images during the creation and transmission process. The following is the sequence of steps we do follow:

- Convert the image from RGB to YCbCr and only take the Y component
- Apply a median filter
- Apply contrast stretching and intensity normalisation
- Recover the YCbCr
- Convert the YCbCr back to the RGB.

Step 2: Detection and feature extraction of the red and bright lesion

- Red and bright lesion detection.

  Apply adaptive segmentation with a sensitivity value of 0.15 with dark foreground polarity and a sensitivity value of 0.85 with bright foreground polarity for red and bright lesions, respectively.
  - Obtain the segmented properties, extends, and aspect ratio
  - Filter out the segmented image as per extent and aspect ratio
  - Apply area filtering and morphological closing
- Red and bright lesion feature extraction
  - Obtain all-region properties
  - Obtain the number of regions (1st), mean area (2nd), mean perimeter (3rd) and mean solidity (4th) feature
  - Stack all the features together.

Step 3: Fusion

This segmentation step is employed after the red and the white lesion feature extraction process is completed.

- The red lesion features and bright lesion features are appended lexicographically.

Step 4: Training algorithm

This segment describes the essential steps of the image construction (training) and set of targets and the following sequence of steps we follow:

- For $i = 1$, to the number of training images
- Obtain the ith image from the database
- Apply to preprocess and extract the features
- Stack the feature results to the training image array
- Assign a target class based on the severity of the dataset.

Step 5: Train classification

This sub-segment defines the required steps to train the selected classifier, and the following classifiers are chosen to perform the operations.

- Train the Improved SVM-RBF, SVM-Polynomial, SVM-Linear, KNN, and DT classifier.

Step 6: Testing algorithm

This segment defines the overall processing steps to predict the results through the feature extraction process by the given classifier and the following sequence of steps, we follow:

- Obtain the desired test image
- Apply image preprocessing and extract the features
- Predict the Improved SVM-RBF, SVM-Polynomial , SVM-Linear, KNN, and DT classifier features.
- Obtain the model of five prediction results.

5.2.5 Training and testing

Figure 5.3 details with the training and testing. The preprocessing stage of the whole training is the first step in the training process, after which the features are collected from the exudates and haemorrhage zones. The target classes each provide one of the three classifiers with the training features extracted from each picture. The classifiers are then set aside for testing. Similarly, the preparation phase also starts the testing procedure. Additionally, the pattern of votes is determined, and each classifier's prediction results are regarded as a vote. The greater vote may later determine categorisation.

5.2.6 Novel ISVM-RBF

Taking the maximal margin C into account allowed for enhancements to be performed in the hyperplane, and the gamma (γ) values and categorisation were adjusted correspondingly. The generic classifier can be stated as follows:

$$f(s) = \sum_{i=1}^{n} \propto_i \; k(s, s') \tag{5.1}$$

The hyperplane and kernel function is denoted by function $f(s)$ and $\propto_i$ respectively. Additionally, the feature vector may well be described as

$$k(s, s') = exp\left[-\gamma||s - s'||_2^2\right]$$

i.e.,

$$k(s, s') = \exp\left[-\frac{||s - s'||^2}{2\sigma^2}\right] \tag{5.2}$$

where $s - s'^2$ and σ represent the distance in Euclidean space and the free parameter, respectively. The output of the RBF classifier is

$$f(s) = \sum_{i=1}^{n} \propto_i \exp\left[-\frac{||s - s'||^2}{2\sigma^2}\right] \quad (5.3)$$

5.3 Results and discussion

All the testing was carried out in the coding environment of MATLAB®. We utilised a 16 GB RAM plus 1TB SSD system with an Intel Core i7 CPU. In this part, we emphasise the main outcomes of the image processing as well as classifier results. A comparison between the suggested work and traditional procedures is also shown.

5.3.1 Dataset

We evaluated this suggested study using the Indian diabetic retinopathy imaging dataset (IDRiD), which was formally the first database from India. Figure 5.2 shows the samples used in this research, which uses an 8:2 ratio of 80% photos for training and 20% for testing.

5.3.2 Image processing results

In this part, we compare the images obtained during the preprocessing stage and those obtained in the final output of the classification process. The findings of a research study for condition grade 4 are effectively achieved in the detecting zones, as depicted in Figure 5.3. No metrics are measured for such segmentation results since the illness grading database used here does not give any ground truth again for segmentation.

Table 5.2 displays a comparison of several classifier findings. It is noteworthy that the combined model outperforms the separate models. The ISVM-RBF performs better than KNN, DT, SVM-P, as well as SVM-L across all classifiers [13–15].

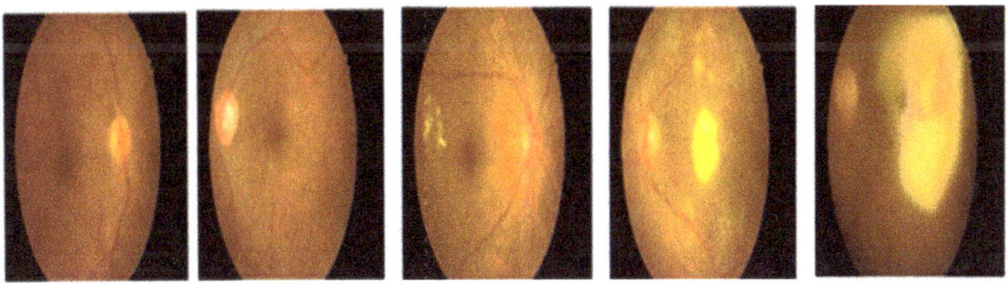

(a) (b) (c) (d) (e)

Figure 5.2 Sample images from the IDRiD dataset: (a) normal, (b) mild, (c) moderate, (d) severe, and (e) PDR

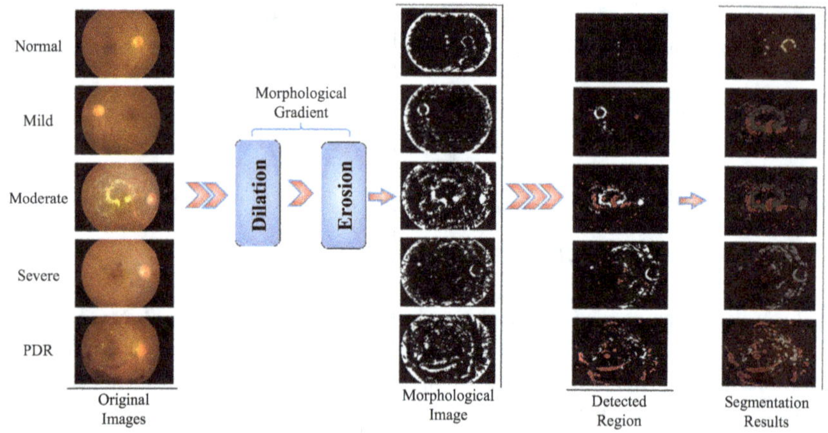

Figure 5.3 Image processing steps: (a) original DR image, (b) enhanced image, (c) detected region (bright and red), and (d) segmentation results

Table 5.2 Various classification results scored in this work

Model	Severity threshold	Acc	Sen	Spec	F1-Sc
KNN	1	86.93	85.81	90.21	8573
	2	92.24	90.52	92.69	8297
	3	94.83	86.53	94.43	8732
	4	95.26	80.29	93.26	9211
Binary trees	1	84.10	94.47	88.52	9563
	2	94.35	93.14	73.63	9121
	3	91.62	87.27	84.62	9031
	4	92.54	83.10	92.46	8298
SVM-polynomial (mixed models)	1	91.10	93.40	92.82	9299
	2	91.52	94.28	88.16	9283
	3	94.59	81.84	98.56	8987
	4	96.43	78.45	100.00	8637
SVM-linear (mixed models)	1	92.33	92.93	92.40	9043
	2	92.82	93.56	87.69	9203
	3	93.52	83.29	95.02	8446
	4	96.51	72.24	97.53	7276
ISVM-RBF (mixed models)	1	96.93	95.89	97.72	9335
	2	95.63	97.92	97.40	9017
	3	96.90	92.60	97.56	9194
	4	98.89	85.76	100.00	9212

The results showed that ISVM-RBF techniques outperformed other approaches in terms of results. Figure 5.4 illustrates how the ISVM-RBF strategy predicts more accurately than other techniques. Figures 5.4–5.6 show the comparative analysis of accuracy, sensitivity, and specificity, respectively. Higher illness severity

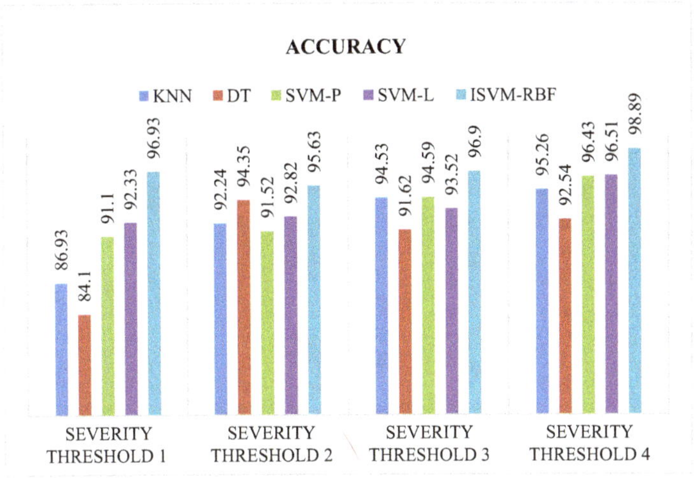

Figure 5.4 ISVM-RBF, KNN, DT, SVM-P, and SVM-L accuracy comparison

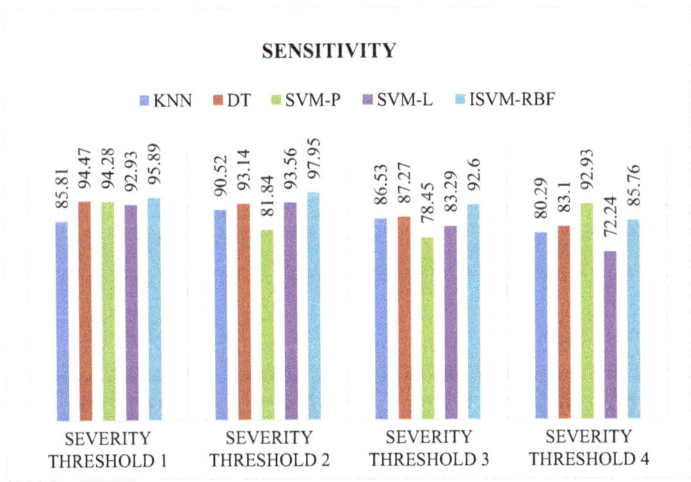

Figure 5.5 ISVM-RBF, KNN, DT, SVM-P, and SVM-L sensitivity comparison

translates to, on average, greater performance. The characteristics look more discriminating due to the sharper illness manifestation at increasing severities. As a result, it improves the performance of the categorisation process. The combined models (dashed red line) outperformed other classifiers in performance. Nevertheless, each item in this table represents the percentile of all the samples included in the test set. The mixed model's categorisation has an overall prediction rate of 92.3%.

5.3.3 Comparison with the state-of-the-art studies

The suggested study and typical research, in which the authors used several databases to describe their findings, are contrasted in Table 5.3. Since few studies consider the accuracy and sensitivity parameters and other research only considers accuracy, the parameters utilised in this comparison represent a trade-off. The outcomes of the suggested work were more accurate and specific. However, the suggested study has a higher sensitivity than most research.

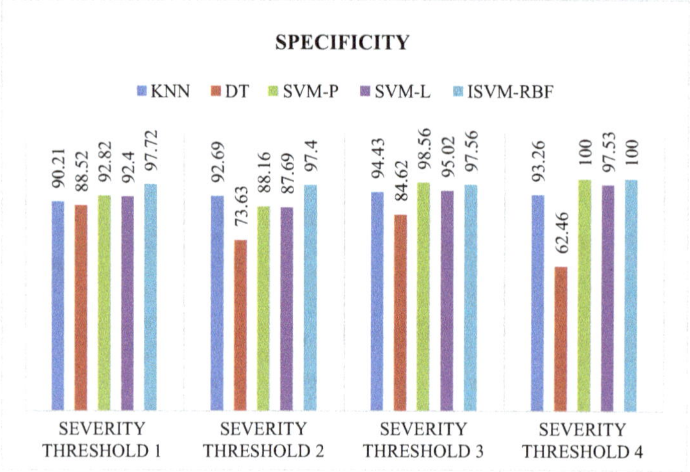

Figure 5.6 ISVM-RBF, KNN, DT, SVM-P, and SVM-L specificity comparison

Table 5.3 A comparison between the proposed work and state-of-the-art

Author	Method	Year	Dataset	Acc (%)	Sen (%)	Spec (%)
[16]	CNN + handcrafted features	2019	IDRiD	90.70	-	-
[17]	CANeT	2020		92.60	-	-
[18]	CNN			90.29	88.75	96.89
			MESSIDOR	90.89	88.75	96.30
[19]	RSNET		IDRiD	86.33	-	-
[20]	CNN		IDRiD	81.00	-	-
			Kaggle			
[21]	Fine KNN	2021	IDRiD	94.00	-	-
			MESSIDOR	98.10	-	-
[22]	CNN-Inception V3	2022	MESSIDOR	97.92	-	97.44
Proposed work	**ISVM-RBF mixture model**	**2023**	**IDRiD**	**98.89**	**85.76**	**100.00**

5.4 Conclusion

The methodology for identifying and categorising red and brilliant lesions is presented in this research. The approach in this study implements three classifiers together with a combined voting mechanism after preprocessing feature extraction and classification processes. The suggested work's accuracy and specificity were higher than the state-of-the-art levels of 98.89% and 100%, respectively. However, as the whole method solely relies on the preprocessing and feature extraction process, the outcomes are always a trade-off between the necessary parameters. The voting mechanism for the three classifiers strengthens the traits that increase dependability. Future research may include implementing deep learning approaches into the algorithm to compare the outcomes with the existing work.

Funding

This research was funded by the Foreign Young Talents Programme of the State Bureau of Foreign Experts Ministry of Science and Technology China. No. QN2023034001

References

[1] OMS, "*Global Report on Glaucoma*," Isbn, 2016. https://sci-hub.si/https://apps.who.int/iris/handle/10665/204874%0Ahttps://apps.who.int/iris/bitstream/handle/10665/204874/WHO_NMH_NVI_16.3_eng.pdf?sequence=1%0A http://www.who.int/about/licensing/copyright_form/index.html%0Ahttp://www.who.int/about/licens.

[2] Abramoff, M. D., Niemeijer, M., and Russell, S. R., "Automated detection of diabetic retinopathy: barriers to translation into clinical practice," *Expert Rev. Med. Devices*, vol. 7, no. 2, 2010, doi: 10.1586/erd.09.76.

[3] Pan, X., Jin, K., Cao, J., *et al.*, "Multi-label classification of retinal lesions in diabetic retinopathy for automatic analysis of fundus fluorescein angiography based on deep learning," *Graefe's Arch. Clin. Exp. Ophthalmol.*, vol. 258, no. 4, 2020, doi: 10.1007/s00417-019-04575-w.

[4] Qummar, S., Khan F. G., Shah S., *et al.*, "A deep learning ensemble approach for diabetic retinopathy detection," *IEEE Access*, vol. 7, 2019, doi:10.1109/ACCESS.2019.
2947484.

[5] de la Torre, J., Valls, A. and Puig, D., "A deep learning interpretable classifier for diabetic retinopathy disease grading," *Neurocomputing*, vol. 396, pp. 465–476, 2020, doi: 10.1016/j.neucom.2018.07.102.

[6] Zeng, X., Chen, H., Luo, Y. and Ye, W., "Automated diabetic retinopathy detection based on binocular siamese-like convolutional neural network," *IEEE Access*, vol. 7, pp. 30744–30753, 2019, doi: 10.1109/ACCESS.2019.2903171.

[7] Zhang, W., Zhong J., Yang S., *et al.*, "Automated identification and grading system of diabetic retinopathy using deep neural networks," *Knowl.-Based Syst.*, vol. 175, pp. 12–25, 2019, doi: 10.1016/j.knosys.2019.03.016.

[8] Bilal, A., Sun, G., Li, Y., Mazhar, S. and Khan, A. Q., "Diabetic retinopathy detection and classification using mixed models for a disease grading database," *IEEE Access*, vol. 9, pp. 23544–23553, 2021, doi:10.1109/ACCESS.2021.3056186.

[9] Bilal, A., Sun, G., Mazhar, S., Imran, A. and Latif, J., "A transfer learning and u-net-based automatic detection of diabetic retinopathy from fundus images," *Comput. Methods Biomech. Biomed. Eng. Imaging Vis.*, vol. 10, no. 6, pp. 1–12, 2022, doi: 10.1080/21681163.2021.2021111.

[10] Rekhi, R. S., Issac, A. and Dutta, M. K., "Automated detection and grading of diabetic macular edema from digital colour fundus images," In *2017 4th IEEE Uttar Pradesh Section International Conference on Electrical, Computer and Electronics, UPCON 2017*, IEEE: New York, NY, USA, 2017, vol. 2018 January. doi: 10.1109/UPCON.2017.8251096.

[11] Bilal, A., Sun, G. and Mazhar, S., "Diabetic retinopathy detection using weighted filters and classification using CNN," In *2021 International Conference On Intelligent Technologies CONIT 2021*, IEEE: New York, NY, USA, 2021, doi: 10.1109/CONIT51480.2021.9498466.

[12] Bilal, A., Sun, G., Mazhar, S. and Imran, A., "Improved grey wolf optimization-based feature selection and classification using CNN for diabetic retinopathy detection," *Lect. Notes Data Eng. Commun. Technol.*, vol. 116, pp. 1–14, 2022, doi: 10.1007/978-981-16-9605-3_1.

[13] Sharma, A. K., Nandal, A., Dhaka, A., *et al.*, "Enhanced watershed segmentation algorithm based modified ResNet50 model for brain tumor detection," *BioMed Res. Int., Hindawi*, 2022.

[14] Sharma, A. K., Nandal, A., Dhaka, A., *et al.*, " HOG transformation based feature extraction framework in modified Resnet50 model for brain tumor detection biomedical signal processing and control," In *Biomedical Signal Processing & Control*, Elsevier, vol. 84, p. 104737, 2023.

[15] Sharma, A. K., Nandal, A., Dhaka, A., Dixit, R., "A survey on machine learning based brain retrieval algorithms in medical image analysis," In *Health and Technology*, Springer, vol. 10, pp. 1359–1373, 2020.

[16] Harangi, B., Toth, J., Baran, A. and Hajdu, A., "Automatic screening of fundus images using a combination of convolutional neural network and hand-crafted features," in *Proceedings of the Annual International Conference of the IEEE Engineering in Medicine and Biology Society, EMBS*, IEEE: New York, NY, USA, 2019, doi:10.1109/EMBC.2019.8857073.

[17] Li, X., Hu, X., Yu, L., Zhu, L., Fu, C. W. and Heng, P. A., "CANet: cross-disease attention network for joint diabetic retinopathy and diabetic macular edema grading," *IEEE Trans. Med. Imaging*, vol. 39, no. 5, pp. 1483–1493, 2020, doi:10.1109/TMI.2019.2951844.

[18] Elswah, D. K., Elnakib, A. A. and El-Din Moustafa, H., "Automated diabetic retinopathy grading using resnet," in *National Radio Science Conference, NRSC, Proceedings*, IEEE: New York, NY, USA, 2020, vol. 2020-September. doi:10.1109/NRSC49500.2020.9235098.

[19] Saranya, P. and Prabakaran, S., "Automatic detection of non-proliferative diabetic retinopathy in retinal fundus images using convolution neural network," *J. Ambient Intell. Humaniz. Comput.*, vol. 15, pp. 1–10, 2020, doi: 10.1007/s12652-020-02518-6.

[20] Alcalá-Rmz, V., Maeda-Gutiérrez, V., Zanella-Calzada, L. A., Valladares-Salgado, A., xCelaya-Padilla, A. and Galván-Tejada, C. E., "Convolutional neural network for classification of diabetic retinopathy grade," in *Lecture Notes in Computer Science (Including Subseries Lecture Notes in Artificial Intelligence and Lecture Notes in Bioinformatics)*, Springer Nature Switzerland AG 2020, vol. 12468 LNAI. doi:10.1007/978-3-030-60884-2_8.

[21] Bhardwaj, C., Jain, S. and Sood, M., "Hierarchical severity grade classification of non-proliferative diabetic retinopathy," *J. Ambient Intell. Humaniz. Comput.*, vol. 12, no. 2, 2021, doi: 10.1007/s12652-020-02426-9.

[22] Bilal A., Zhu L., Deng A., Lu H. and Wu N. "AI-based automatic detection and classification of diabetic retinopathy using u-net and deep learning," *Symmetry,* vol. 14, no. 7, pp. 1427, 2022.

Chapter 6

A survey image classification using convolutional neural network in deep learning

Fathima Ghouse[1] *and Rashmika*[1]

The processing of image data for transmission and machine perception, in addition to providing enhanced visual data for human simplification, makes image processing an always fascinating field. Digital images can be made better through the use of image processing. Grayscale conversion, image segmentation, edge detection, feature extraction, and classification are all methods of image processing. The current deep learning convolutional neural network-based image classification system is discussed in detail in this chapter. The fundamental motivation behind the work introduced in this chapter is to think about and examine the various models involving convolutional brain network in profound learning for picture grouping. DenseNet, MobileNet, VGGnet, GoogleNet, ResNet, ImageNet, NasNet, and AlexNet were among the standard models that the algorithm was applied to. The most famous convolutional brain network is utilized for object discovery and item classification groupings from pictures are AlexNet, GoogleNet, and ResNet. Generally, the chapter gives definite information on picture handling and order strategies.

6.1 Introduction

If we take into account a grayscale image, the values of each pixel in the matrix will range from 0 to 255—an image is a matrix of pixel values. In the event that we consider an RGB image, the combined values of R, G, and B will be present in each pixel. In order to extract useful information from an image, it is possible to convert it into digital form. The process takes the image as an input and use effective algorithms to provide outcomes that could be an image, data, or features related to that image. Image rectification, image enhancement, image classification, images fusion, and other fundamental operations are all part of image processing. The method for classifying images consists of training the framework and then testing it. The preparation cycle takes the trademark properties of the pictures and

[1]Adhiyamaan College of Engineering, India

structures an overall extraordinary model for a specific class. Regardless of whether the classification problem is binary or multi-class, the procedure is carried out for all classes. Using the system's trained generalized model, the test images are categorized in the testing step into various classes. The division of the training features into classes serves as the basis for class assignment.

An artificial neural network (ANN) is a type of network that is based on a biological neural network and is similar to the neurons in the human brain in that it is interconnected with other networks in a similar way. Nonlinear dynamic modeling and identification of systems as well as the solution of complex problems are made possible by ANNs. Convolutional neural networks (CNN) are the most effective at solving computer vision problems. CNN is a more powerful and accurate method for solving classification problems in general. Recurrent neural network (RNN) can also be known as an ANN that uses the time-series data. Natural language processing (NLP) and speech recognition is a specialty of RNNs. They are both distinct in their mathematical approaches, which allows them to solve specific problems more effectively. While CNNs are superior for dealing with spatial data like images, RNNs are excellent for dealing with temporal or sequential data. Despite the fact that sparsity is introduced and the same neurons and weights are reused over time in both RNN and CNN, both models operate in a similar manner. A benefit of using CNNs is that they can create an internal representation of a 2D image. When working with images, this enables the model to learn position and scale in various data structures.

CNNs are a special type of network design for deep learning algorithms that are utilized for tasks like image recognition and pixel data processing. Although there are various forms of deep learning neural networks, CNN is the network design of choice for applications involving image identification, classification, and computer vision. CNN has a number of hidden layers that aid with information extraction from images. Convolutional, ReLU, polling, and fully linked layers are the four crucial CNN layers. Multiple filters work together to execute the convolution process in a convolutional layer. After performing an element-by-element operation, the ReLU layer sets all of the negative pixels to 0. A downsampling technique called polling layer decreases the feature's dimensionality.

6.2 Deep learning

Profound learning is a subset of artificial intelligence (AI) where a three-or more-layered brain network is utilized. Deep learning is used by many AI applications and services to automate more physical and analytical tasks without the need for human intervention. An ANN, a type of advanced machine learning algorithm, provides support for the majority of deep learning models. As a consequence of this, deep learning is also referred to as deep neural learning or deep neural networking. There are a few kinds of brain organizations, including repetitive brain organizations, convolutional brain organizations, fake brain organizations, and feedforward brain organizations, each with benefits for explicit use cases.

However, they all function in a somewhat similar manner: data are fed into the model, and the model determines whether it has made the right decision or interpretation regarding a particular data element.

6.2.1 Artificial neural network

The biological neural networks that shape the human brain are the inspiration for the ANN. Like the human brain, ANNs have layers of neurons that are interconnected to one another. In the field of computerized reasoning, a fake brain organization endeavors to copy the organization of neurons that makes up the human mind so PCs can figure out things and pursue choices in a human-like way. In the ANN, computers are programmed to behave similarly to interconnected brain cells. An affiliation point for every neuron is somewhere in the range of 1,000–100,000. The human mind stores information in a way that makes it more likely to spread, and we can recall more than one piece of it at once if it is fundamental. The human brain is made up of amazing parallel processors, as we can say [1]. Three layers make up this ANN: The input layer receives external data, the hidden layers process and compute information, and the output layer provides the final result or prediction based on the computations in the hidden layers.

6.2.2 Recurrent neural network

An ANN that works with sequential or time-series data is known as a RNN [2]. RNN is utilized in profound learning and the formation of models that copy neuron action in the human mind. All of the calculations are stored in RNNs' "memory." Because it generates the output for all inputs or hidden layers, it uses the same parameters for each input. In contrast to other brain organizations, this lessens boundary intricacy. The brain organization's feedback is received and handled by the info layer x prior to being given to the center layer. In the center layer h, there are a few secret layers, each with its own enactment works, loads, and predispositions. An intermittent brain organization can be utilized on the off chance that the different boundaries of various secret layers are not impacted by the first layer, for example the brain network has no memory. The intermittent brain organization will normalize the different enactment works, loads, and inclinations, guaranteeing that each secret layer has similar properties. Instead of making different secret layers, it will just make one and circle over it however many times depending on the situation.

6.2.3 Feed forward neural network

It is type of ANN in which nodes cannot be formed as loops. It can also be known as multilayer neural layer because the data or information is moving forward. During the flow of data, the input layer will receive the data and travels through the hidden layer, does some operation in it, and exists in the output layer. It consists of a two-layered network with the hidden neurons and trained with the help of LMBP algorithms.

6.3 Convolutional neural network

One of the most well-known deep neural networks is the CNN. In design recognition, the desire to use stowed away layers has outperformed conventional methods. Profound learning has ended up being a useful asset since it has capacity to deal with enormous measure of information. Image classification, similarity detection, and object recognition are the primary applications of a type of ANN known as a convolutional neural network (also referred to as a CNN). It can perceive face, individual, sign, etc. CNNs, which are primarily used for image classification, are now used in every field that has a problem that requires classification and in a number of tasks that perform exceptionally well across a variety of applications. One of the first applications in which CNN architecture was successfully implemented was handwritten digit recognition. The network has improved with the addition of new layers and various computer vision techniques since CNN's creation. The CNN is depicted in Figure 6.1 as a list of layers that convert the 3D image volume's width, height, and depth into a 3D output volume. The input convolution and pooling layers, fully connected (FC) layers, and the output layer for classification make up the CNN architecture that Jinglan Zhang [3] describes. Each of these layers serves a distinct purpose. The completely associated layer comes after the convolutional layer, making the FC layer more mind boggling. As an image's complexity increases, the CNN will be capable of successively becoming aware of larger quantities and more complex capabilities until it identifies the item in its entirety [4]. When compared to the others, the final fully linked layer activation feature frequently stands out. The ideal activation feature must be selected for each interest. A neuron's need to be activated is determined by

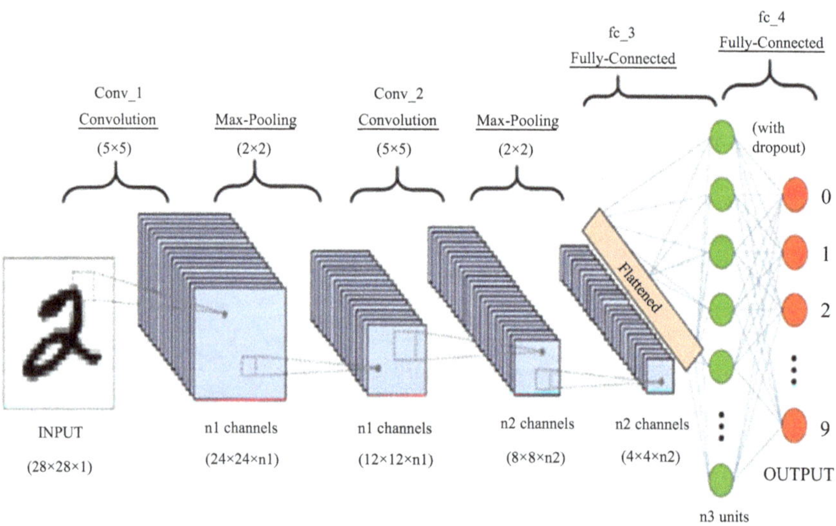

Figure 6.1 Convolutional neural network

its activation function. This indicates that it determines whether or not the neuron's entry into the network is significant, necessitating the use of a simpler mathematical operation.

Newer architectures investigate novel ways of constructing convolutional layers to improve learning efficiency, whereas traditional network architectures were only made up of stacked convolutional layers. For a wide range of computer vision issues, these architectures offer general architectural recommendations that practitioners of machine learning can adapt. Rich features can be extracted using these architectures for advanced tasks like image classification, object identification, and segmentation. By improving the model's accuracy, CNN architecture's performance can be improved. The following is a list of some methods for increasing accuracy.

6.3.1 Convolutional layer

The convolutional layer is one of the structure blocks of CNNs as a greater part of the calculation is finished in this layer which is utilized to separate the different elements from the information pictures. When the image is black and white, it is entered into the convolutional layer as a 2D array with values between 0 and 255, where 0 indicates that the image is entirely white and 255 indicates that it is completely black. If the image is variety, it is a 3D RGB cluster with a range of 0–255. In Figure 6.2, the numerical activity is carried out in a channel of a particular size between the information pictures. The spot item between the channel and the pieces of the info picture that are of a similar size as the channel is taken by sliding the channel over the pictures. The output tells us about the image's edges and corners, among other things.

6.3.2 Pooling layer

The pooling layer, which is followed by the convolution layer, is used to shrink the dimensions. Because of this, the spatial length of the representation can be reduced,

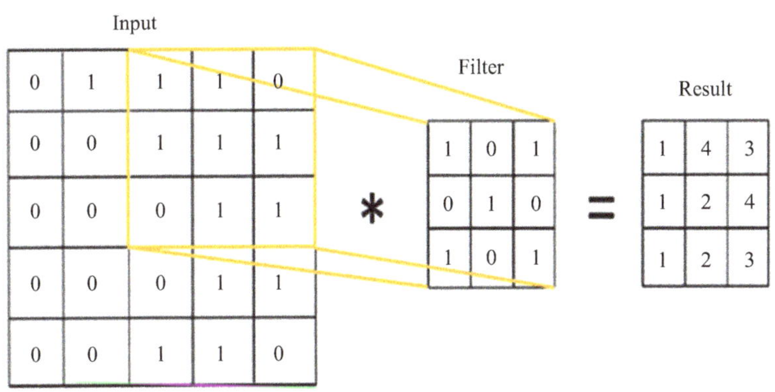

Figure 6.2 Convolutional layer

which reduces the need for a lot of computation and weights. The feature map stores the precise positions of features in the input, so if an object in an image has moved slightly, it might not be recognized by the convolutional layer. Translational invariance is provided by the pooling layer, which means that CNN input remains translational. The highlights will be perceived of the info picture with the assistance of CNN. Pooling can take one of two forms: average pooling and maximum pooling (Figure 6.3). Max pooling is utilized to find the most extreme worth in the pixel from a part of the picture covered by the portion (Figure 6.4). The common value from the kernel-protected portion of the image is returned by average pooling.

6.3.3 Fully connected layer

The layer ends with the fully connected (FC) layer. It consists of the neurons between two distinct layers, as well as the weights and biases. Due to the interconnected nature of each neuron, it is referred to as the FC layer. This layer makes use of activation features like ReLU and Sigmoid to carry out the type manner. In the output layer of neural network models that use the softmax function to forecast a multinomial probability distribution. For problems involving multiple classes with more than two labels, softmax is used as the activation function. Figure 6.5 shows that the input layer, the FC layer, and the output layer are the three layers.

6.3.4 Dropout layer

When the model overfits, the dropout layer is used to get rid of the unwanted layer that caused the overfitting. As a result, unwanted neurons from FC layers must be removed at a rate of 0.5 to prevent this layer from overfitting. Because they prevent overfitting on the training data, dropout layers are useful in CNN training. In the

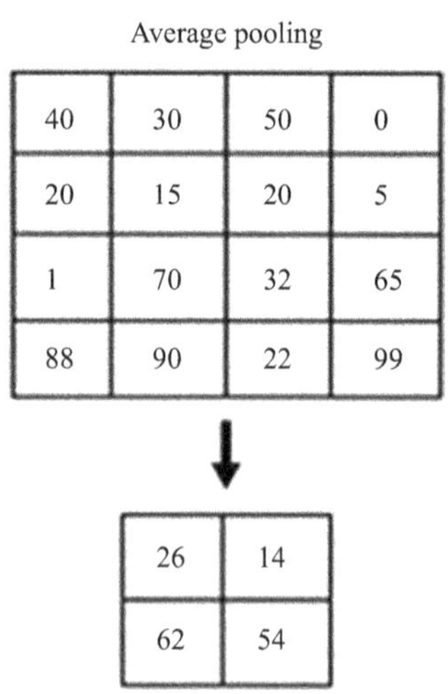

Figure 6.3 Average pooling

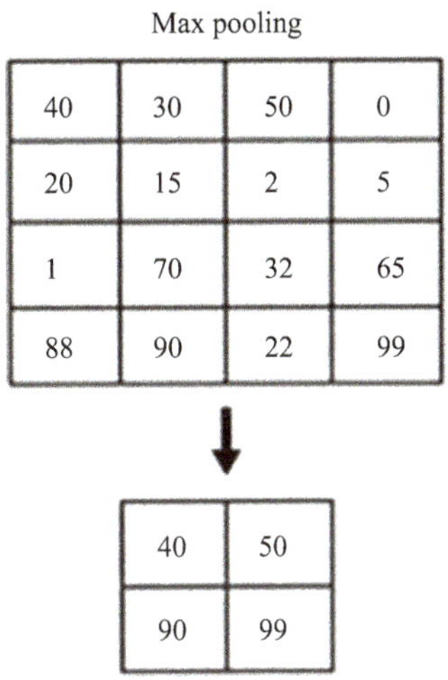

Figure 6.4 Max pooling

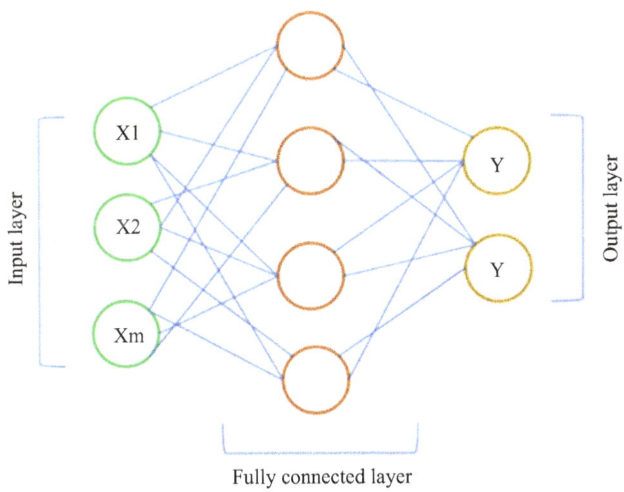

Figure 6.5 Fully connected layer

absence of the initial batch of training samples, learning is disproportionately affected. As a result, it would be impossible to learn features that are only present in subsequent batches or samples. Figure 6.6 does not include a dropout layer and does not reduce overfitting, Figure 6.7 has a dropout layer that skips some layers.

6.3.5 Softmax layer

This softmax layer is related to the cross-entropy function. In a CNN, the softmax layer tests the model's reliability by utilizing a loss function and a cross-entropy function for optimal performance. It is utilized to change the unnormalized yield

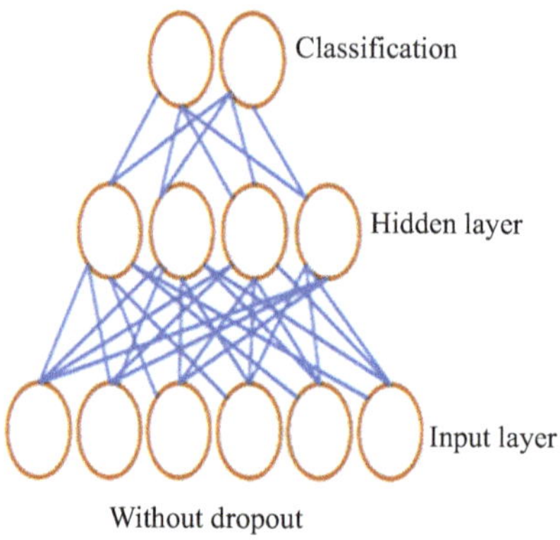

Figure 6.6 Without dropout layer

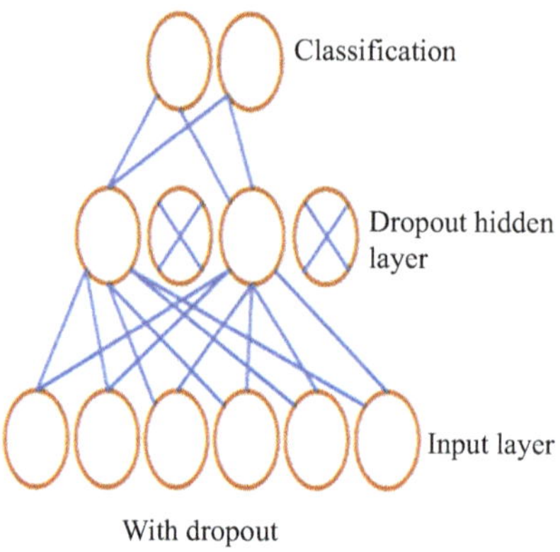

Figure 6.7 With dropout layer

into a standardized result, which is addressed as the vector of k components. The CNN has no use for the softmax layer if probabilities are not required.

6.4 CNN models

6.4.1 VGGnet

The multilayered deep CNN architecture of the VGG, which was developed in 2014 and was rewarded as one of the best excellent vision models, stands for visual geometry group. In order to improve model performance, this VGGnet model was created to increase CNN's depth. VGG is a kind of brain network design that is used for picture object grouping and discovery. VGG, a novel object recognition model with up to 19 layers, is shown in Figure 6.8. Unlike ImageNet, VGG, which was designed as a powerful CNN, outperforms baselines on numerous tasks and datasets. VGG is currently one of the most widely used image-recognition architectures.

The neural network with the 16 layers and weights is represented by the number 16 in the name VGG. It has an input layer, a hidden layer, a convolutional layer, and a FC layer. Thirteen of the 16 layers in VGG are convolutional, and three are FC. In ImageNet, this model is accurate to 92%, and the dataset contains 14 million images. As a result, VGG16 is a substantial network with roughly 138 million parameters. Indeed, even by current principles, it is a gigantic organization. However, what attracts users to the network is the VGGNet16 architecture's simplicity [5]. The height and width are reduced by a pooling layer after a few convolution layers. There are approximately 64 filters that can be used, which can

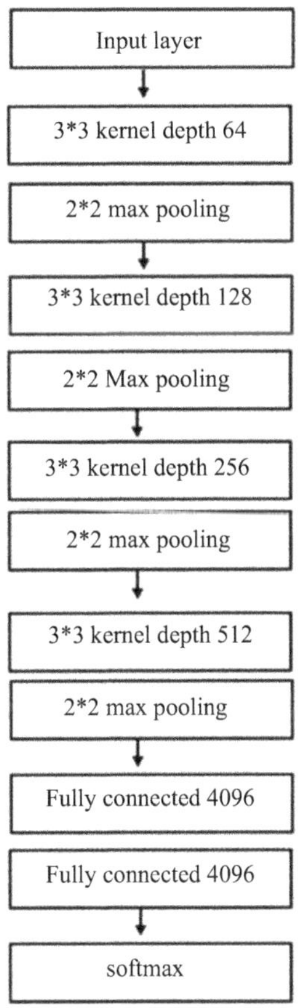

Figure 6.8 VGG architecture

be doubled to approximately 128 and then 256. The quantity of channels will be multiplied in each step by utilizing the convolutional layer.

VGG19 is a modern CNN with pre-prepared layers and a strong comprehension of how shape, variety, and construction characterize a picture. VGG19 is an extremely profound brain network that has been prepared utilizing complex grouping undertakings on large number of pictures. There are 19 layers in the VGG-19 CNN. A pretrained version of the network that was trained on more than one million images can be loaded into the ImageNet database [5]. A total of 14 million images are organized into 1,000 categories on ImageNet. Images of keyboards, mice, pencils, and various animals can be categorized by the pretrained network. As a machine learning model, the DVM classifier and the deep learning models VGG16, VGG19, and InceptionV3 were utilized. The classification performance of the DVM model is 71.06%, while that of the VGG16 model is 96.44%, that of the VGG19 model is 97.96%, and that of the InceptionV3 model is 72.08%.

6.4.2 AlexNet

AlexNet was the first convolutional brain organization to utilize GPU to further develop execution and was created in year 2012 by Alex Krizhevsky. The AlexNet architecture has eight layers: five convolutional layers, three FC layers, and some intermediate layers like pooling and the activation layer, which can speed up the training process (see Figure 6.9). AlexNet has 60 million learning parameters, and the main problem is overfitting. Each convolutional layer employs the ReLU nonlinear activation function [6]. To lessen the quantity of element maps, the AlexNet network has executed cushioning. The information size is fixed due to the presence of completely associated layers. The input size is stated to be 224×224×3, and it comes in the form of RGB images with 96 kernels. These kernels are connected with four strides to the previous layer of kernels, which results in a size of 227×227×3 due to padding. A pretrained version of the network that was trained on more than one million images can be loaded into the ImageNet database [6]. Images of keys, mice, pencils, and a variety of animals can be categorized by the pretrained network into one of the 1,000 object categories.

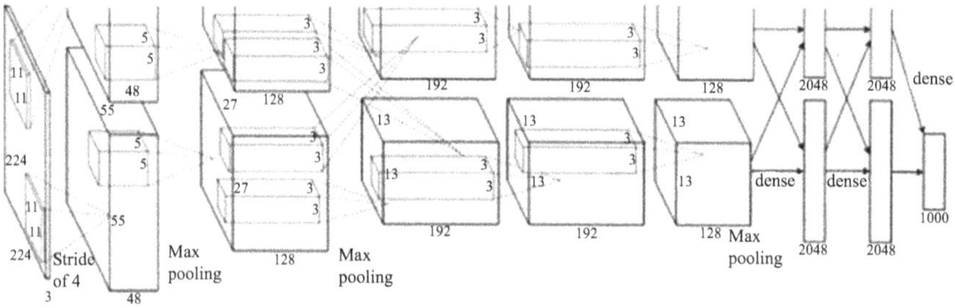

Figure 6.9 AlexNet architecture

The final layer generated a probability distribution for each of the 1,000 classes using a 1,000-node softmax for the ImageNet-required classification of 1,000 labels. The significance of the depth of the network to its performance was one of AlexNet's most important findings. That depth produced a large number of parameters, rendering CPU-based training either impractically slow or completely impractical. Using a GPU could cut down on training time. However, the 3GB of memory that high-end GPUs had at the time was insufficient to train AlexNet.

6.4.3 GoogleNet

GoogleNet is a 22-layer profound convolutional brain network that is proposed in 2014 by scientists at Google. GoogleNet is now used for face detection, adversarial training, object detection, and image classification, among other computer vision tasks. When compared to AlexNet and ZF-Net, the error rate on GoogleNet has decreased. It had a main five mistake pace of 6.67%. This was quite similar to human performance, which was the challenge that researchers had to assess. The picture you need to order should be of a similar size as the organization's feedback size. The first component of GoogleNet's layers property is the image input layer. The organization input size is characterized by the picture input layer's feedback size property.

GoogleNet developed an inception module with a typical dense structure that resembles a sparse CNN. Because only a few neurons are effective, the number of convolutional filters with a given kernel size is kept small. The size of the input image architecture is fixed at 224×244, and the image is preprocessed. The Google Net engineering has 22 defined layers, which are convolutional layers and completely associated layers. In the event that we incorporate non-defined layers like max pooling, the GoogleNet Model has a sum of 27 layers and there are 9 beginning modules in Figure 6.10.

The CNN's beginning organization is quite possibly the main development in brain network research. There are three versions of the inception network: version one, version two, and version three. GoogleNet is the name of the initial version of this network, which is in charge of classification and detection in the ILSVRC. Overfitting this network, which has a lot of deep layers, could cause many issues. GoogleNet architecture is being developed as a solution to these issues. The network will become wider rather than deeper by incorporating filters of various sizes that can operate on the same level. In order to speed up computations and reduce the size of the network, this convolutional layer includes input filters 1×1, 3×3, and 5×5.

6.4.4 DenseNet

CNN which is a densely connected network is known as DenseNet. It was developed primarily to increase accuracy due to the high level neural network's diminishing gradient caused by the distance between the input and output layers. The input is given as an image and it goes through the network where some operations are done and finally output will be predicted. Aside from the basic convolutional and pooling layers, DenseNet has two important blocks. They are the transition

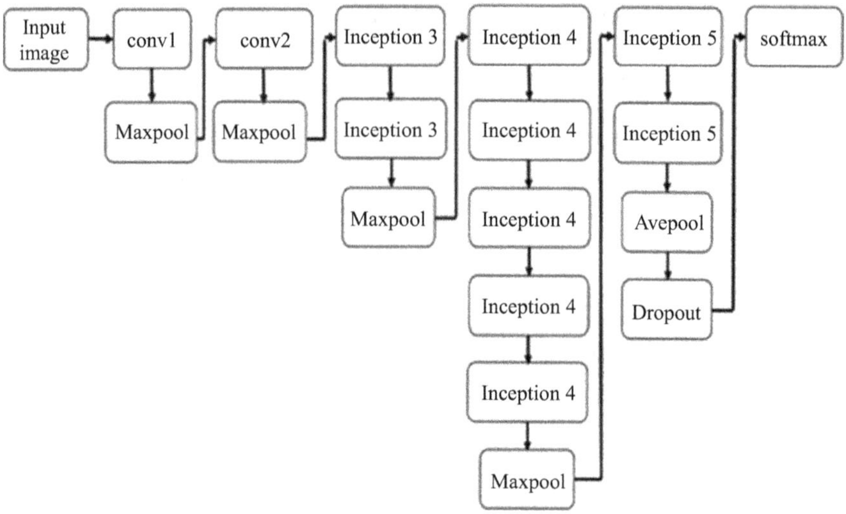

Figure 6.10 GoogleNet architecture

layers and the dense blocks. The first convolution block has 64 7×7 filters with a stride of 2 in Figure 6.11. Following that is a max pooling layer which has the stride 2 and the size is 3××3 [7].

Densely connected convolutional network comes from the fact that every layer in a DenseNet architecture is connected to every other layer. For example, if there are L layers in the network, there are L(L + 1)/2 direct connections. The feature maps of all preceding layers are used as inputs for each layer, and its own feature maps are used as input for each subsequent layer. DenseNet links every layer to every other layer, which is as straightforward as it may appear. This is the central idea, and it has a lot of power. DenseNet takes the concatenation of feature maps from previous layers as the input for a layer.

6.4.5 MobileNet

The tensor flow first mobile computer vision model of the MobileNet model was developed for use in mobile applications and on edge devices. In order to reduce the number of parameters, it makes use of depth-wise separable convolutions. The profundity wise comprises two layers that is profundity wise convolution and point wise activity. The profundity wise layer plays out the lightweight sifting by applying the convolutional channel. Through computing input channels, it is utilized for the construction of new features in point-wise layers. This model has fewer parameters and a high accuracy to help MobileNet improve this dense network block [8]. TensorFlow's first mobile computer vision model is the MobileNet model, as the name suggests, designed for use in mobile applications. Depth-wise separable convolutions are used in MobileNet. When compared with the organization with normal convolutions of similar profundity in the nets, it altogether

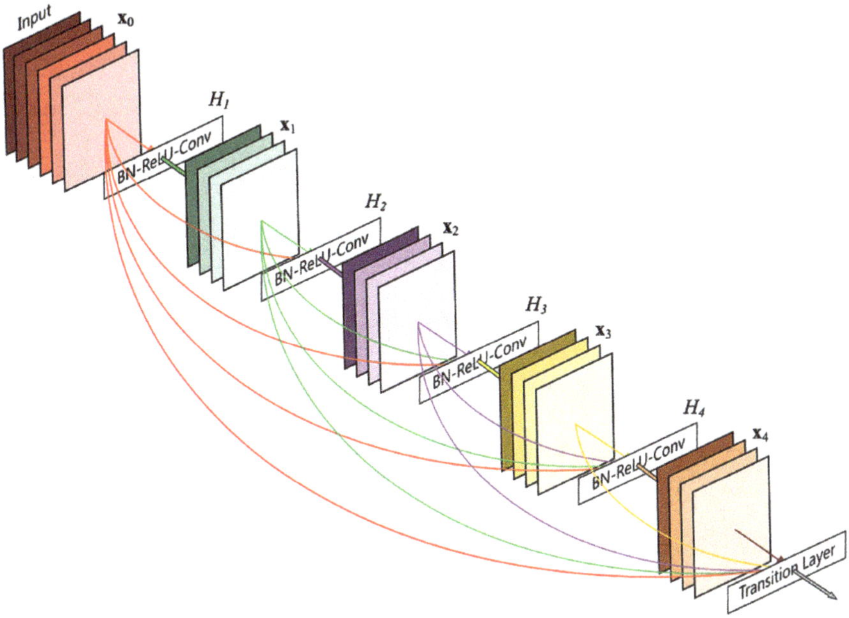

Figure 6.11 DenseNet architecture

decreases the quantity of boundaries. As a result, lightweight deep neural networks are created. MobileNet is an effective model for mobile and embedded vision applications thanks to its simplified architecture and its use of depth-wise separable convolutions to construct lightweight deep CNNs.

When compared to the conventional network, which has the same depth, this model is used to reduce the parameters. There are two kinds of blocks in MobileNet v2: a downsizing block with stride 2 and a residual block with stride 1. In Figure 6.12, MobileNet and MobileNet v2 have three layers: the first is depth-wise convolution, the second is convolution without nonlinearity, and the third is convolution with ReLU.

6.4.6 ResNet

ResNet can solve the problem of vanishing gradients. It utilizes the skip association, which frames the remaining block by skirting a few layers between the initiation layer and the subsequent layer. The two terms "residual neural network" and "artificial neural network" are synonymous. It is a gateless or open-gated variant of the Expressway Net, which was the first functioning extremely deep feedforward brain network with many layers that was significantly deeper than previous brain networks. Profound remaining organizations, for example, the famous ResNet-50 model, are convolutional brain networks with 50 layers. Residual neural networks, also known as ResNets, are ANNs that construct

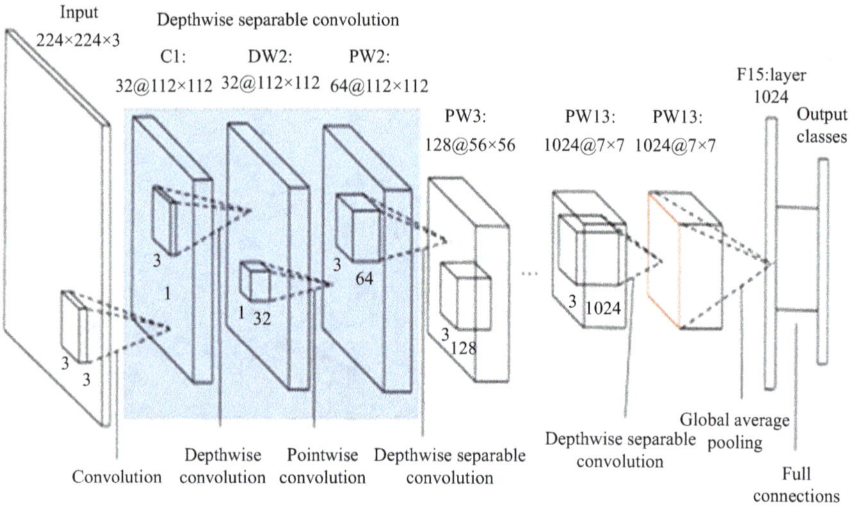

Figure 6.12 MobileNet architecture

networks by stacking residual blocks on top of one another [9]. Computer vision makes use of the deep learning model known as residual network. There are hundreds or thousands of convolutional layers in this architecture of a CNN.

There are million images in the ImageNet database which can be served as training grounds for the Inception-ResNet-v2 CNN [9]. The organization has 164 layers and can characterize pictures into 1,000 item classes, including consoles, mice, pencils, and different creatures. The convolutional neural architecture known as Inception-ResNet-v2 is an extension of the Inception family that includes residual connections. We assume that the learning-derived desired underlying mapping will serve as input for the top activation function. The mapping must be directly understood by the portion on the left of the dotted-line box. The residual block gets its name from the fact that the portion contained within the dotted-line box on the right must learn the residual mapping. Assuming the personality planning is the ideal fundamental planning, the lingering planning is and hence more straightforward to learn; we just have to push the upper weight layer's loads and inclinations. With remaining blocks, data sources can engender quicker across layers because of the lingering associations.

In Figure 6.13, ResNet-18 is a CNN with 18 layers. The pretrained network is capable of classifying images of keyboards, mice, pencils, and a variety of animals into one of the 1,000 object categories. As an immediate outcome of this, the organization has gained exact element portrayals for a great many pictures. A 224×224 image input is accepted by the network. On the ImageNet dataset, ResNet 34 is a pretrained image classification model. This PyTorch execution depends on the design depicted in the Torch Vision bundle's paper "Profound

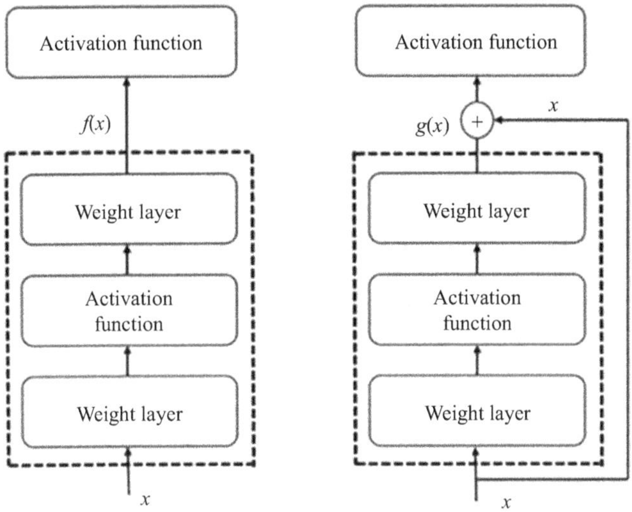

Figure 6.13 ResNet architecture

Lingering Learning for Picture Acknowledgment." A single image in the RGB order of 1, 3, 224, 224 makes up the model input blob. Uploading your dataset to Roboflow is the first step, assuming you already have one. Organize your images into folders based on the names of your classes before loading a classification dataset. After that, before clicking "Create New Dataset," visit roboflow.com and sign up for a free account there.

ResNet50 is a proactively prepared Profound Learning model. A model which is been pretrained can be used as a good starting point because it was trained on a different task than the current one. The features learned on the previous task are applicable to the new task. ResNet-50 has 50-layer convolutional brain network with one MaxPool layer, 48 convolutional layers, and one normal pool layer. Counterfeit brain networks known as remaining brain networks are developed by stacking lingering blocks into networks. ResNet-101 is a huge convolutional mind network with 101 layers. A pretrained version of the network that was trained on more than one million images can be loaded into the ImageNet database [5]. Images of keyboards, mice, pencils, and a variety of animals are among the 1,000 object categories that the pretrained network is able to classify.

6.4.7 NasNet

In 2012, the neural search architecture network known as NasNet was developed by the Google Brain team. The normal cell, which defines the map size in terms of the same dimensions, and the reduction cell, which reduces the map size based on its height and width, are the two primary uses for this Nasnet. The NasNet-Large CNN will be applied to larger datasets after being trained on a smaller set of data. NasNet is utilized for better execution and for planning brain network which includes a ton of experimentation which can be tedious and costly. The network can classify images into 1000 different object categories, including keyboards, mice, pencils, and various animals. The authors of NasNet do not predefine the blocks or cells, despite the fact that the entire architecture is predefined as shown below. Instead, the search method based on reinforcement learning is used to locate them. The number of initial convolutional filters and the number of motif repetitions N are therefore free parameters that can be scaled. Normal cells and reduction cells are the names given to these cells. Convolutional cells that produce a component map with a similar aspect as their feedback are known as expected cells. A component map with a two-overlap decrease in level and width is created by convolutional cells. A RNN's controller RNN only investigates the normal and reduction cell structures.

NasNet applies its operations to the small dataset first before moving its blocks to the large dataset in order to achieve a higher mAP. When Nasnet uses the scheduled drop path, a modified drop path, for effective regularization, it performs better. The number of cells utilized in the initial Nasnet engineering is not predetermined, and only normal and reduction cells are utilized, as depicted in Figure 6.14. The size of the feature map is determined by normal cells, whereas reduction cells return a two-fold smaller feature map in terms of height and width. Nasnet utilizes a control engineering in view of repetitive brain organizations to foresee the whole design of the organization in light of the two starting secret states.

6.4.8 ImageNet

ImageNet is a WordNet-based image database with hundreds of thousands of images representing each node of the hierarchy. The creation of ImageNet was intended to serve as a resource for encouraging research into new computer vision techniques. The ImageNet dataset contains 14,197,000 annotated images belonging

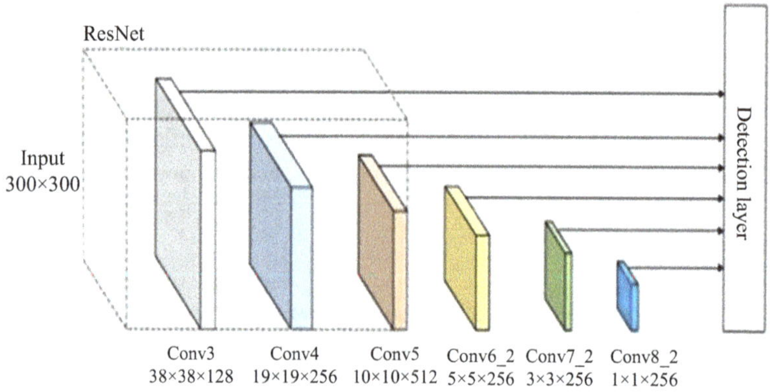

Figure 6.14 NasNet architecture

to the 20,000 subcategories of the WordNet hierarchy. Additionally, they offer the bounding box for approximately one million images, which can be utilized for further localization. The venture has helped with the headway of PC vision and profound learning research [10]. Since its inception, ImageNet has furnished scientists with a typical arrangement of pictures to use as benchmarks for their models and calculations. Increased research into machine learning and deep neural networks has made it easier to classify images and complete other computer vision tasks. A substantial visual database designed for use in the creation of software for visual object recognition is the ImageNet project. The project has manually annotated over 14 million images to identify the objects depicted, and bounding boxes are included in at least one million of the images.

6.5 Image classification

The process of categorizing and labeling agencies of pixels or vectors within an image based on specific rules is known as image category. According to Deepika Jaswal *et al.* [11], a computer will analyze an image as an array of pixels, with the size of the matrix dependent on the image resolution. Image classification is the process of determining what an image represents. With images, the model will be well-trained to recognize various image classes. The definition and representation of the characteristics that appear in an image as a distinct gray level in terms of the object that these characteristics reflect is the goal of image classification. The most important aspect of image evaluation or processing is the image category. Supervised image classification techniques and unsupervised image classification techniques are the two main types of image classification methods. Numerous examiners utilize a combination of directed and solo classification strategies to increment absolute last result assessment and sorted maps. However, due to its usefulness for high-decision facts, object-based totally category has gained more traction.

Education data is not used in the fully automatic unsupervised classification method. This suggests that hidden layers or facts agencies are found by machine learning algorithms to analyze and cluster unlabeled datasets without the need for human intervention. Using the appropriate algorithm, the system determines the specific function of the image through image processing.

Directed picture arrangement utilizes reference tests that have previously been characterized to prepare the classifier and afterward order new and unclassified information. The procedure of outwardly choosing tests for preparing information in the picture and relegating them to pre-chosen classifications is known as directed characterization strategies. This is finished to give factual measurements that will be utilized to work on the general picture.

6.6 Literature survey

The paper by Muhammad [6] is an automatic system for identifying and classifying fish species using the Alexnet model in deep learning. They have compared with two model VGGnet and Alexnet where VGGnet achieves 90.48% and Alexnet achieves 86.65%. Tourists are interested in the identification and description of freshwater and pond fish species. The dataset has been collected from QUT datasets which has 3,960 images of fishes. The AlexNet model is preferred over others because it has fewer layers and has a training and validation accuracy of more than 90%.

Soybean crops are heavily afflicted by disease, resulting in significant losses in the agricultural economy. The most common diseases, such as bacterial blight, frogeye leaf spot (FLS), and brown spot, can cause significant damage to crops in the field. To overcome this, Scahin *et al.* [13] proposed an effective deep learning-based CNN-based method for identifying soybean disease. So, the pretrained model AlexNet achieves 98.75% and GoogleNet achieves 96.25%. The dataset is collected from soybean field which contains 649 for AlexNet and 550 for GoogleNet.

The system is used in the work of Mrugendra *et al.* [14] to identify snake species based on their visual characteristics in order to provide the appropriate treatment and prevent subsequent deaths. To accomplish the goal, this system employs image processing, CNN, and deep learning techniques. Erroneous snake species identification from perceptible trends results in a high death toll from snake bites, making it difficult to identify snake species.

The research conducted by Nur liyana *et al.* [15] is used to identify snake species based on human descriptions—both verbal and visual—of the species. Unstructured text was used to present the human descriptions. NLP, which includes pre-processing, feature extraction, and classification, is used to identify the snake species. It will be beneficial for medical professionals to be able to identify the snake species that patients describe in order to treat them.

Ayad saad and Hakan [16] proposed a computerized butterfly species ID model utilizing profound convolutional brain organizations. The butterfly species had been identified using CNNs. Examination and evaluation of the exploratory results received the utilization of three unique local area frameworks as directed. Using a transfer learning approach, the training and testing achieves 80% of success, despite the images' issues with occlusion, background complexity, butterfly distance, shooting angle, and butterfly position.

A novel image classification approach for bird species identification was proposed in the work of Siva Krishna *et al.* [17]. Using novel preprocessing and data augmentation techniques, they are extracting bird features and have trained a CNN on the largest publicly accessible dataset. In this organization, design accomplishes a mean typical accuracy score of 95% while foreseeing the principal types of each picture document and scores 96%, close to 100% possibilities of bird species from the dataset.

The work of Vijayalakshmi and Joseph Peter [18] is the classification and recognition of fruits, which is a difficult task because they have different types of fruits. A five layer of CNN is used to identify banana fruits. Five different types of fruits are analyzed and their features were extracted from the algorithm in deep learning. Finally, this fruit identify process is carried out by the two-algorithm random forest and K-nearest neighborhood (k-NN). They achieved 96.98% accuracy rate for the deep feature random forest classification algorithm.

Punyanuch *et al.* [19] presents a model for using face images to identify dog breeds. This model is developed for identifying which type of dog breed with the help of recognizing the faces. This model uses the technique as transfer learning approach which has been pretrained by the help of CNN and it achieves an accuracy of 89.92%.

A novel two-step deep learning classifier is proposed by Hazem Hiary [20] for identifying species-specific flowers. In the first place, the blossom area is consequently portioned to permit confinement of the base jumping box around it. The proposed bloom division methodology is exhibited as a matched classifier in an absolutely convolutional network framework. Second, to distinguish between the various flower species, they create a potent classifier based on CNNs. They proposed the original strides during the preparation stage to guarantee that it is strong, precise for continuous arrangement. Their characterization results surpass 97% on all datasets.

The model proposed by Nagifa Ilma [21] uses CNNs in deep learning to classify snakes into venomous and non-venomous categories. The characteristics, head shape, body shape, physical appearance, skin texture, and eye color of snakes are used to differentiate between venomous and non-venomous species. The information has been separated into preparing, approval, and testing fragments in three stages. With fivefold cross validation, the model is able to classify snake images at 91.30%, while the model without cross validation achieves 90.50%.

Kazi Md *et al.* [22] proposed a model which is used for identifying the bird's species individually. The bird's images are taken from the diverse scenarios which are seen in different sizes, shapes, and colors from the human point of view. A pretrained CNN achieves better accuracy for given image. They proposed the deep learning model which is capable of identifying the individual bird images from the given input images and the pretrained ResNet model has been used with the CNN networks to encode the images. The ResNet model is used as a pretrained CNN network as the base model to encode the image. It achieves the accuracy of 97.98% in the bird species classification.

Mansi Chalawari *et al.* [23] proposed a model utilized for arrangement of canine variety with the assistance of move learning strategy in convolutional brain organization. It consists a combination of multiple CNN models for dog breed which can be classified for improved performances. This model can be used as mobile app or net app for real-world and user-defined image. Once they give the image as input, it will analyze the image and identify it has dog breed. This model classifies the animal on both generic and fine-grained levels. It uses the pretrained ResNet50 with transfer learning technique to achieve an accuracy of 86%.

Nur Nabila [24] proposed a picture handling procedure and convolutional brain network for distinguishing butterfly species. Butterfly is important for health, education, environment, and aesthetics. The significance of the model is to comprehend the impact of propensity misfortune and climate changes. The images of butterfly obtained from different sizes undergo through the resize, dimensional, and cropping to classify images and produce final outcomes. It uses a pretrained model of GoogleNet that achieves an accuracy of 97.5%.

A deep CNN, according to Thi Thanh *et al.* [25], is used to identify plant species from flower images. It shows that GoogleNet achieves the best accuracy among another model. The image preprocessing is done usually with complex background. A KDES descriptor technique is used for extracting features of processed organ images through three levels, that is, pixel level, patch level, and image level. So, this extracted three levels are concatenated to form a KDES vector. Finally, the model of GoogleNet achieves an accuracy of 66.60%.

Rinu and Manjulu [26] in their work focus on the detection of plant diseases with the help of CNN. Agriculture has a huge impact on life and economics of humans. The decline in productivity performance and crop losses are occurring as farmers are unable to identify the diseases in the initial stage. So, this model is helpful to identify if the plant is infected by disease or not. The pretrained model of VGG16 is used for detecting and classifying the plant images and achieves an accuracy of 94.8%.

6.7 Discussion

In the work of Muhammad [6], the QUT fish dataset consisted of 3,960 images that were collected from the unique setting. The four parameters—batch length, dropout layer, and the number of convolutional and fully linked layers—are taken into account. It uses a condensed version of the AlexNet model, which has two FC layers and four convolutional layers. In contrast to the original AlexNet model, which has 86.65% testing accuracy, the proposed and improved AlexNet model has 90.48% testing accuracy. First, they deployed a batch size of ten and noticed that the proposed system's validation and testing accuracy obtained were 94.23% and 87.35%, respectively. The accuracy of the validation and checking out has risen once the batch length was increased to 20. The accuracy of testing and validation was 96.28% and 88.52%, respectively. This work's disadvantage was that it could only classify freshwater fish species quickly; for underwater species that must contend with background noise, muddy water, and environmental obstacles, it cannot classify the fish species.

In the work of Scahin *et al.* [13] the pretrained model has been used; in particular GoogleNet and AlexNet models have been applied for soybean diseases in Scahin *et al.*'s work [13] through transfer learning methodology. The proposed and trained GoogleNet CNN architectures have been used with the preprocessed pictures. To categorize the four types of goods from the provided disease information set, the proposed models have undergone retraining. The fourth layer, which is close to the described range of classes, was reconfigured as the final layer. The study's four disease classes—bacterial blight, brown spot, FLS, and one healthy magnificence—and one healthy class are observed. For the purpose of identifying three soybean illnesses, the the AlexNet model is trained with 649 images that achieves an accuracy of 98.75% and the GoogleNet model is trained with 550 images and achieves an accuracy of 96.25% of damaged and healthy soybean leaves, respectively. The limitations of the work include the possibility of improving the model's performance rate by adjusting the minibatch size, bias learning rate, and weight.

Mrugendra *et al.*'s work [14] measured the fine-tuned and optimized models primarily in terms of the overall performance metrics and training results. For training and validation, a total of 3,050 images are divided into 28 species. The device is fed a random set of snake images after the images have been trained and tested, and the likelihood that the label is accurate is saved and observed. DenseNet, VGG16, and MobileNet are the three models used to train images. The accuracy of the DenseNet model validation and test are 78% and 72%, respectively. The VGG16 model approval exactness and test precision are 62.7% and 58.65%, respectively. The accuracy of the MobileNet model validation and test is 17.2% and 12.28%, respectively. The limitation of this work is the point at which the huge information base with assortment of pictures is given it will not be able to perceive the species.

Nur liyana *et al.* [15] proposed a model that involves a collection of text-based description of snake's species provided based on snake images through questionnaire strategies. Then, important functions have been extracted through the use of term frequency-inverse file frequency (TF-IDF), and those capabilities have been supplied to system through transfer learning technique to study and predict the snake species with the help of Weka tool. The dataset has been collected from the 60 responders with the help of 180 text-based approach for classification and testing. The stop phrase elimination, stemmer, and tokenizer features extraction techniques were then applied. The final step in determining the weight of each phrase in each file was the completion of the TF-IDFT remodel. Here, four machine learning algorithms have been compared: naive Bayes achieves 61.11%, k-nearest neighbor achieves 55.56%, support vector machine achieves 68.33%, and decision trees J48 achieves 71.67% were used during the entire process. Consequently, with 71.67%, J48 has the highest percentage. The overall results in this chapter demonstrate that J48 is the best choice for the text classification task. The drawback of the work is that the respondents are unable to describe the snake image, so the snake species cannot be identified.

They gathered 44,659 images dataset from the website of the Butterflies Monitoring & Photography Society of Turkey in 104 categories for the study by Ayad Saad [16]. The butterfly images are the focus of a refined transfer learning strategy. There are training elements (80%) and testing elements (20%) in each class. The normal exactness values for 100 years and the normal misfortune values were repeated for each dataset in the models because the preparing and testing records needed to be changed in every run. The butterfly images were classified using the three deep learning models, that is, VGG16, VGG19, and ResNet50. VGG16 train exactness 80.4% and test precision 79.5% VGG19 train precision 77.6% and test precision 77.2% ResNet train exactness 84.8%, and test exactness 70.2%. For our dataset, VGG16 and VGG19 produce approximately equivalent results. In the training phase, the ResNet version performs better than VGGNET. The drawback of this work is if the images have problems such as the position of butterflies, the shooting angle, butterfly distance, occlusion and background complexity the prediction of image is difficult.

A dataset consisting of 275 distinct bird species classes is included in the current study by Siva Krishna [17]. The dataset will be used for training, validation, and testing of three different algorithms. Given that it was important to ensure that the classification model will perform correctly in real-world circumstances, it may be necessary to maintain a complete separation between the testing set and the training set. The classification model could then be trained with this information. The training set received more than 70% of the records, while the testing set and validation set received the remaining records. They selected the training dataset for setting a random fine-tuning method. This deep learning model employs a CN network to identify the species of bird. There are 275 categories for the 39,364 images that make up the collection. The dataset has been trained with a pretrained model of ResNet 152 v2. The loss function was 15.06% and the observed test accuracy was 95.71%. The work's limitations include the inability to improve the recognition rate by adding additional layers.

There are three stages in the proposed method. In the first stage, RGB images are taken as input and resized for further processing. Using a CNN, the level of feature extraction is completed. If the banana fruit is present, the level results are sent to KNN and Random Forest, two extraordinary classifiers, to distinguish it from the other fruits in the database. In the paper by Vijayalakshmi and Joseph Peter [18], random forest and K-NN classifying algorithms are used to identify fruits. Comparing the exhibit lists of the proposed CNN-based classifiers and the current Pig-based extraction, it is discovered that the CNN-based classifiers achieved 96.98% exactness with KNN and 97.11% precision with irregular woods. The constraints of this work are hard to recognize single organic products while different organic product pictures are introduced.

The experiments performed by Punyanuch *et al.* [19] consisted of three main phases: data information, training, and testing. The step of preparing the data is important because we concentrate on dog face images. After that, it is divided between the testing technique and the training technique. Using a CNN, the training model produced a classification of dog breeds. The variant is utilized for breed classification and model assessment. They looked at MobilenetV2, InceptionV3, and Nasnet as three CNN models. The training data for each model includes image augmentation, rotation, translation, and random noise. The Nasnet model with a preparation set containing pivot pictures accomplishes the best exactness of 89.92%. With over 80% classification accuracy across all settings, the proposed method has the potential to perform well. The paper depicts the use of face recognition to identify dog breeds; however, when the dog face is side posed, it does not recognize the dog face in the image.

The method proposed by Nagifa Ilma [21] makes use of CNN in deep learning to distinguish between the two species of snake. The dataset, which consists of 1,766 images of snakes divided into venomous and non-venomous categories and obtained from Kaggle, is used to implement the proposed model thanks to the neural network. Here each picture is reformatted into 224×224 pixels. The snake images can be classified using a fivefold cross validation model with the highest accuracy of 90.50%. For improving the accuracy process, they are utilizing the transfer learning technique. The proposed model is contrasted with different models, for example, Origin Net, VGG19, Resnet50, Xception Net, Mobilenet v2, Beginning Resnet-v2, VGG16 with 82.38%, 43.75%, 81.81%, 82.94%, 82.35%, 89.62%, and 62.50% individually. The proposed model has been outflanked with 90.50% precision. The accuracy can be improved by including additional images of snakes against a variety of backgrounds.

The proposed profound convolutional brain network is utilized to order the bird species and other pretrained ResNet design models are utilized. This model is trained with 1.2 million images on ImageNet dataset which contains 1,000 different species. The dataset is separated into three categories: validation, testing, and training. Here, the base model is used with different variants of ResNet models. Whereas the ResNet18 achieve the top five test accuracy of 96.71%, ResNet34 achieves 97.40%, ResNet50 achieves 97.83%, and ResNet101 achieves 97.98%. So, the ResNet101 shows less error and is significantly higher than the other three models.

The proposed model is used to identify dog breed with the help of ResNet50 model along with transfer learning technique in CNN. They used two variety of datasets such as dog and human, so the total images of dog are 8,351 and total human images are 13,243. This VGG16 model is used to classify the dog and human images, which gives less accuracy. So, ResNet50 achieves the best accuracy of 86% and with the help of transfer learning technique it even achieves better accuracy. It was trained with 133 breed of dog images and 40 epochs with GPU support.

The study of Nur Nabila *et al.* [24] is used for identifying butterfly images with the help of CNN techniques. To classify the butterfly images, a pretrained GoogleNet is used and after the classification, accuracy is calculated with the help of confusion matrix. The dataset consists of four butterfly species and was collected from online and some are captured using a camera and a total of 120 images are captured. The data is divided into two processes: one is 80% of training and the other 20% of testing. So, the classification of butterfly species achieves an accuracy of 97.5%.

The work of Thi Thanh *et al.* [25] is used for identifying plant species from flower images using CNN. To select the best model, it is compared with AlexNet, CaffeNet, and GoogleNet and it is concluded that GoogleNet has performed better compared to other models. The dataset is taken from PlantCLEF and with different backgrounds so that CNN can be applied directly on different images. The database consists of 1.2 million images with 1,000 different flowers. From the flower dataset, 27,975 images are used for training and 8,327 images are used for testing. This model achieves an accuracy of 50.60% of AlexNet, 54.84% of Caffenet, and 66.60% of GoogleNet. It shows that GoogleNet achieves the best accuracy among all models.

This examination is finished on the discovery of plant illnesses and grouping utilizing convolutional brain network in profound learning in the work of Rinu and Manjula [26]. The VGG16 model does two tasks, which performs better detection of diseases. Here, the graphical user interface will permit the user to use the images from the datasets and the result will be displayed on the user interface. They can detect the object in the image and the next classifying process is done. In Level 0, the user will choose the plant disease image that will detect and recognize the type of plant diseases. In Level 1, the CNN model will take the image from the pre-trained dataset and recognize the type of plant diseases. Level 2 can be used for recording the necessary details in the system function. The dataset collected from the Kaggle website has images of diseased and healthy plants. The training dataset consists of 54,305 images of 14 different types of plant and 38 different classes of plant diseases. The VGG16 model achieves an accuracy of 94.8%.

6.8 Conclusion

In this chapter, comparison and analysis of different models with its accuracy using CNN is performed. Various models of CNNs, such as AlexNet, GoogleNet, and ResNet along with their accuracy are obtained. The previous image classification algorithm took a long time to develop and lacked improved performance. However, the utilization of CNN calculation has fundamentally expanded the exhibition of picture arrangement framework as well as diminished the time prerequisite. Based on the literature review and analysis, it can be said that the deep learning method in image processing is faster and more accurate than using traditional methods.

References

[1] Mhatre, M. S., Siddiqui, D., Dongre, M. and Thakur, P., "A review paper on artifical neural network: a prediction technique," *International Journal of Scientific and Engineering Research*, vol. 8, no. 3, pp. 1–3, 2017.

[2] Kaur, M. and Mohta, M., "A review of deep learning with recurrent neural network," *International Conference on Smart System and Inventive Technology*, pp. 460–465, 2019.

[3] Zhang, J., Humaidi, A. J., Ai-Dujaili, A., *et al.*, ''Review of deep learning concepts, CNN architectures, challenges, applications, future directions,'' *Journal of Big Data*, vol. 8, no. 53, 2021.

[4] Albawi, S., Mohammed, T. A. and Al-Zawi, S., "Understanding of a convolutional neural network," In *International Conference on Engineering and Technology (ICET)*, pp. 1–6, 2017.

[5] Sharma A. K., Nandal A., Zhou L., *et al.*, "Brain Tumor Classification Using Modified VGG Model-Based Transfer Learning Approach," *New Trends in Intelligent Software Methodologies*, vol. 337, pp 338–550, 2021.

[6] Muhammad Ather Iqbal, Zhijie Wang, Zain Anwar Ali, and Shazia Riaz, "Automatic fish species classification using deep convolutional neural networks," *Wireless Personal Communications: An International Journal*, vol. 116, no. 2, pp. 1043–1053, 2021.

[7] Sharma, A. K., Nandal A., Dhaka A., *et al.*, "A survey on machine learning based brain retrieval algorithms in medical image analysis," *Health and Technology*, Springer, vol. 10, pp. 1359–1373, 2020.

[8] Khasoggi, B. and Ermatita, S. "Efficient mobilenet architecture as image recongition on mobile and embedded devices," *The Indonesian Journal of Electrical Engineering and Computer Science*, vol. 16, no. 1, pp. 389–394, 2019.

[9] Sharma, A. K., Nandal A., Dhaka A., *et al.*, "HOG transformation based feature extraction framework in modified Resnet50 model for brain tumor detection biomedical signal processing and control," In *Biomedical Signal Processing & Control*, Elsevier, vol. 84, 2023.

[10] Krizhevsky, A., Sutskever, I. and Hinton, G. E., "Imagenet classification with deep convolutional neural network," *Communications of the ACM*, vol. 60, no. 6, pp. 84–90, 2017.

[11] Jaswal, D., Sowmya. V. and Soman, K. P., "Image classification using convolutional neural networks," *International Journal of Scientific & Engineering Research*, vol. 5, no. 6, pp. 1661–1668, 2014.

[12] Jadhav, S. B., Udupi, V. R. and Patil, S. B., "Identification of plant diseases using convolutional neural networks," *International Journal of Information Technology,* vol. 13, pp. 267–275, 2020.

[13] Sachin B. Jadhav, Vishwanath R. Udupi, Sanjay B. Patil, "Convolutional Neural Networks for Leaf Image-Based Plant Disease Classification," *International Journal of Artificial Intelligence*, vol. 8, No. 4, pp. 328–341, 2019.

[14] Vasmatkar, M., Zare, I., Kumbla, P., Pimpalkar, S. and Sharma, A., "Snake species identification and recognition," In IEEE Bombay Section Signature Conference (IBSSC), pp. 1–5, 2020.

[15] Izzati Rusli, N. L., Amir, A., Hanin Zahri, N. A., Ahmad, B., "Snake species identification by natural language processing," *The Indonesian Journal of Electrical Engineering and Computer Science*, vol. 13, no. 3, pp. 999–1006, 2019.

[16] Almryad, A. S. and Kutucu, H. "Automatic identification for field butterflies by convolutional neural networks," *Engineering Science and Technology, An International Journal*, vol. 23, no. 1, pp. 189–195, 2020.

[17] Siva Krishna Reddy, A. V., Srinuvasu, Dr. M. A., Manibabu, K., Sai Krishna, B. V. and Jhansi, D., "Image based bird species identification using deep learning," *International Journal of Creative Research Thoughts*, vol. 9, no. 7, pp. 319–322, 2021.

[18] Vijayalakshmi, M. and Joseph Peter, V., "CNN based approach for identifying banana species from fruits," *International Journal of Information Technology*, vol. 13, pp. 27–32, 2021.

[19] Borwarnginn, P., Kusakunniran, W. and, Karnjanapreechakorn, S., "Knowing your dog breed: identifying a dog breed with deep learning," *International Journal of Automation and Computing*, vol. 18, no. 1, pp. 45–54, 2021.

[20] Hiary, H., Saadeh, H., Saadeh, M. and Yaqub, M., "Flower classification using deep convolutional neural network," *IET Computer Vision*, vol. 12, pp. 155.6, 2018.

[21] Ilma, N., Rezoana, N., Shahadat, M., Ul Islam, R. and Andersson, K., ''A CNN based model for venomous and non-venomous snake classification,''

Communications in Computer and Information Science book series (CCIS), vol. 1435, pp. 216–231, 2021.

[22] Md Ragib, K., Shithi, R. T., Ali Haq, S., Hasan, M., Mohammed Sakib, K., Farah, T., "PakhiChini; automatic bird species identification using deep learning,'' In *Fourth World Conference on Smart Trends in System, Security and Sustainability (WorldS4)*, pp. 1–6, 2020.

[23] Chalawari, M., Koli, M., Deshmukh, M. and Pathak, P., "Dog breed classification using convolutional neural network and transfer learning," *International Journal for Research Trends and Innovation*, vol. 6, no. 4, pp. 99–103, 2021.

[24] Kamaron Arzar, N. N., Sabri, N., Mohd Johari, N. F., Shari, A. A., Mohd Noordin, M. R. and Ibrahim, S., "Butterfly species identification using convolutional neural network," *IEEE International Conference on Automatic Control and Intelligent Systems (I2CACIS)*, vol. 3, no. 5, pp. 221–224, 2020.

[25] Nhan Nguyen, T. T., Le, V. T., Le, T. L., Natapon Pantuwong, H. V. and Yagi, Y., "Flower species identification using deep convolutional neural network," *Regional Conference on Computer and Information Engineering*, 2016.

[26] Rinu R. and Manjula S. H., "Plant disease detection and classification using CNN," *International Journal of Recent Technology and Engineering*, vol. 10, no. 3, pp. 52–56, 2021.

Chapter 7

Text recognition using CRNN models based on temporal classification and interpolation methods

Sonali Dash[1], Priyadarsan Parida[2], Ashima Sindhu Mohanty[2] and Gupteswar Sahu[3]

Analysis and recognition of text by hand is a very challenging task and is almost needed in every field either through technical or non-technical process. These days there is a huge demand and need for both digitalization of handwritten text and minimizing the use of hardware such as keyboard for text entry. The digitalization of handwritten text has various purposes like storage, searching, modification and sharing of data while the need to recognize and read the data by writing in the air is needed to support the growing technical advancements such as Augmented Reality (AR), Virtual Reality (VR), etc. One simple way to cater to these needs is to do it manually but it becomes really difficult to get it done when there is either a lot of data or not a normalized structure. Due to these challenges, there is a need to present a fully independent model. While there have been various attempts at to get to a solution some of them are not efficient and the others are not generalized. In this chapter, a structure is presented to recognize text from handwritten data and air-writing using a combination of recurrent and convolutional neural networks (CRNN) along with connectionist temporal classification (CTC) and interpolation methods.

7.1 Introduction

In the last decade, new technologies are being integrated into people's daily life and we are becoming more accustomed to using it in various ways. People are using the technology to access the internet or to talk to their friends or just to record

[1]Department of Electronics and Communication Engineering, Chandigarh University, India
[2]Department of Electronics and Communication Engineering, GIET University, India
[3]Department of Electronics and Communication Engineering, Raghu Engineering College, India

something. This rise has provided us solutions to problems like storing and conversion of hard handwritten text to a digitalized one but nothing comes without a few downsides. With the advent of technology and our dependence on it, we need to physically carry a device at all times. This calls for a solution and one possible way to deal with this is using air-writing.

Recognition of handwritten text is the task of recognizing and transcribing the handwritten text to some form of digital text. This problem can be further categorized into online and offline recognition [1]. Since online recognition is done when the input text is recorded using a pressure sensitivity device, the temporal and geometric information regarding the text is available for the recognition model [2]. This makes it easier in comparison to offline recognition where the text is written and is made available using a scanner and the resulting images are then used to predict the text. The transcription of the documents makes them more accessible and easier to use [3]. The problem with the offline data is that the input length and the clarity of the text vary from source to source, word to word and this causes problems when we need or trying to make a generalized text recognition model.

The use of NN has been revived due to the great success of deep convolutional neural networks (DCNN), which are helpful in different computer vision tasks. However, the vast majority of current deep neural network research has focused on detecting or classifying object categories [4] that generally predict and give result in the form of a single label.

In the real world, the text either from a scene or handwriting usually occurs in the form of sequences; these generally require the machine to predict a succession of object labels, instead of a single label. Therefore, object identification can naturally treat as a sequence recognition problem, with numerous classes to be worked on at the same time. The varying length of sequence-like objects is another unique property suited to them [5]. So, the traditional DCNN models cannot be applied upon these sequence objects as DCNN generally works with objects of fixed dimensions while these sequences fail to produce variable-length label sequence [4].

The need to carry a physical device at all times is imposing a new burden on us. The next-generation technology can help us eliminate the necessity to carry a physical device all the times. VR and AR help us project the output directly into the user's smart glasses [6]. Speech recognition modules help us communicate with the technology in a better way. However, it does not fulfil all the requirements needed for the communication with the technology [7]. Another method that has received a lot of interest as a communication method is gesture recognition. Accelerometers, photo sensors and camera-based technologies are coming forth as new mediums of interaction and they often rely on gestures as a suitable alternative to traditional touch screen or keyboards [8] which are not really compatible with AR, VR and mixed reality. So, to be in sync with the current technology we would need some advancement in the way we enter texts. Among other options writing in air (WiTA) seems to be a promising solution.

The art of writing something in the free space using hand or figure movements is known as air-writing [9]. It can be considered as a unique and more detailed way of using gestures. However, recognition of characters written in air is not an easy task [9]. The characters are recognized and distinguished from general gestures using fine grain movements and different people can write each character in their own style [7]. In old conventional ways, the writings, alphabet, numbers and words are recorded using a pen and paper in a multistroke way. This means that one may pick up their pen number of times while writing something. Movements caused by movement of pen is quite difficult to perceive as the user is not actually touching something and they may feel disoriented at times which may result in bad inputs [6,10]. To add to this, the traditional WiTA systems rely on traditional statistical models, which further restricts their performance eventually restricting their application in day-to-day life [11]. Despite all these shortcomings, the improvements made in this field are promising [7].

With the advent of smartphones with sensors, it is easy for users to record their gestures using a mobile application for collecting data and this data can then be processed as per user requirements [6]. Normally, during the collection of the data for recognition, one will use a motion sensor such as gyroscope to obtain better results [6,10]. But on the other hand, there is a bigger issue at hand. The data given by the sensors that is to be used for recognition is of variable length, which is quite evident from the fact that different people may take different amount of time to write something [6]. So, we cannot work with deep neural networks like convolutional neural networks which only take a fixed length input. But this issue can be solved by truncating/removing the starting and ending of the signal [12]. This seems a fairly simple process but it results in a lot of data loss, and therefore, some important features are lost in the process. On the contrary, using an interpolation method can be helpful. The statistical approach or technique used for the prediction of unknown values for the given values is known as interpolation [7,13]. It covers the full length of the signal as it maps the signal data to a predefined fixed time without removing any part, thus making data loss minimal.

There are a lot of different interpolation techniques that are available and studied with respect to image processing [7,13,14]. Furthermore, a fine-tuned CNN can make the training process easy, which in turn improves the efficiency. The model presented in [7] will be reviewed along with some techniques that are already present.

The rest of this chapter is organized in the following structure. Section 7.2 contains more information about the past works that have been done to solve the respective problems. Section 7.3 gives an in-depth summary of the datasets that are used to prepare the model presented in the chapter. Section 7.4 proposes and discusses the model used in the chapter while presenting the evaluation matrix used to measure the efficiency of the models. Section 7.5 discusses the results achieved by the models proposed and discussed in this chapter and Section 7.6 concludes the theory presented in the chapter.

7.2 Related works

For a specific sequence-like object, some attempts have been made to overcome this problem (e.g., scene text). The algorithms presented in Bissacco *et al.* [15] first detect individual characters, then use DCNN models trained on labelled character pictures to recognize these identified characters. These methods often require the development of a powerful character detector which can work on the original word image to recognize and crop each character present in the image. Other systems handle scene text recognition as a problem of image categorization, with each English word being assigned a class label. To summarize, DCNN alone are insufficient for the task and cannot be used for text recognition directly.

Many attempts have been suggested to solve this problem using conventional methods rather than CNN as mentioned in [16]. It is observed that even though these suggested methods are giving promising results but they are underperformed the CNN methods.

An effective solution to this problem is using recurrent neural networks (RNN) with DCNN and a loss function to get better results [5,17].

Another major part of the deep neural network family is RNN models that were created primarily to handle sequences. One of the benefits of using RNN is that in both training and testing, it does not require the position of each element in a sequence object image. Only a pre-processing step is required to turn the input image into a sequence of image features. Figure 7.1 shows the components of the ANN used for HTR [5].

The ANN is given a grey-value image containing text as an input. This is passed through multiple CNN layers that are trained to get the relevant features [2]. These CNN layers output a 1D or 2D matrix (commonly known as sequences) and this is passed to RNN which then creates a matrix that have scores for each character in the sequence element. Decoding is the process of determining the most probable labelling from a matrix, and it can be done with the help of a CTC output layer [17].

Some research and models presented were based on the fact that a user will be using their fingers to write something in air. The recognition system will then follow and interpret the finger movement to get an idea of the character. For capturing finger movement, different types of sensors can be used. One category is on-body sensors like smart watches which will be worn by the user [18,19] or using custom designed sensors [20]. But the problem with these sensors is that they limit the usage as you need to carry the sensor everywhere. To overcome this, some researchers used off-body sensors which work on encoding the action for each word [21]. Accordingly, users will need to work in accordance with the encodings which in turn degrade usability. In a similar fashion, some researchers used motion [22] or Kinect sensors [11].

The study by Liu *et al.* [23] focused on the orientation of the surface on which users were to write, as well as the use of hand's stabilization point, the rotation injection approach for data augmentation based on rotation matrix [7,23].

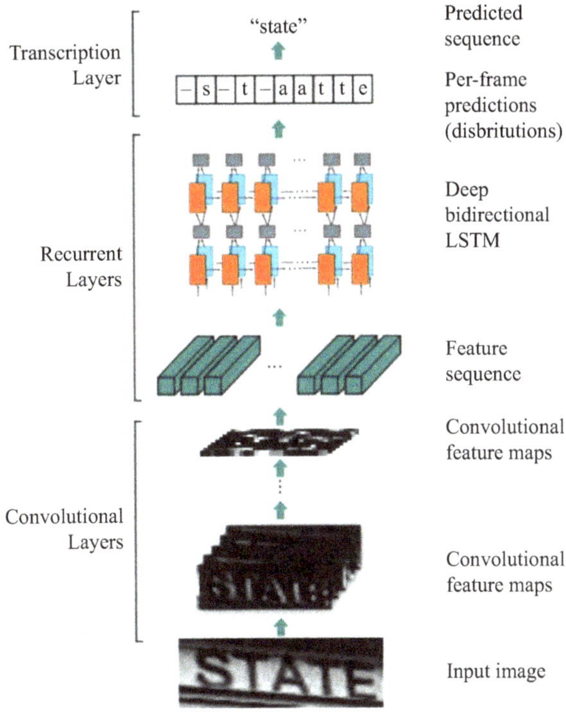

Figure 7.1 The CRNN network architecture for transcribing medieval handwritten texts and offline text recognition uses CNN layers followed by two layers of LSTM layers, and a final transcription layer for scene text recognition

Using a machine learning approach, they were able to distinguish 64 characters with a 99.99% accuracy. Kim *et al.* [24] proposed their WiTA dataset containing air-written Korean and English characters using RGB cameras and using Stereo-Temporal Convolution for the model. Alam *et al.* [25] developed a finger-joint tracking character recognition system [7]. The identification of character, alphabet, number, or special key is done by calculating the distance between the thumb and the finger joint. When one hand was invoked at a time, they got 91.95% accuracy, and they got 91.85% accuracy when both hands were used at the same time [7].

We have used the model proposed by Abir *et al.* [7] in which they have applied several interpolation techniques on publicly available multiple datasets [6,10,26,27]. In the research, they used a 2D-CNN model that followed standard practices in the study [7,28,29]. These measures resulted in an outstanding performance, outperforming all the user-dependent and user-independent models on air-writing [7].

7.3 Datasets

For this chapter, we will use IAM [30] and CVL [31] datasets. This version of the IAM dataset given by Marti and Bunke [30] has over 16,000 unique words which are given by approximately 1,539 forms filled by 657 writers as listed on IAM website. The CVL dataset has English and German words and it contains handwriting from 7 texts written by 311 writers. Table 7.1 shows the distribution of words and lines in the IAM and CVL datasets.

There are seven publicly available datasets [6,10,26,27] on air-writing used in the chapter [7]. There are varying number of classes, characteristics, subjects, and data gathering methods in each dataset. All of the datasets are user-dependent, meaning that samples from all users were utilized for both testing and training, whereas five of the datasets are user-independent, meaning that one users' data was used for training, while the data of the other users was used for testing [7]. Table 7.2, given in Abir *et al.* [7], presents a concise summary of the datasets.

Table 7.1 Summary of the IAM [30] dataset used for handwritten text recognition part

Dataset	**#Chars**	**#Unique words**			**#Lines**		
		Train	**Valid**	**Test**	**Train**	**Valid**	**Test**
IAM [30]	79	11,242	2,842	3,707	10,244	1,144	1,965
CVL [31]	54	383	275	314	11,438	652	1,350

Table 7.2 Summary of the datasets used for the air-writing part

Datasets	**No. of users**	**No. of features**	**No. of classes**	**No. of samples**	**Training principle**	
					User independent	**User dependent**
RTC [27]	10	3	26	30,000	C	✓
RTD [10]	10	2	10	20,000	C	✓
6DMG-digit [26]	6	13	10	600	✓	✓
6DMG-lower [26]	6	13	26	1,470	✓	✓
6DMG-upper [26]	25	13	26	6,500	✓	✓
6DMG-all [26]	25	13	62	8,570	✓	✓
Smart band [6]	55	6	26	21,450	✓	✓

7.4 Model and evaluation matrix

This section contains the method by which the respective models for handwritten text and air-writing were built. This section explains the data pre-processing required for optimal performance, followed by the architecture of the models used in the chapter and finally, an evaluation matrix describing the process and method to calculate the results is discussed.

7.4.1 Process of data pre-processing

7.4.1.1 Handwritten text recognition

The proposed model consists of three major components: a CNN, the long short-term memory neural network (LSTM) which was proposed to avoid the problem of vanishing gradient with RNN [32,33], and CTC as the loss and decoding function [34]. Input data for this part is an image of text to be recognized. This can be a scan of the original document or an image containing the data. The input image is then transformed into a size of 128 × 32 pixels. If the image does not match the specification, then we will need to resize it (without distorting it) until it is either 128 pixels wide or 32 pixels high. Then, the image is moved or copied to a target image with dimensions 128 × 32, as shown in Figure 7.2.

Data augmentation for the same can be easily done by duplicating the image to an arbitrary point rather than aligning it to the left or right or scaling it randomly [35].

7.4.2 Air-writing recognition (writing in air)

This chapter expands on experimentation of interpolation methods on air-writing sensor data with a fixed signal length given by Abir *et al.* [7] As the datasets are balanced, no further processing was needed.

7.5 Description and working of the model

7.5.1 Handwritten text recognition

The handwritten text recognition model proposed in this chapter works on a network of different layers of deep neural networks along with a CTC algorithm that works as an encoding and loss function and a decoding algorithm. In this chapter, we have discussed and compared some of the most prominent decoding algorithms that can be used for the same.

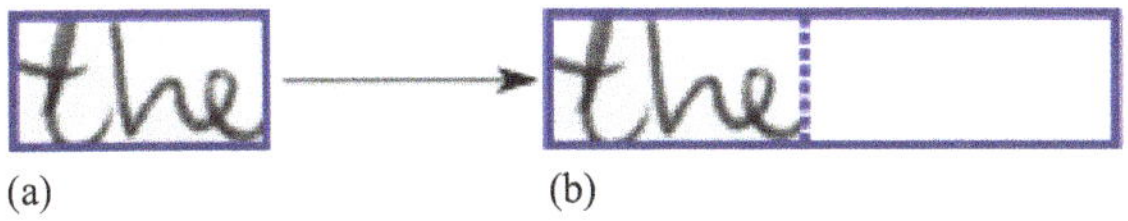

Figure 7.2 (a) Random image with an arbitrary dimension, (b) white colour is used to fill the empty space

7.6 Convolutional neural network

To extract relevant features from the input image, a CNN network is used. This reduces the sample size of the input so that it can be processed further.

The key difference between a multi-layer perceptron (MLP) and CNN is that the latter has at least one convolution layer instead of matrix multiplication [36]. There are five layers to the proposed CNN. The working of the CNN is divided into three parts, as shown in Figure 7.3.

1. Convolution operation. This is the first part of the CNN. The first two layers have a 5×5 kernel mesh and a 3×3 kernel mesh in the final three layers. The convolution of input and kernel is used to compare the activity of the neuron [17].
2. Non-linear RELU: This is applied to the activation of a neuron. It is given by

$$y' = h(a) = \max(0, a) \tag{7.1}$$

   where a is the activation of the neuron. This zeroes out all the negative parts while keeping the positive part unchanged.
3. Pooling layer: This layer is responsible for giving a scaled down version of the given input by replacing a certain place in the output by the summary statistic of nearby outputs.

CNN layers produce a 32-character sequence with 256 characteristics per layer. The calculation of all the features per time stamp is shown in Figure 7.4.

The LSTM block is shown in Figure 7.5. For the LSTM block , there are two recurrences: inner and outer [37]. The inner recurrence is a self-loop of the state cell also known as constant error carousel (CEC) [33]. The information flow inside a LSTM is managed by gates. The amount of effect on the state cell by data is controlled by the input, and the output controls the effect of inner state on the outer state [38]. The problem of vanishing gradient is solved by CEC. As a result, a vanilla RNN can bridge 5–6 time steps between the input and target event, as compared to an LSTM which can bridge 1,000 time steps [38]. The feature sequence for the LSTMRR used in this chapter has 256 features for each time step. The output from the RNN is mapped to a 32×80 matrix. There are two layers of RNN that are stacked on top of each other with 256 units each. There are 79 characters in the IAM dataset. Along with these, one more character is added for

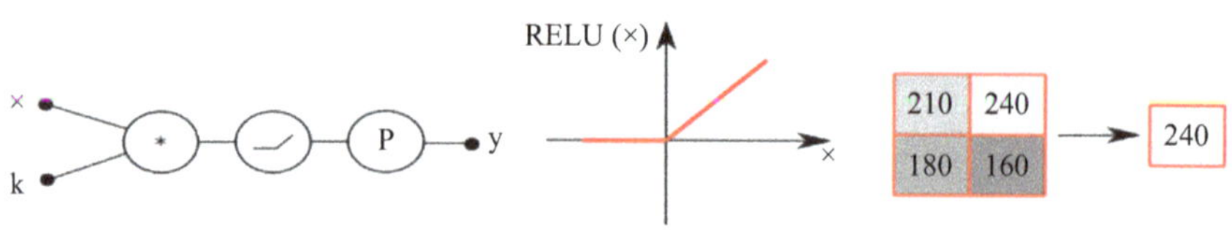

Figure 7.3 (a) Convolutional layer network, kernel k convolves picture x, (b) plot of ReLU, (c) and max-polling on a 2×2 matrix

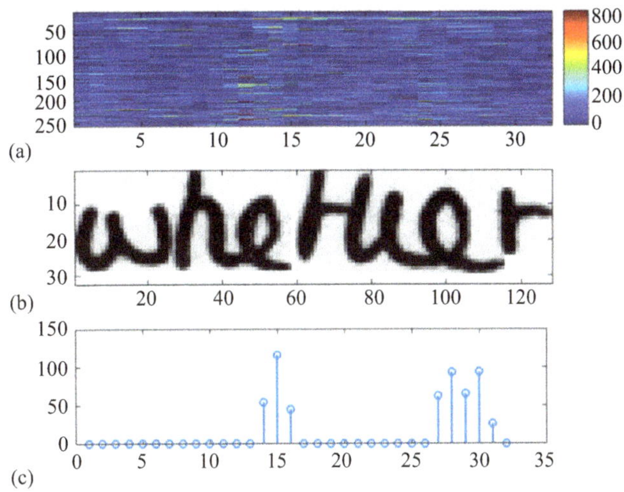

Figure 7.4 (a) Computation of 256 features by the CNN layers per time step and (b) input image used. (c) The plot of the 32nd feature shows a high correlation to the presence of character 'e'.

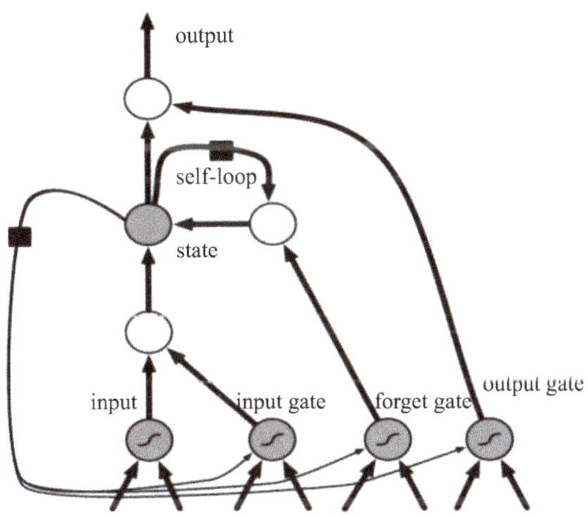

Figure 7.5 Illustration of the working of an internal cell and the gating neuron of an LSTM block

the CTC operation. As a result, each of the 32 time steps has 80 entries. This creates a bidirectional RNN where the given input is traversed all the way from front to back and back to front. This results in two output sequences that produce a feature of size 32×512 when concatenated. The resultant square matrix is then mapped and

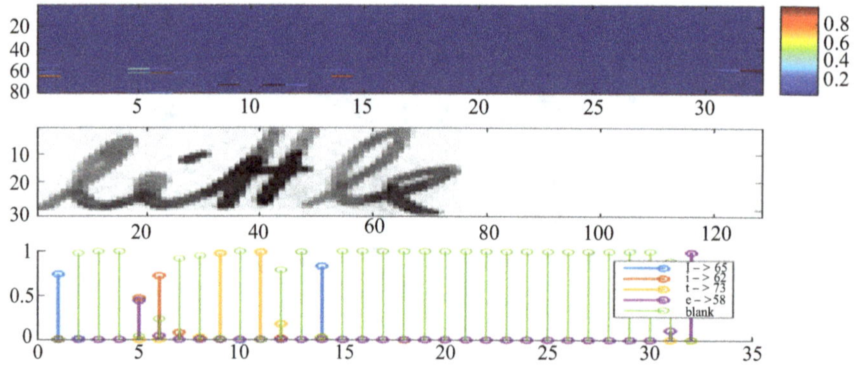

Figure 7.6 Example of RNN that generates scores for the word 'little'

fed to CTC [35]. Figure 7.6 shows the output matrix for the input word 'little'. This matrix gives us scores for all characters, which includes the CTC blank label as its 80th element [35]. The entries in the matrix correspond to the elements in this order '!"#&'()*+,-./0123456789:;?
ABCDEFGHIJKLMNOPQRSTUVWXYZabcdefghijklmnopqrstuvwxyz'.

The characters are now predicted as to where they appear in the image. With this we can sort out the scores of the predicted elements as shown in the bottom-most graph and take the most probable characters to get the best path which can then be decoded to get the final result.

7.7 Connectionist temporal classification

This is used for removing redundant duplication of characters that take up more than one time frame. To take care of this, it merges the duplicates into one. Also, it introduces a ' - ' to be placed between duplicates to get different possible encodings [39]. These are then used to train CRNN. The scores of ground truth are summed up to obtain the loss assignments. The score of a particular path is calculated by multiplying the character scores together [39]. The score of the path 'a—' is $0.4 \times 0.7 \times 0.6 = 0.168$ and the score of the path 'aaa' is $0.4 \times 0.3 \times 0.4 = 0.048$, as shown in Figure 7.7 [39]. The scores of all the paths to the corresponding text are summed up to generate the score that corresponds to the given ground truth. Loss for a prediction is given by taking the negative sum of all the log-probabilities. The gradient to the loss with respect to different neural network parameters (like weights of the convolutional kernels used) is computed and used to update the parameters in order to train the neural network [35].

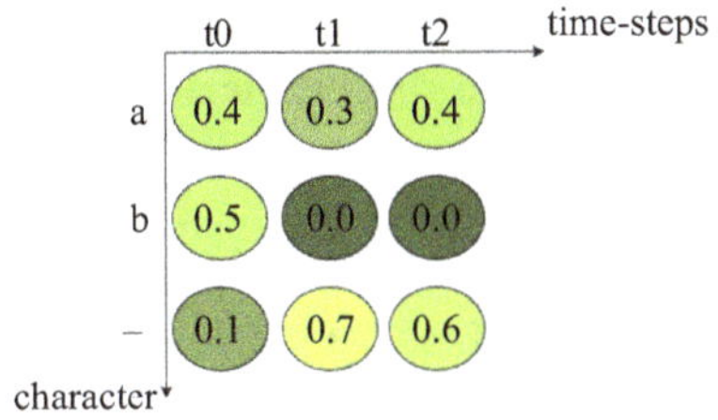

Figure 7.7 *Graphical model of best-fit decoding algorithm*

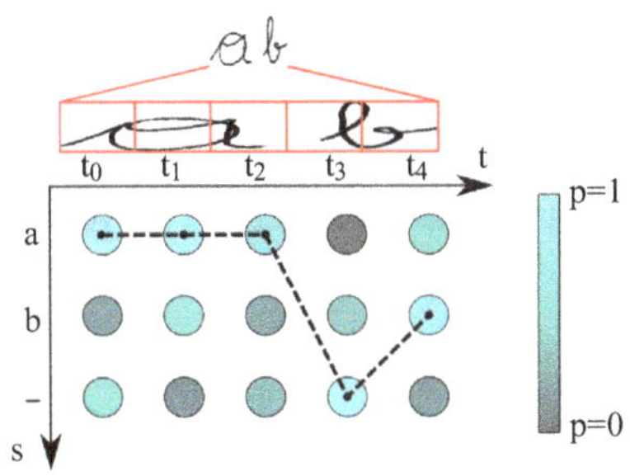

Figure 7.8 *An illustration of the time-step matrix which is used to find the ideal solution*

7.8 Decoding

Now, we want to get the best text. So, we can either examine the output matrix of every potential match or can use 'best path' algo [40]. The best path decoder determines the optimal path by taking into account the character that has the highest probability at each time step. The creation of the actual text by deleting the duplicate characters and blanks is done in this step [39]. The characters are 'a', 'b' and '-' (blank) along with five time steps [35].

By using the best path decoder to solve this problem, we find out that the best fit character at time step t0 is 'a', which is the same for both time step t1 and time step t2. At t3, the blank character has the highest probability. At the final time step t4, character 'b' has the highest probability. Thus, we are given 'aaa-b' as the path. Now, the final output is determined by removing the duplicate characters, which results in 'a-b' and then removing the blank spaces results in 'ab' which is returned as the recognized text. Figure 7.8 shows the working of best fit decoding algorithm [17]. The other decoding methods like token passing [40] is used when the words that are to be recognized are known to the system beforehand [40]. Beam search decoding works by calculating the probability of multiple labelling candidates and word beam searching algorithm [41] was proposed to improve on the Vanilla Beam Search and the running time of token passing [41]. The words in the recognized text are constrained by a prefix tree that uses a dictionary. The experiment outcomes that are given in [41] clearly show that word beam search (WBS) outperforms vanilla beam search, token passing and other decoding methods. Thus, it is best to use it as the decoding algorithm. For this chapter, we will be using best fit, token passing and WBS and give a comparative result for both of them when worked upon IAM dataset.

The WBS decoding method used in this chapter can have different parameters and can be implemented in different settings. Some of the different parameters for the WBS are as follows:

1. The branch width: This specifies the maximum width of each branch and the algorithm will try to stick within the given range. The more the branch width, the more will be the branches formed during decoding. But increasing the BW increases the time taken to recognize the word.
2. Word mode or non-word mode: To make the algorithm work for both characters in the dictionary words and the non-characters in between words, this algorithm has a feature to switch between the word mode and non-word mode state. These two states are interlinked and the transition can be done in between them.
3. Language model mode: The language model LM can have different modes. There can be four modes depending the type of LM and dataset:
   (a) Words: In this, there is no LM but only a dictionary.
   (b) N-grams: In this, the beam labelling gets a score by LM each time a transition is made from word to non-word state.
   (c) N-grams + Forecast: In this, every time a beam is extended using a word character, all the possible next words are queried using the prefix tree. All the beams with possible words are scored by LM and summed up.
   (d) N-grams + Forecast + Sample: This is the worst case where there is no match and all the words in the dictionary are used for the forecast.

For this chapter, we have used the word mode of the LM. The dictionary is made up of training set and testing test given by Tr and Te, respectively. The model discussed in this chapter to recognize the text from air-writing consists of finding the optimal frame length for the inputs, selecting the proper interpolation method that will give us the ideal fixed length and a network of CNN layers that will be used to make the predictions. In this chapter, we will also discuss about all the possible combinations that can be there for the interpolation methods and choose the best one according to the results [42].

7.9 Optimal fixed length

Maintaining a fixed length is really important as deep learning architectures work on a fixed length signal for training and predictions. Due to the nature of data collection and different user data, the signal length likely differs and an operation to fix them is needed.

To get the fixed length: take the mean of the lengths as fixed length. This divides the data into two parts: the parts which fall short of the fixed length are up-sampled, while the other half is down-sampled. As loss is occurring in downsampling, a further increase in the fixed length can be done so that data loss is minimized all the while keeping the fixed length manageable. This is in accordance with Abir *et al.* [7].

7.10 Using different interpolation techniques for finding the ideal fixed frame length signals

The statistical approach or technique used for the prediction of unknown values for the given values is known as interpolation [7,42]. It is widely used for reshaping images without damaging their qualities. For the discussion in this chapter, we used the concept of interpolation for calculating the fixed-length signals that are to be fed to the model, thus minimizing data loss [7]. The use of interpolation on images has shown that different interpolation algorithms have different effects on images. Among all the available interpolation algorithms, the widely used ones are nearest neighbour, bicubic, Lanczos and bilinear interpolation methods [14]. A brief on the interpolation methods are as follows:

Nearest neighbour: It is the simplest method [43]. The output is produced using the input's closet sample point, resulting in a discontinuous interpolated data [37].

The interpolated point Y_j is calculated as follows:

$$Y_i = \begin{cases} Y_B; j < \dfrac{a+b}{2} \\ Y_A; j >= \dfrac{a+b}{2} \end{cases} \tag{7.2}$$

where the indices y_A and y_B are represented by a and b where $a < j < b$ [7].

Bicubic: This interpolation method is built as an improved version of the cubic interpolation method in the 2D regular grid [7]. To determine the grey level, the closet 16 pixels to the specified input coordinates are used along with a polynomial or cubic convolution technique [7,44]. Equation (7.3) defines the bicubic interpolation kernel $Y(x)$ [7,44],

$$Y(x) = \begin{cases} (a+2)|x|^3 - (a+3)x^2 + 1; |x| <= 1 \\ 1a|x|^3 - 5a|x|^2 + 8a|x| - 4a; |x| < 2 \\ 0 \end{cases} \tag{7.3}$$

where a is generally between -0.50 and -0.75 [7]

Lanczos: It is used to give a smooth interpolation of the signal values between samples. This is done by mapping each sample to a scaled and translated copy of the kernel [7]. The kernel $L(x)$ is given by

$$L(x) = \begin{cases} 1; x = 0 \\ \dfrac{a \sin(\Pi x) \sin \dfrac{\Pi x}{a}; -a <= x < a, x \neq 0}{0; \text{otherwise}} \end{cases} \tag{7.4}$$

where a is the size of the kernel.

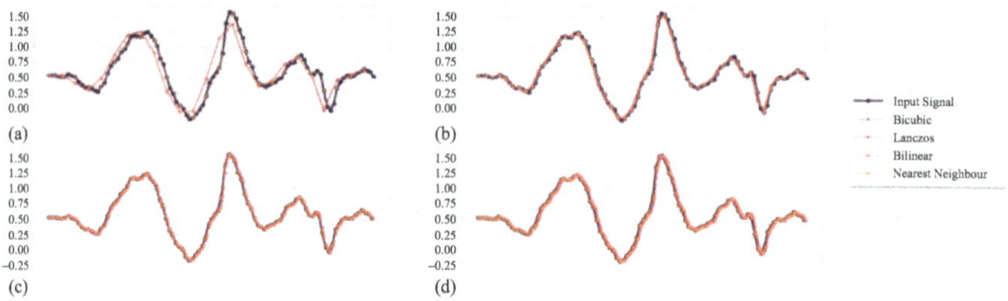

Figure 7.9 Example of demonstration of the impact of various interpolation techniques used to upsample and downsample time-series data. (A) 1/4 downsampling for l between 100 and 25, (B) 1/2 downsampling for l between 100 and 50, (C) 2× upsampling for l between 100 and 200 and (D) 4× upsampling for l between 100 and 400. Data used for the input signals is coming from smart-band dataset.

The number of lobes in a Lanczos kernel is $2a - 1$. In these $2a - 1$ lobes, there are a positive lobes that are present at the centre and the other $a - 1$ lobes that alternate between positive and negative are found on both sides.

The interpolated value $S(x)$ at an arbitrary real argument x, for a one-dimensional signal is achieved by convolving the samples discreetly with the Lanczos kernel [7].

$$\sum_{i=\lfloor x\rfloor-a+1}^{\lfloor x\rfloor+a} s_i L(x-i) \tag{7.5}$$

where the kernel size is defined in [7].

Bilinear: Linear interpolation is the process of generating new data points using linear polynomials over a set of given known points. By performing linear interpolation in one direction we can achieve bilinear interpolation in the other [7]. In bilinear interpolation, the weighted average is applied on the four closet values, which gives us the value of any random position. This makes the output smooth as one interpolation operation is done in a particular direction while the other one is done in the direction perpendicular to that [7].

$$H(x) = \begin{cases} 0; |x| > 1 \\ 1 - |x|; |x| < 1 \end{cases} \tag{7.6}$$

where the distance between any two points is given by x.

The effects of the above-discussed interpolation methods for downsampling and upsampling on time-series data signals are shown in Figure 7.9 [7].

7.11 CNN architecture

These are the types of neural networks that are inspired by the design of the visual cortex of animals [7,45]. Using a combination of convolution and pooling layers, we can automatically learn the spatial hierarchy of the features. They are being used for various tasks related to object detection, classification and segmentation. But in recent times, they have been used in human activity recognition like gestures [46]. This chapter uses the CNN model proposed by Abir *et al.* [7] to recognize air-writing using the time-series data collected from different sources [7]. Aside from the input layer, the CNN is made up of four sets of layers. The first three layers are responsible for feature extraction, and they are made of maxpooling, dropout layers and a pair of 2D convolution layers with ReLU being the activation function [7]. The third group's output is flattened in the fourth layer, and the prediction layer is given by a dense layer with softmax activation function that is accompanying the dropout layer. The network receives a tensor with the format: $L{\times}f \times 1$, as its input. Here, f denotes the number of features (time signals) and l denotes the signal length [7]. The tensor is constructed using a widely used structure [4,28] of conv-conv-pooling-dropout. This process of removing the pooling layers between the convolution layers and keeping two convolution layers instead of one with a higher filter size reduces the number of trainable parameters [28] all the while making the decision function more discriminative. The dropout fully connected layers and maxpooling provide the best performance [7,29]. So, in this chapter, dropout is used after every maxpooling and convolution layer. Table 7.3 presents the discussed CNN model.

7.12 Evaluation matrix

7.12.1 Handwritten text recognition

Evaluation is done using the IAM dataset. The goal is to compare the performance of best path and WBS decoding methods while using the same NN. Word error rate (WER) and character error rate (CER) are the most common error measures used for handwritten text recognition [47].

7.12.2 Air-writing recognition

The model's performance for both user-independent and user-dependent principles is calculated using the accuracy formula presented in [7]. An accuracy of ten-fold cross-validation was reported for the RTC, RTD and smart-band datasets in the user-dependent category for the comparison between the proposed model in [7] and the existing methods. It's worth noting that the datasets are evenly distributed in terms of sample per letter and per subject [7]. Thus, using accuracy as the sole parameter for the evaluation of the classification, performance is justified [7]. The formula used for the accuracy is shown in (7.7).

$$Accuracy = (Number\ of\ Correct\ Predictions)/(Total\ number\ of\ Predictions) \tag{7.7}$$

Table 7.3 Network architecture for a 2D-CNN structure based upon Smart-Band dataset

Operation group	Name of the layer	Size of filter	Number of filters	Size of the stride	Padding	Activation function	Size of output	Number of parameters
	Input Layer	NA	NA	NA	NA	NA	200 × 6 × 1	0
	Conv 1 of 1	2×2	32	1×1	1×1	ReLU	200 × 6 × 32	160
Group 1	Conv 2 of 1	2×2	32	1×1	1×1	ReLU	200 × 6 × 32	4,128
	MaxPool1	2×2	1	2×2	0		100 × 3 × 32	0
	Dropout1			p = 10%			100 × 3 × 32	0
	Conv 1of 2	2×2	64	1×1	1×1	ReLU	100 × 3 × 64	8,256
Group 2	Conv 2 of 2	2×2	64	1×1	1×1	ReLU	100 × 3 × 64	16,448
	MaxPool2	2×2	1	2×2	0		50 × 2 × 64	0
	Dropout2			p = 20%			50 × 2 × 64	0
	Conv 1 of 3	2×2	128	1×1	1×1	ReLU	50 × 2 × 128	32,896
Group 3	Conv 2 of 3	2×2	128	1×1	1×1	ReLU	50 × 2 × 128	65,664
	MaxPool3	2×2	1	2×2	0		25 × 1 × 128	0
	Dropout3			p = 20%			25 × 1 × 128	0
	Flatten						3,200	0
Group 4	Dense					ReLU	512	16,38,912
	Dropout4			p = 50%			512	0
	Dense					Softmax	26	13,338
							Total	17,79,802

The output size and the number of parameters vary depending on the dataset used as they are based on the signal length, l and number of features. The smart-band dataset has six features and a signal length of 200. As a result, we get the 2D-CNN network as shown in [7].

7.13 Results and discussion

7.13.1 Handwritten text recognition

The neural network used in this chapter is in accordance with the CRNN presented in [5] and implemented using TensorFlow [35]. In this model, there are seven convolution layers followed by two RNN layers and a final CTC layer in this model. The neural network's output signal has a length of 100 time steps [41]. The IAM dataset is made up of 79 unique characters but 1 more character '-'is added to the list, making the size of the output matrix of the RNN as $T \times (C + 1) = 100 \times 80$. Four scoring modes are evaluated for the WBS decoding [41]. The language model (LM) used to predict upcoming words and give probabilities [48] can be trained by either rudimentary LM using the text given in the training set which is then made more extensive by concatenating it with a wordlist (denoted as Te+L) having 370,099 words or an ideal LM which only contains the texts given in the test set (denoted as Te) [41]. The resulting dictionary sizes for the rudimentary and ideal LM for IAM dataset are as follows: 373,412 and 3,707 unique words, respectively [41]. For the CVL dataset, WBS performs better than the best path and token passing algorithms as all of the words in the CVL can be used for the dictionary needed by the LM. The small size of the CVL dataset enables us to use all of the words given in the dataset, thus outperforming the rest. It is shown and proved in the paper presented by [41] that WBS outperforms best fit decoding on both ideal LM and rudimentary LM. The results of the comparison of WBS with best fit and other decoding are presented in Table 7.4. There are some cases where token passing outperforms the WBS with regard to WER. This is due to the fact that WBS uses the word mode, while token passing is using N-grams mode. These modes are explained in Section 4.2.1.

7.13.2 Air-writing recognition

After extensive testing on a smart-band dataset with a suitable length l and several interpolation methods, the best value is selected among them [6]. Afterwards, the

Table 7.4 Comparison of different decoding algorithms on IAM and CVL datasets

Decoding algorithm	IAM	CVL
Best path	8.94/29.47	1.94/5.22
Token passing	10.04/11.72	1.35/1.32
WBSTe	4.86/10.15	1.07/1.45
WBSTr	9.77/27.2	1.11/1.68

The results are given in CER/WER format. CER and WER stand character error rate and word error rate, respectively. WBSTr and WBSTe stand for WBS trained from the training set and the testing set, respectively.

findings were compared to the best interpolation method and existing signal length unification methods such as truncating and padding. Finally, the methodology was applied to the remaining six datasets [10,26,27] and the results are confirmed using the same methods as presented in [7].

7.13.2.1 Finding optimal length

This is the process of selecting the best fit fixed signal length from all the signal lengths. The input to the 2D-CNN model is of fixed length, so we need to find a fixed length signal; hence, this is an important aspect of data pre-processing. Figure 7.10 illustrates a histogram of the sample length from the smart-band dataset [6].

Since each sample has a time-series data of six time series, the size of the matrix for each sample of the letter is fixed at $l \times 6$ where the fixed signal length of the signal is defined as l. Figure 7.10 shows the signal length distribution for the smart-band dataset and through observation we can confer that selecting a constant signal length is a question of choice (see Section 4.2.2). So, we can select the fixed length for any dataset in the same manner.

We took into account the following two aspects based on statistical features. First, we calculated and analysed the round mean by upsampling half of the length distribution data and downsampling the other half of the length distribution data. However, downsampling by an interpolation method may result in the loss of important data, while upsampling by an interpolation method densely populates the signal. Second, we may have some data redundancy if we upsample the majority of the data, but data loss is often avoided. As a result, in our experiments signal lengths of 100 and 200 are used, with 100 being used to balance interpolation for upsampling and downsampling and 200 being used to up-sample the majority of the data to minimize data loss. Hence, the shape of the matrix containing all of the samples will be $21,450 \times l \times 6$, where l is either 100 or 200 [7].

7.13.2.2 The effects of different interpolation methods

At $l = 100$, different methods were considered as almost half of the data needed to be up-sampled or down-sampled. As a result, we obtained 20 different arrangements of the interpolation methods (see Table 7.5). Also, working on 20 different

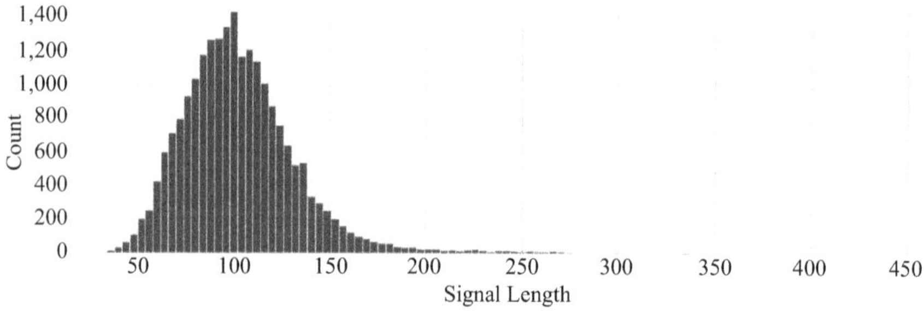

Figure 7.10 A histogram representing the distribution of signal length in the smart-band dataset

Table 7.5 Comparison of different interpolation methods for different upsample and downsample settings [7]

Method for upsampling	Method for downsampling	Length of signal, l	Accuracy	
			Average (%)	Standard dev.
Bicubic	Bicubic		87.35	0.41
	Lanczos	100	87.21	0.59
	Bilinear		87.76	0.45
	Nearest neighbour		87.04	0.21
Lanczos	Bicubic		87.54	0.18
	Lanczos	100	86.5	0.7
	Bilinear		86.84	0.46
	Nearest neighbour		86.73	0.27
Bilinear	Bicubic		86.91	0.77
	Lanczos	100	86.77	0.63
	Bilinear		86.25	0.8
	Nearest neighbour		87.38	0.09
Nearest neighbour	Bicubic		87.38	0.37
	Lanczos	100	87.32	0.76
	Bilinear		86.67	0.58
	Nearest neighbour		87.08	1.24
Bicubic	Bicubic		88.54	0.31
Lanczos	Lanczos	200	87.35	0.31
Bilinear	Bilinear		88.46	0.19
Nearest neighbour	Nearest neighbour		88.08	0.59

The table shows the effects of choosing different combinations of interpolation methods for the process of upsampling and downsampling [7].

combinations for $l = 200$ made little sense because then there was so little data left to be down-sampled. The same interpolation method was used for both upsampling and downsampling [7].

From Table 7.5, we can see that the performance for all the possible combinations of $l = 100$ is marginally poor as important data is lost while downsampling. For $l = 200$, the bicubic, performed the best. The temporal data loss was minimized because we up-sampled most of the data for $l = 200$ [7]. As a result, we must choose the best length. In a similar manner, for a given data signal, the comparative analysis is carried out using conventional methods such as padding and truncation as illustrated in Table 7.6. The signal's sampling rate is modified accordingly to upsampling and downsampling, while the signal's main characteristics are preserved, with truncating removing the extra part of the signal to make it fit and

Table 7.6 Comparison of traditional truncating and padding methods vs the bicubic interpolation [7]

Approach Used	Signal length, l	Padded or up-sampled samples	Truncated or down-sampled samples	Flops	Inference time (ms)	Accuracy	
						Avg. (%)	Std.
Pre-sequence	50	212	21,238	5,99,177	1,648	60	1
	100	10,892	10,558	9,92,393	1.492	84.57	0.3
Padding and truncation	200	21,161	289	17,78,825	1.709	86.58	0.37
	400	21,449	1	34,17,225	2.056	86.38	0.24
Post-sequence	50	212	21,238	5,99,177	1.429	48.15	0.44
	100	10,892	10,558	9,92,393	1.843	80.25	0.23
Padding and truncation	200	21,161	289	17,78,825	1.77	85.84	0.42
	400	21,449	1	34,17,225	2.519	85.63	0.35
Bicubic Interpolation	50	212	21,238	5,99,177	1.475	84.98	0.35
	100	10,892	10,558	9,92,393	1.533	87.35	0.41
	200	21,161	289	17,78,825	1.687	88.54	0.31
	400	21,449	1	34,17,225	2.43	88.02	0.48

The table compares the best interpolation method on Smart-Band dataset (see table 7.5) with the traditional padding and truncating method as presented in [7].

padding using zeros to fill the extra empty slots. The terms post and pre refer to the parts of the sequence where padding by zero or truncation is done [7].

The inference time and the number of floating-point operations of the model increase when we increase the signal length to 400; yet the model's performance declined in both situations of the padding and truncation methods. Meanwhile, at signal length 200 (which is also the most computationally optimal), the bicubic algorithm did substantially better. Because the accuracy of bicubic interpolation is highest at $l = 200$, it is chosen as the fixed length. Table 7.7 shows how the ideal fixed length is chosen for the other databases using similar principles [7].

7.13.2.3 Result and discussion

The air-writing is evaluated in this literature based on two principles [6,49]: (1) input dependent on the user and (2) input independent of the user. The method presented in this literature and proposed in [7] clearly outperforms the previous methods [6,9,10,27,49] and other methods described in [7].

Table 7.8 shows the comparison of our proposed model applied upon RTC, RTD and smart-band datasets with the existing approaches. From the table, we can

Table 7.7 The selected signal length, l *[7]*

Dataset	Minimum	Maximum	Length of signal, *l*
Smart-band	34	438	200
RTC	21	173	125
6DMG-digit	29	218	175
6DMG-upper	27	412	250
6DMG-lower	27	163	150
6DMG-all	27	412	250
RTD	18	150	125

The minimum and maximum values for all the datasets [7].

Table 7.8 Performance of user-independent and dependent principles on RealSense trajectory based and smart-band datasets [7]

Training principle	Approach used	Accuracy		
		RTC	RTD	Smart band
User-dependent principle	KNN-DTW based [3]	-	-	89.2
	2D-CNN based [20]	97.29	-	-
	LSTM based [5]	-	99.17	-
	CNN-LSTM fusion [33]	98.74	99.63	-
	Discussed	99.63	99.76	91.34
User-independent principle	1D-CNN [3]	-	-	83.2
	Discussed	-	-	85.59

concur that our model outperforms the existing approaches by a wide margin in terms of accuracy for both user independent and user dependent principles [7].

7.14 Conclusion

Advancement of technology has helped us to improve on a lot of things while it may present some problems like hardware dependability. It is useful in solving some of the most challenging and tedious tasks like text recognition. This chapter focuses on both these aspects of technology. In this research, we explore how to effectively automate text recognition from variable-length handwritten sources using CRNN models and connectionist temporal classification algorithms, which serves as both an encoder and a loss function. This chapter also discussed the WBS decoding algorithm in comparison to the traditional decoding algorithms and found that WBS clearly overperforming the other methods.

As for the solution to hardware dependability and the issue of being in synchronization with current technology like AR and VR, this chapter discussed a way to solve the method of 'writing in the air'. For this, we have used CNN models with interpolation methods to improve efficiency. A brief comparison of several interpolation algorithms is also presented, with the findings being stated.

References

[1] Plötz, T. and Fink, G. A. (2009). Markov models for offline handwriting recognition: a survey. *International Journal on Document Analysis and Recognition (IJDAR)*, 12(4), 269–298.

[2] Pacha, A. (2019). Self-learning optical music recognition. In *Vienna Young Scientists Symposium*, Institut für Information Systems Engineering: Vienna, Austria, pp. 34–35.

[3] Sánchez, J. A., Romero, V., Toselli, A. H. and Vidal, E. ICFHR2014 competition on handwritten text recognition on transcriptorium datasets (HTRtS). In *2014 14th International Conference on Frontiers in Handwriting Recognition*. IEEE, 2014, pp. 785–790.

[4] Sharma, A. K., Nandal, A., Dhaka, A. and Bogatinoska, D.C., "Brain tumor classification via UNET architecture of CNN technique." *International Conference on Cyber Warfare, Security (SpacSec2021)*, Springer, 9–11 Dec 2021, Manipal University Jaipur, 2022, pp. 18–33.

[5] Shi, B., Bai, X. and Yao, C. (2016). An end-to-end trainable neural network for image-based sequence recognition and its application to scene text recognition. *IEEE Transactions on Pattern Analysis and Machine Intelligence*, 39(11), 2298–2304.

[6] Yanay, T. and Shmueli, E. (2020). Air-writing recognition using smart-bands. *Pervasive and Mobile Computing*, 66, 101183.

[7] Abir, F. A., Siam, M., Sayeed, A., Hasan, M., Mehedi, A. and Shin, J. (2021). Deep learning based air-writing recognition with the choice of proper interpolation technique. *Sensors*, 21(24), 8407.

[8] Zabulis, X., Baltzakis, H. and Argyros, A. A. (2009). Vision-based hand gesture recognition for human-computer interaction. *The Universal Access Handbook*, 34, 30.

[9] Chen, M., AlRegib, G. and Juang, B. H. (2015). Air-writing recognition—part i: modeling and recognition of characters, words, and connecting motions. *IEEE Transactions on Human-Machine Systems*, 46(3), 403–413.

[10] Alam, M., Kwon, K. C., Abbass, M. Y., Imtiaz, S. M. and Kim, N. (2020). Trajectory-based air-writing recognition using deep neural network and depth sensor. *Sensors*, 20(2), 376.

[11] Mohammadi, S. and Maleki, R. (2019). Real-time kinect-based air-writing system with a novel analytical classifier. *International Journal on Document Analysis and Recognition (IJDAR)*, 22(2), 113–125.

[12] Dwarampudi, M. and Reddy, N. V. (2019). Effects of padding on LSTMs and CNNs. arXiv preprint arXiv:1903.07288.

[13] Aly, H. A. and Dubois, E. (2005). Image up-sampling using total-variation regularization with a new observation model. *IEEE Transactions on Image Processing*, 14(10), 1647–1659.

[14] Roy, R., Pal, M. and Gulati, T. (2013). Zooming digital images using interpolation techniques. *International Journal of Application or Innovation in Engineering & Management (IJAIEM)*, 2(4), 34–45.

[15] Bissacco, A., Cummins, M., Netzer, Y. and Neven, H. (2013). Photoocr: reading text in uncontrolled conditions. In *Proceedings of the IEEE International Conference on Computer Vision*, IEEE: USA, pp. 785–792.

[16] Rodriguez-Serrano, J. A., Gordo, A. and Perronnin, F. (2015). Label embedding: a frugal baseline for text recognition. *International Journal of Computer Vision*, 113(3), 193–207.

[17] Scheidl, H. (2018). Handwritten text recognition in historical documents, (Doctoral dissertation, Wien). reposiTUm. https://doi.org/10.34726/hss.2018.43931

[18] Yin, Y., Xie, L., Gu, T., Lu, Y. and Lu, S. (2019). AirContour: building contour-based model for in-air writing gesture recognition. *ACM Transactions on Sensor Networks (TOSN)*, 15(4), 1–25.

[19] Moazen, D., Sajjadi, S. A. and Nahapetian, A. AirDraw: leveraging smart watch motion sensors for mobile human computer interactions. In *2016 13th IEEE Annual Consumer Communications & Networking Conference (CCNC)*. IEEE: Las Vegas, NV, USA, 2016. pp. 442–446.

[20] Sakuma, K., Blumrosen, G., Rice, J. J., Rogers, J. and Knickerbocker, J. Turning the finger into a writing tool. In *2019 41st Annual International Conference of the IEEE Engineering in Medicine and Biology Society (EMBC)*. IEEE: Berlin, Germany, 2019. pp. 1239–1242.

[21] Markussen, A., Jakobsen, M. R. and Hornbæk, K. Vulture: a mid-air word-gesture keyboard. In *Proceedings of the SIGCHI Conference on Human Factors in Computing Systems*, Association for Computing Machinery, New York, NY, USA and Toronto, ON, Canada, 2014. pp. 1073–1082.

[22] Chen, M., AlRegib, G. and Juang, B.-H. (2015). Air-writing recognition—part i: modeling and recognition of characters, words, and connecting motions. *IEEE Transactions on Human-Machine Systems*, 46(3), 403–413.

[23] Lin, X., Chen, Y., Chang, X. W., Liu, X. and Wang, X. (2018). Show: smart handwriting on watches. *Proceedings of the ACM on Interactive, Mobile, Wearable and Ubiquitous Technologies*, 1(4), 1–23.

[24] Kim, U. H., Hwang, Y., Lee, S. K. and Kim, J. H. (2021). Writing in the air: unconstrained text recognition from finger movement using spatio-temporal convolution. arXiv preprint arXiv:2104.09021.

[25] Alam, M. S., Kwon, K. C. and Kim, N. (2021). Implementation of a character recognition system based on finger-joint tracking using a depth camera. *IEEE Transactions on Human-Machine Systems*, 51(3), 229–241.

[26] Chen, M., AlRegib, G. and Juang, B. H. 6dmg: a new 6d motion gesture database. In *Proceedings of the 3rd Multimedia Systems Conference*, Association for Computing Machinery, New York, NY, USA, 2012, pp. 83–88.

[27] Alam, M. S., Kwon, K. C. and Kim, N. Trajectory-based air-writing character recognition using convolutional neural network. In *2019 4th International Conference on Control, Robotics and Cybernetics (CRC)*. IEEE: Tokyo, Japan, 2019, pp. 86–90.

[28] Simonyan, K., and Zisserman, A. (2014). Very deep convolutional networks for large-scale image recognition. arXiv preprint arXiv:1409.1556.

[29] Wu, H. and Gu, X. (2015). Towards dropout training for convolutional neural networks. *Neural Networks*, 71, 1–10.

[30] Marti, U. V. and Bunke, H. (2002). The IAM-database: an English sentence database for offline handwriting recognition. *International Journal on Document Analysis and Recognition*, 5(1), 39–46.

[31] Kleber, F., Fiel, S., Diem, M. and Sablatnig, R. Cvl-database: an off-line database for writer retrieval, writer identification and word spotting. In *2013 12th International Conference on Document Analysis and Recognition*. IEEE: Washington, DC, USA, 2013. pp. 560–564.

[32] Hochreiter, S. and Schmidhuber, J. (1997). Long short-term memory. *Neural Computation*, 9(8), 1735–1780.

[33] Hochreiter, S. (1991). Untersuchungen zu dynamischen neuronalen Netzen. Diploma, Technische Universität München, 91(1), 31.

[34] Graves, A., Fernández, S., Gomez, F. and Schmidhuber, J. Connectionist temporal classification: labelling unsegmented sequence data with recurrent neural networks. In *Proceedings of the 23rd International Conference on Machine Learning*, Association for Computing Machinery, New York, NY, USA, 2006, pp. 369–376.

[35] Sharma, A. K., Nandal, A., Dhaka, A., *et al.* HOG transformation based feature extraction framework in modified Resnet50 model for brain tumor detection. In *Biomedical Signal Processing & Control*, Elsevier, vol. 84, 2023 p. 104737.

[36] Goodfellow, I., Bengio, Y. and Courville, A. *Deep Learning*. MIT Press: USA; 2016.

[37] Gnauck, A. (2004). Interpolation and approximation of water quality time series and process identification. *Analytical and Bioanalytical Chemistry*, 380(3), 484–492.

[38] Sharma, A. K., Nandal, A., Dhaka, A., and Dixit, R. A survey on machine learning based brain retrieval algorithms in medical image analysis. In *Health and Technology*, Springer, vol. 10, 2020. pp. 1359–1373.

[39] Bansal, S. *Explaination of CTC*. Available from https://sid2697.github.io/Blog_Sid/algorithm/2019/10/19/CTC-Loss.html. [Accessed 23 Mar 2022], 2019.

[40] Graves, A. Supervised sequence labelling. In *Supervised Sequence Labelling With Recurrent Neural Networks*. Berlin, Heidelberg: Springer; 2012. pp. 5–13.

[41] Scheidl, H., Fiel, S. and Sablatnig, R. Word beam search: a connectionist temporal classification decoding algorithm. In *2018 16th International Conference on Frontiers in Handwriting Recognition (ICFHR)*. IEEE, 2018. pp. 253–258.

[42] Davis, P. J. (1975). *Interpolation and Approximation*. Courier Corporation, Dover Publications: New York.

[43] Amin, J., Sharif, M., Haldorai, A. K., Yasmin, M., and Nayak, R. S., Brain tumor detection and classification using machine learning techniques. RAMSACT 2022, Springer, Accepted May 2022.

[44] Keys, R. (1981). Cubic convolution interpolation for digital image processing. *IEEE Transactions on Acoustics, Speech, and Signal Processing*, 29(6), 1153–1160.

[45] Sharma, A. Kumar., Nandal, A., Ganchev, T., and Dhaka, A. Breast cancer classification using CNN extracted features: a comprehensive review. In *Application of Deep Learning Methods in Healthcare and Medical Science (ADLMHMS-2020)*. Taylor and Francis: CRC Press; 2022.

[46] Duffner, S., Berlemont, S., Lefebvre, G. and Garcia, C. 3D gesture classification with convolutional neural networks. In *Proceedings of the 2014 IEEE International Conference on Acoustics, Speech and Signal Processing (ICASSP)*, Florence, Italy, 4–9 May 2014; pp. 5432–5436. 49.

[47] Bluche, T. (2015). Deep neural networks for large vocabulary handwritten text recognition, Doctoral dissertation, Paris 11.

[48] Jurafsky, D. and Martin, J. H. *Speech and Language Processing: An Introduction to Natural Language Processing, Computational Linguistics, and Speech Recognition*. Prentice Hall: Upper Saddle River, NJ, USA, 2000.

[49] Xu, S. and Xue, Y. Air-writing characters modelling and recognition on modified CHMM. In *2016 IEEE International Conference on Systems, Man, and Cybernetics (SMC)*. IEEE: Budapest, Hungary, 2016, pp. 001510–001513.

Chapter 8

Microscopic *Plasmodium* classification (MPC) using robust deep learning strategies for malaria detection

Rapti Chaudhuri[1], Shuvrajeet Das[1] and Suman Deb[1]

Pathogenic microbes cause harm to human lives. Very specific mention is *Plasmodium* species which belongs to such category of pathogen, biologically known as malaria parasite, the devastating cause of life loss. The presence of pathogenic *Plasmodium* microbes are done by optical blood sample analysis through microscope. Manual identification of microscopic pathogen is a challenging and time-consuming task with respect to their similar structural formation. Even through digital microscope, identification of such pathogenic structures depends on several complex factors which is always dominated by human limitation over a long period of sample scanning. Machine-driven identification of such pathogenic microbes will be an additive benefit to speedy identification of probable presence of *Plasmodium* in human RBC. The machine-driven classification, segmentation and identification by applying deep learning techniques can be incorporated for obtaining near-perfect identification solution. The closed approximation inference has been carried out by applying convolutional neural network models for proper classification followed by precise identification of variant *Plasmodium* species in a single slide. Classification models such as SE_ResNet, ResNeXt, MobileNet and XceptionNet are studied extensively and applied on taken dataset after data preprocessing, augmentation and regularization, with state of art comparison. Resultant analysis has been done graphically and numerically as well for attaining desired parametric conditions. The aforesaid models are mainly considered here for their confirmed reliability and consistency in producing saturated results relative to the concerned data type and constrained parameterized structure. During experiment the proposed methodology has resulted in the identification of pathogenic *Plasmodium* microbes in an optimum amount of time and classifying the type of *Plasmodium* parasite to its exact class, working as a state of decision support system in medical pathology.

[1]Department of Computer Science and Engineering, National Institute of Technology Agartala, India

8.1 Introduction

Manual slide analysis for microscopic pathogen is a very tedious task. In various work [1–4] the complexity of this analysis has been reported. With advent of technology the present deep learning techniques are agile enough to mimic human ability to identify the microscopic pathogen with a great range of precision with in challenging time-bound. But while doing so the analysis of the blood samples require considerable care for confusion on close structural identification among the similar species of microscopic pathogens like *Plasmodium*. Blood slides to be first identified are read on the basis of presence of four classes of *Plasmodium* species, *Plasmodium falciparum*, *Plasmodium vivax*, *Plasmodium malariae* and *Plasmodium ovale*. The process has been carried out applying convolutional neural network (CNN) models for precise classification and it has been followed by bounding box identification for more efficient and accuarte prediction of *Plasmodium* species to detect type of malaria. The type of input with resultant output is illustrated briefly in Figure 8.1.

8.1.1 Classification of Plasmodium *using CNN*

CNN is known to be the foremost deep learning classification model extensively used for image classification. Therefore inspite of existence of various heuristic as well meta heuristic approaches like naive Bayes classifier, random forest, decision tree, genetic algorithm, and particle swarm optimization. CNN has been considered over these in simulating the current unsolved research problem. Convolution helps in reduction of input image with the usage of filter of different sizes. The procedural steps of convolution includes primarily multiplication of an input matrix of image pixel to a considerable sized filter or kernel. With this concept, the present work has used CNN

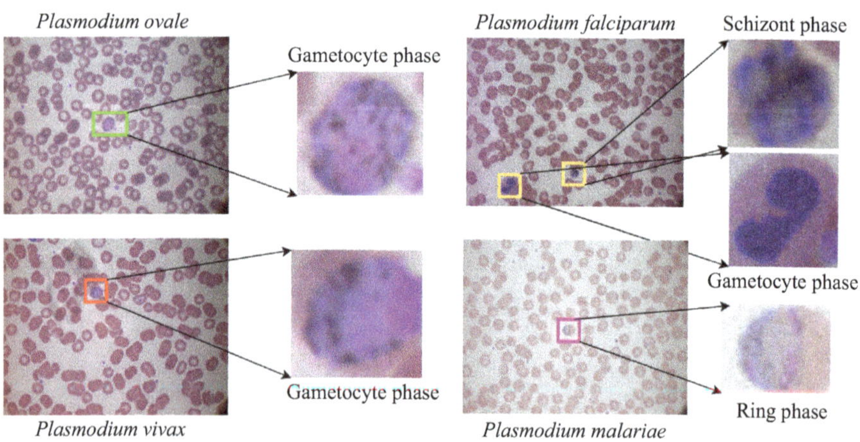

Figure 8.1 Different variants of Plasmodium *species taken as input resulting in output of respective detected phases through classification followed by bounding box identification*

models extensively in case of classifying *Plasmodium* genus with its variable four species which are prone to spread malaria in human body. Basic architectural structure of CNN consists of five layers namely, the convolutional layer, pooling layer, fully connected layer, dropout and activation layer. The brief introduction to each of these functional layers has been done in next phase of introduction.

Convolution layer in CNN architecture: Being the first layer of CNN architecture it tends in feature extraction process of input images. Numerically the procedure has been analyzed in the form of multiplication of filter sized $n \times n$ to image of size $M \times M$. The output is obtained in the form of edge and corner extraction by sliding mechanism applied on filter over the input image.

Pooling in CNN architecture: The convoluted image is then passed to the pooling layer where the respective dimension of image gets decreased. Pooling can be of various types like Max pooling, Average pooling which come into existence as per need. The pooled image is then passed to the fully connected layer for undergoing flattening.

Fully connected layer in CNN architecture: The pooled input images are fed to one or more fully connected layer, and here the actual process of classification takes place. After flattening sometimes these are again passed towards the dropout layer according to the requirement of environmental demands.

Dropout in CNN architecture: For the dropout layer, two main concept of data balance bias and variance are needed to be considered extensively. Bias depicts numerically the difference between the average prediction of the algorithm and the absolute original value of the data, that is, it calculates the error in prediction. High bias results in underfitting situation, giving a poor category of recognition both in training as well as test data. Variance denotes the variability of model prediction for a concerned data point (Figure 8.2). The algorithm works more than well on training data results in high variance possessing the chance of producing high error in recognizing new test dataset. For predicted result taken as N, input function $f(m)$, the output of predicted data points obtained as a straight line is represented as shown in (8.1).

$$N = f(m) + e \tag{8.1}$$

e is the error term.

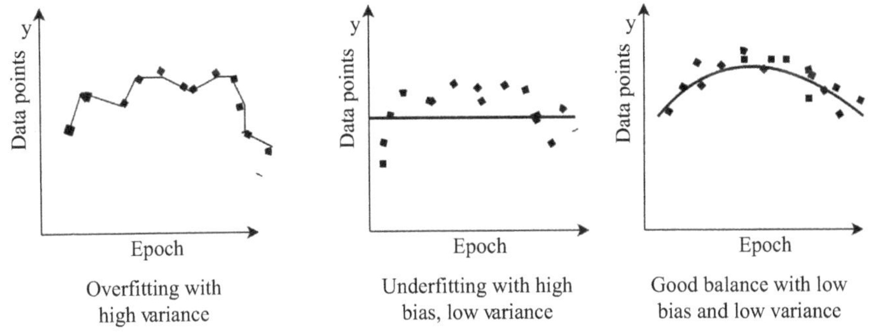

Figure 8.2 Illustration of variable data points representation in overfitting, underfitting and good balance situation

Corresponding predicted value $\hat{f}(m)$ is subtracted from obtained value. So the expected squared error can be presented as (8.2), (8.3) and (8.7):

$$E(m) = E\left[(N - \hat{f}(m))^2\right] \tag{8.2}$$

$$E(m) = (E\left[\hat{f}(m)\right] - f(m))^2 + E\left[(\hat{f}(m) - E\left[\hat{f}(m)\right])^2\right] + \sigma_e^2 \tag{8.3}$$

Error (E) results from the sum of bias square, variance and irreducible error.

$$E(m) = bias^2 + variance + irreducible\_error \tag{8.4}$$

In case of reduction of overfitting conditions primarily, dropout technique has been introduced. A few neurons are dropped or pruned randomly at the time of training so that the size of the respective model gets reduced.

Activation layer in CNN architecture: This layer is the main component which takes decision of firing the selected information in forward direction. Most commonly used Activation functions are ReLU, Softmax, tanH and the Sigmoid functions. The softmax function is numerically presented in (8.5).

$$n_i = \frac{e^{m_i}}{\sum_j e^{m_j}} \tag{8.5}$$

Specifically sigmoid function is used for binary classification and softmax for multi-class classification of input images.

Different models of CNN for classification: Models are distinguished according to their occupied components, their structural orientation and application including convolutional layers, pooling layers, dropout, data augmentation, addition of filters, fully connected layer and types of activation layer for producing the classified result.

Different literature survey [5] as well as experimental results provides four discussed models to be better relative to other existing techniques. Four different existing models of CNN have been analyzed in an extensive manner with respect to accurate prediction of malaria parasite in sample RBC. The basic model of CNN is illustrated in Figure 8.3. Precision and recall are used to numerically express the confidence of a particular prediction problem.

Equations for Precision, Recall and F_1 score

Mathematically, precision, recall and F_1 score are expressed as:

$$Precision(P) = \frac{TruePositives}{(TruePositives + FalsePositives)} \tag{8.6}$$

$$Recall(R) = \frac{TruePositives}{(TruePositives + FalseNegatives)} \tag{8.7}$$

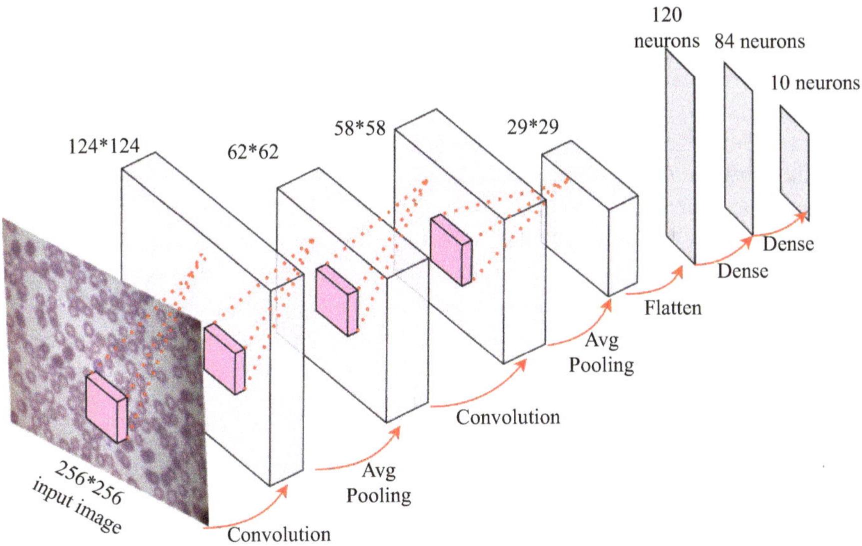

Figure 8.3 Fundamental model architecture of basic CNN for analyzing malaria parasite Plasmodium *species in taken RBC cell image slide*

$$F_1 = \frac{2 * PR}{(P + R)} \tag{8.8}$$

Architectural illustrations have been done for proper analysis of respective performances executed by different models on input RBC images with *Plasmodium* species.

8.2 Related works

The chief goal of this research piece is identification which follows classification of respective *Plasmodium* species based on their stages in standard microscopic slides. This proposed automated computer-aided optimum time procedure has been analyzed on the basis of certain carried out experiments. Some of the related works have been illustrated here with extensive study and simulation analysis.

Shekhar *et al.* [6] proposed a new robust ML-assisted CNN model for automated classification and prediction of infected cells with in standard microscopic slides equipped with thin blood smears. The aforesaid work applied three CNN models including Basic CNN, VGG-19 Frozen CNN and VGG-19 Fine Tuned CNN for measuring comparative performance for their accuracy.

Guo *et al.* [7] presented a smartphone based end-to-end platform to detect multiple DNA of malaria. The research work made use of low-cost paper based microfluidic test, accompanied with deep learning techniques along with blockchain technology for secure data connectivity. Performance accuracy has been evaluated with perfect illustration.

Krishnadas *et al.* [8] used two object detection models YOLO v5 and scaled YOLO v4 for respective classification of stages and detection of type of malaria parasite. Seventy-nine (approx.) accuracy has been achieved overall in identification.

Rahman *et al.* [9] proposed and analyzed a Deep Learning structure for end-to-end arrangement accompanied with both feature extraction and classification applied on raw patches of RBC. An accuracy of approx 97.7% has been achieved in this particular scenario.

Kunwar *et al.* [10] presented and proposed a new image processing system for *Plasmodium* parasite identification in blood smear slide. The work also adds determination of infected cell types based on extracted features.

Kapoor *et al.* [11] made the use of two deep convolutional neural networks, namely, VGG-19 and ResNet-50 models to carry out analysis on malaria dataset in human RBC. The model accuracy has been studied based on precision, recall and F1 score. The result analysis showed 97% accuracy of ResNet-50 structure outperforming the performance of VGG-19. Main demerit lies in more consumption of training and execution time along with more data collection executed for better performance. It serves as a challenge in this regard to be carried out with other constrained parameterized environment.

Pattanaik *et al.* [12] illustrated automatic segmentation of RBC in microscopic blood smear phone images by the usage of ML-assisted techniques. A deep ResNet module (multi-magnificent) has been proposed for automated binary classification (infected and non-infected cell detection). The aforesaid proposed network resulted with 98% accuracy in precise prediction. The basic limitation of this chapter lies in choosing phone images instead of standardized microscopic slide input. This might resulted in erronious grab of pixels, perfect resolution etc., and the second limitation is execution of binary classification which itself reduces the chance of generating challenges in this regard. This might not be suitable procedure to be carried out in comparatively complex environment types, resulting in low accuracy of prediction process.

Jameela *et al.* [13] utilized CNNs and other DL-assisted structures such as image processing for evaluation of parasitemia in microscopic blood slides. Models like ResNet 50, ResNet 34, VGG-16 and VGG-19 are implemented on same collected dataset. VGG-19 outperforms the other taken models with an accuracy of 0.973 (approx) along-with application of fine tuning. For training process epoch is taken more resulting in comparatively more execution time. Other DL-assisted models could be implemented for proper result analysis.

State-of-art comparison is elaborated in Table 8.1.

8.3 Methodology

A concept of different methods adapted for experimentation are gathered and analyzed briefly to generate basic idea of the whole procedure. Four models of CNN has been considered in this structural work which outperform other existing models relative to comparative performance accuracy in complex type of

Table 8.1 State-of-art comparison with techniques used in existing related works

Reference	Technique	Accuracy	Pros	Cons
Kumar *et al.* [10]	Image processing, segmentation	No performance accuracy has been mentioned	Segmentation has been applied and feature extraction has been added for better result	No DL-assisted model analysis is carried out which would increase the clarity of detection and accuracy
Rishika *et al.* [11]	VGG-19, ResNet-50	97% Accuracy	Good classification with better accuracy	More training and execution time, more number of data required for elevating performance of model which may not be suitable for other complex environments
Pattanaik *et al.* [12]	MM-ResNet	98% Accuracy	Skip connection and other data preprocessing techniques are done for better result	Phone images are chosen as input instead of standardized microscopic image followed by binary classification reducing most challenges, unsuitable in complex type of environment

environment with optimum executable time. The output has been passed into the next module for perfect recognition of *Plasmodium* species using bounding box identification process.

The models validate the processed input images with an accurate prediction and the respective performances have been illustrated in graphical manner as well as presented in numerical form.

The chief goal of the concerned research piece is to propose a malaria parasite diagnosis using feature extraction model with deep CNN-based optimized network. Automated computer-aided recognition strategies using Deep learning-based techniques would help pathological practitioners to detect the unsolved issue early. The proposed methodology has been classified into five modules: data preprocessing phase, augmentation, regularization using batch normalization, classification based on pattern matching, and identification depending on bounding box formation.

8.3.1 Data preprocessing

The preprocessing stage is compiled with denoising, smoothing, conversion of large sized image to required pixel size for meeting GPU (graphical processing

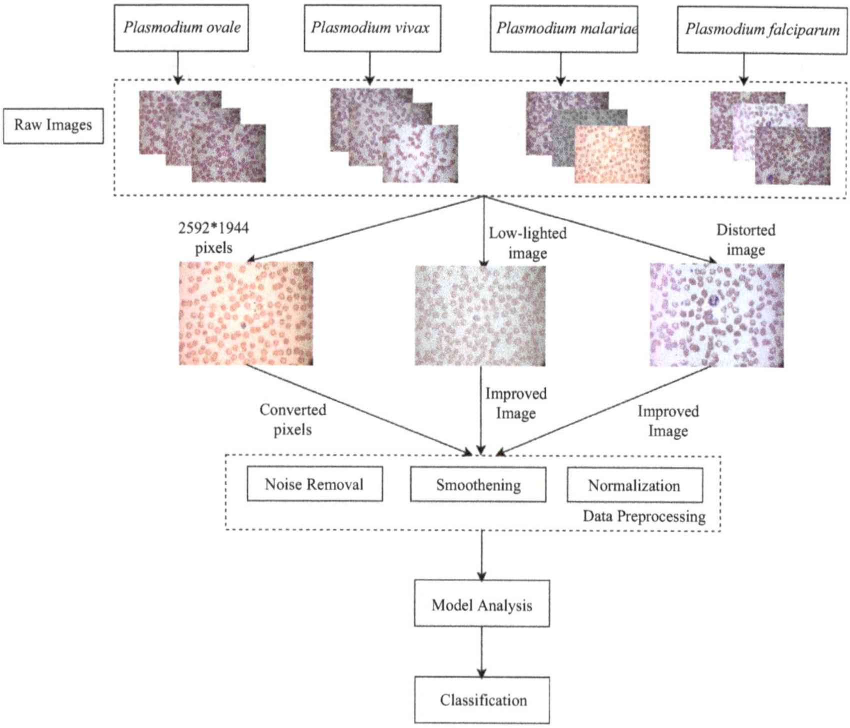

Figure 8.4 Illustration of the entire procedure with mention of step-wise data preprocessing

unit) requirement and enhancing image quality. Medical slide images often get distorted making the evaluation a challenging task. The species of same genus are mostly similar in body structure difficult to extract patterns resulting in more complexity in identification. Therefore preprocessing constitutes a major substep to acquire a better outcome.

A total of 2592 × 1944 pixel sized images of different species of *Plasmodium* are collected from kaggle scientific archive. The collected images are converted to 256 × 256 images by pixel for meet the GPU requirement. Four models each of which are trained taking variable epoch with respect to considered constrained environment and performance tendency on concerned dataset relative to inference time. After training, the models are validated for checking the consistency. The classified images are transferred to next module for proper identification using object detection algorithm. Preprocessing steps are mentioned in Figure 8.4.

8.3.2 Data augmentation

Augmentation has been basically preferred to increase amount of data with slight modifications like shearing, cropping, rotating images resulting in number of already existing data copies. It increases the possibility of training accuracy as well as testing performance. Data augmentation procedure has been illustrated below on the basis of taken *Plasmodium* dataset.

8.3.3 Weight regularization using batch normalization

Smaller values of model weights have been assumed for reducing overfitting problems. Model complications are reduced using weight regularization. Loss function is applied to regularize the matter. Most commonly used L_2 regularization method is illustrated numerically in (8.9).

$$A(\theta; i,j) = B(\theta; i,j) + \frac{\beta}{2}\theta^T\theta \tag{8.9}$$

$B(\theta; i, j)$ is the loss function and $\frac{\beta}{2}\theta^T\theta$ is L_2 regularization term. Normalization results better by the approach of Batch Normalization which uses learnable network layers with α and β. The computation of the illustrated approach is illustrated in (8.10)–(8.13).

$$\lambda = \frac{1}{e}\sum_{i=1}^{e} x_i \tag{8.10}$$

$$\theta^2 = \frac{1}{e}\sum_{i=1}^{e} (x_i - \lambda)^2 \tag{8.11}$$

$$\widehat{x}_i = \frac{x_i - \lambda}{\sqrt{\theta^2 + \phi}} \tag{8.12}$$

$$y_i = \alpha\widehat{x}_i + \beta \tag{8.13}$$

To cope up with convergence speed, a huge initial learning rate is chosen. This rate minimizes the parameters usage and network dependency resulting in enhanced network's generalization. Fine tuning of hyper parameters and loss function reduction optimizes classification strategy.

8.3.4 Classification based on pattern recognition

Based on certain deep neural network models classification has been carried out. Feature extraction and pattern matching pointed out to be primary techniques of basic classification.

8.3.5 Models for multi-class classification

SE_ResNet: Taking citation from the work described in the paper [14], an updated block namely, *Squeeze-and-Excitation (SE)* of ResNet module ResNet has been introduced and analyzed with respect to concerned working area. This work has used 41 layers in the model to meet the respective need of outcome. Figure 8.5 illustrates the basic architecture of SE_ResNet with respect to the concerned working space. The SE block has been constructed according to the principle of transformation function denoted by F_S which maps the input vector $A \in R^{H'*W'*D'}$

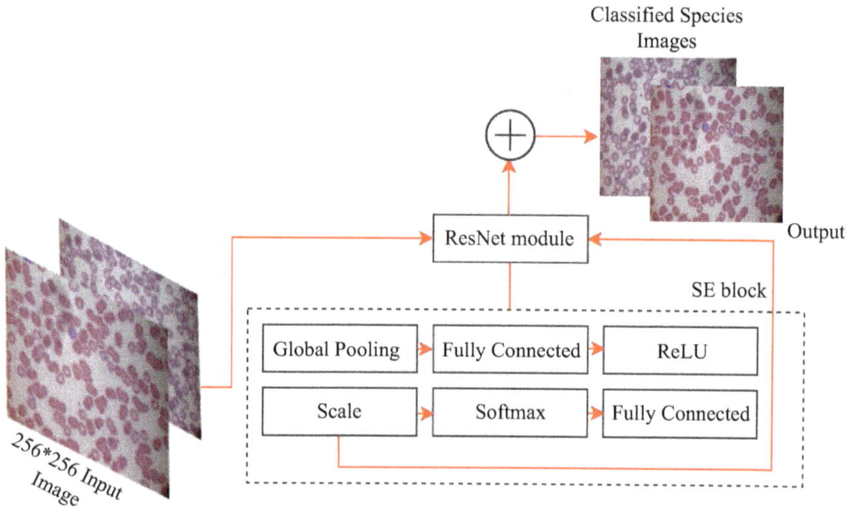

Figure 8.5 Fundamental model architecture of SE_ResNet for analyzing malaria parasite Plasmodium *species in taken RBC cell image slide*

to feature maps $K = [k_1, k_2,...,k_D]$, $Z \in R^{H*W*D}$. It has been numerically expressed in (8.14).

$$k_D = k_D * A = \sum_{i=1}^{D'} v_D^i * A^i \tag{8.14}$$

(H*W) is the spatial dimensions of feature maps resulted in creation after fed through squeeze operation on D^{th} filter. Accumulated feature maps construct a channel descriptor. $Z = [z_1, z_2, ..., z_D]$ presents the learned set of filter kernels, Z_D is the parameter of Dth filter. From global receptive field, the required subsequent information is fed through the above mentioned channel descriptor.

This resultant aggregated feature maps forms collective modulation weights in squeeze-excitation block. The enhanced refined weights are then passed to ResNet module for obtaining required output.

ResNeXt: ResNeXt [15] the very upcoming model version after ResNet is ResNeXt, as the name suggests. ResNet 50 model is considered as the fundamental version containing 50 layers which includes 48 convolutional layers with 1 max pool and 1 average pooling layer. The fundamental architecture of ResNeXt compared to ResNet module is illustrated in Figure 8.6. Here, ResNeXt model uses 55 layers to carry out the respective convolutional processes.

Vanishing gradient constitutes solely an important problem to be handled to retrieve lossless information. Here deep neural network come into action. But the exponential degradation problem of saturated accuracy at the time of convergence is tackled by deep residual network. Identity layers are introduced in ResNet

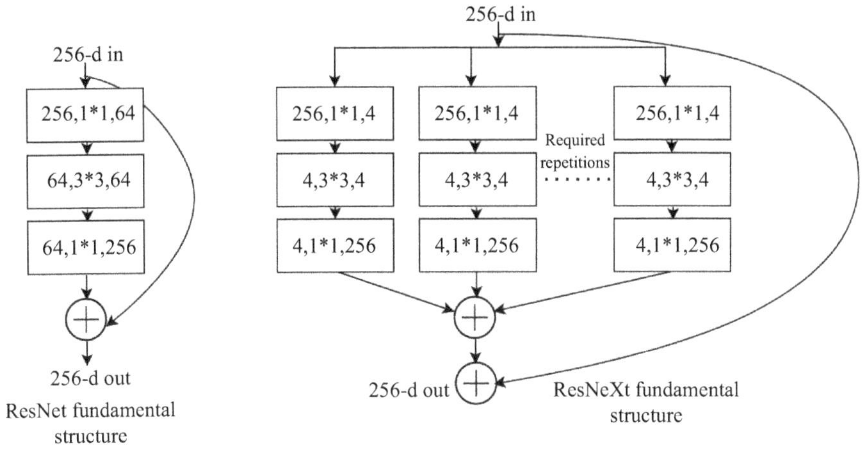

Figure 8.6 Fundamental model architecture of ResNeXt with ResNet module for analyzing malaria parasite Plasmodium *species in taken RBC cell image slide*

structure with no extra change in parameters, thus resulting in considerably optimized computation time. The identity mapping has been briefly structured in numerical way in (8.15).

$$M(x) = G(x) + x \tag{8.15}$$

where $M(x)$ denotes the the layers fitted to residual mapping and $G(x)$ is the non-linear layer introduced to another mapping of x identity layers.

ResNeXt [16] reduces extra required hyperparameters by the use of cardinality concept, which mentions that the transformation size sets work as additional dimension equipped by the width and depth of the ResNet model. The process of conventional transformation seen in ResNet module appears to be repeating to a certain extent in case of ResNeXt. All the transformation are at last concatenated. The network block is splitted which is followed by transformation and merging obtaining desired output with required less time complexity. Applying the fundamental concept of perceptron, the whole structure is numerically expressed in (8.16). g_i considered as respective training weights and h_i are the inputs.

$$\sum_{i=1}^{N} g_i h_i \tag{8.16}$$

N depicts the number of channel input vector to the neuron. The input are set free in the form of (h_1, h_2, ... h_N). It has been followed by transformation and scaling to $h_N g_N$. The scaled up module is assembled by $\sum_{n=0}^{N-1}$ as shown in (8.17).

$$\sum_{n=0}^{N-1} g_N h_N \tag{8.17}$$

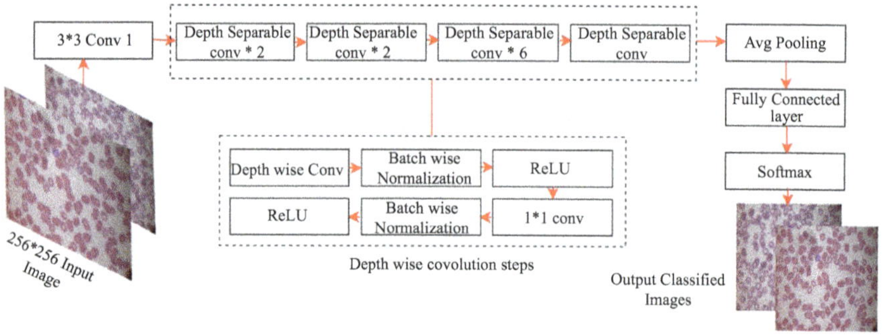

Figure 8.7 Fundamental model architecture of MobileNet for analyzing malaria parasite Plasmodium *species in taken RBC cell image slide*

The whole procedure has been reintroduced in ResNeXt [7] and the assembly has been considered more precisely by a generic function.

D is the cardinality which is not bound to be equal to N. $\lambda_i(h)$ is another arbitrary function which projects h into an embedding. Afterwards the transformation process is carried out.

MobileNet: A special structure of CNN model, MobileNet, is constructed for solving the issues with mobile applications. The model works on the principle of depth wise as well as point wise convolutions [17]. In this present context, MobileNet uses 29 layers to meet the needed requirement. The basic block diagram is shown in Figure 8.7. The usage of separable convolution technique minimizes the subsequent number of parameters, which results in a time-bound execution. The conceptual idea of filter's depth separation and spatial dimension results in idea of depth-wise separable convolution. Execution cost of depth-wise convolution is Df^2*M*Dk^2 where DK^2 means the channel wise spatial convolution, M denotes the depth of subsequent layers and $M*N*Df^2$ is the execution time of point wise convolution where M is the total height of layers and N, the total depth of layers with kernel each of size 1*1.

Depth-wise convolution is presented as channel wise $D_k * D_k$ spatial convolution and it is followed by point-wise convolution strategy. Point-wise convolution denotes to basic application of 1×1 convolution. The prime distinguishable characteristic of MobileNet lies in splitting ability of convolution into 3×3 depth-wise convolution, followed by a 1×1 point-wise convolution, rather than undergoing simple convolution. Table 8.2 defines the basic parameters of architectural CNN models considered for the research piece. Table 8.3 briefly points out the parameterized architectural layers used in the MobileNet model.

XceptionNet: The primary architectural representation of XceptionNet deep neural network includes separable convolutions according to involved deep layers [18]. In this present piece of work, XceptionNet has used a total of 98 layers to carry out the processes. The model has been interpreted as the extreme version of InceptionNet.

The primary architecture is classified into entry, middle and exit flow. The parameterized description is mentioned in Table 8.4. In XceptionNet, the original

Table 8.2 Description of features acquired by constructed four CNN models

Model	Activation function	Pooling	Kernel size	Stride	Padding
SE_ResNet [17]	ReLU, softmax	Global average pooling	3*3	2*2	0*0
ResNext [18]	ReLU	-	1*1, 3*3	1*1	0*0
MobileNet [19,20]	Softmax	Avg. pooling	3*3	2*2	0*0
XceptionNet [21]	tanh, ReLU, softmax	Max. pooling	3*3, 2*2	2*2	0*0

Table 8.3 Architectural parametric illustration of MobileNet [20] model for classification of Plasmodium *species*

Model layers	Filter used	Filter size	Map size	Activation func
Input	-	-	256*256*3	-
Conv1	32	3*3	128*128*32	-
DS Conv × 2	32	3*3	64*64*64	-
DS Conv × 2	32	3*3	64*64*64	-
DS Conv × 6	128	3*3	32*32*256	-
DS Conv	512	3*3	8*8*512	-
Avg. pooling	1024	-	8*8*1024	
Fully connected	-	-	1024	*softmax*

Table 8.4 Architectural parametric illustration of XceptionNet [18] model for classification of Plasmodium *species*

Model layers	Filter used	Filter size	Map size	Activation func
Input	-	-	-	-
Conv 1	32	3*3	2*2	*ReLU*
Conv 2	64	3*3	2*2	*ReLU*
Conv 3	128	3*3	2*2	-
Sep. Conv 1	128	3*3	-	*ReLU*
Sep. Conv 2	128	3*3	-	-
Max pooling 1	3*3	2*2	-	-
Concatenation				
Conv 4	-	3*3	2*2	-
Sep. Conv 3	256	3*3	-	ReLU
Sep. Conv 4	256	3*3	-	-
Max pooling 2	3*3	2*2	-	-
Conv 2	16	5*5	10*10*16	*tanh*
Concatenation				
Conv 5	-	3*3	-	Softmax
Glob. Avg. pooling	-	-	-	-

concept of depthwise separable convolution is modified by the introduction of two major properties of convolution which results in reduced execution time.

- 1×1 convolution is followed by channel wise spatial convolution and it distinguishes itself from the original in order of its processing.
- Intermediate ReLU is absent in XceptionNet for non-linearity, which confirms its presence in InceptionNet.

The architectural configuration includes three modules, namely, entry flow, middle flow and exit flow. Execution of point-wise convolution of size $1 \times 1 \times N$ is done on $K*K*N$ shaped image (K denotes kernel, c denotes the number of channels). This is carried out with depth-wise convolution of $d \times d \times 1$ (d size with one channel) sized filter. Depth-wise convolution procedure is carried out with point-wise convolution technique which decides for reversing in the order of original processing.

8.4 Experimental results and discussion

Python library has been used to carry out the simulation procedure. The efficiency of the proposed methodology has been implemented through four *Plasmodium* species datasets. Dataset are available in the link mentioned:

https://drive.google.com/file/d/1HTb2XNZ0pTBDJVy-74mRFDAV1GwS4VwA/view?usp=share_link

8.4.1 Dataset description

Four species of *Plasmodium* images which are prone to cause malaria, namely, *Plasmodium malariae*, *Plasmodium vivax*, *Plasmodium falciparum* and *Plasmodium ovale*, have been extracted on slides containing RBC, collected from Kaggle archieve. Each species having 1500 images with in and the 2592×1924 pixel sized image have been converted to 256×256 pixel for the sake of training.

8.4.2 Performance measures

This section gives a brief illustration of results obtained starting from data augmentation, enhancing image quality, classification to bounding box microbe identification with simulated analysis of performance and graphical presentation (Table 8.5).

The numerical analysis is done in (8.18).

$$f(h) = \sum_{i=1}^{D} \lambda i(h) \tag{8.18}$$

Augmentation Results Data augmentation involves zooming of image by cropping it, shearing, reflection and rotation of original taken image. It helps in increase in performance measure of model training. Augmentation processing and results are shown in Figure 8.8.

Model Analysis on taken dataset Figure 8.9 shows SE_ResNet having an average amount of validation accuracy 0.52 (approx) with an increasing epoch

Table 8.5 Experimental results obtained by applying different models on Plasmodium *microscopic slide image dataset considering Top-1 accuracy, loss and categorical accuracy with respective parameters*

Model	**Layers used**	**Categorical accuracy**	**Epoch**	**Top-1 accuracy**	**Loss**	**Parameters**
SE_ResNet [14]	41	0.52	30	0.52	2.3	67,000,000
ResNeXt [19]	55	0.60	50	0.59–0.6	1.2	89,000,000
MobileNet [20]	29	0.68–0.7	50	0.68–0.7	0.2	13,000,000
XceptionNet [18]	98	0.25	30	0.25	1.6	-

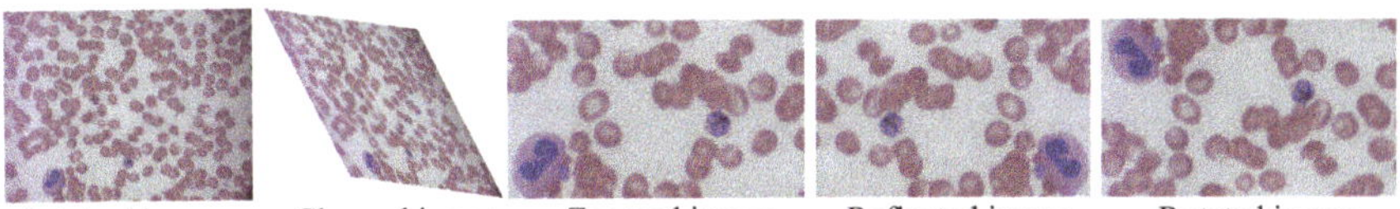

Figure 8.8 Data augmentation processes involving slight modifications for training purpose of each model

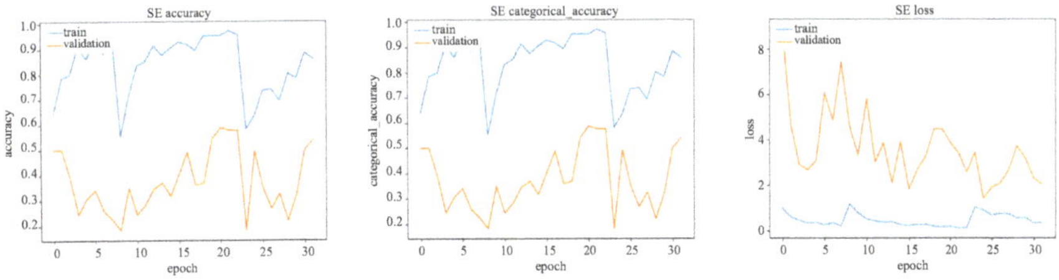

Figure 8.9 Graphical measurements of performance accuracy and losses for both training and validation dataset of the model SE_ResNet

of 30. After epoch 30 the model shows no satisfactory result. The categorical SE accuracy points to 0.52 approximately and with a loss value of 2.3 (approx) relative to 30 epochs. Considering total inference time, SE_ResNet took comparative less time in training as with less epoch, but with not that much satisfactory result.

Figure 8.10 shows ResNeXt having an average amount of validation accuracy 0.59–0.60 (approx.) with an increasing epoch of 50 which is slightly better than average precision. The categorical ResNeXt accuracy points to 0.6 approximately and with a loss value of 1.2 (approx.) relative to 50 epochs. Considering total inference time, SE_ResNet took slightly more time in training and with average amount of epoch, but could be modified for better performance.

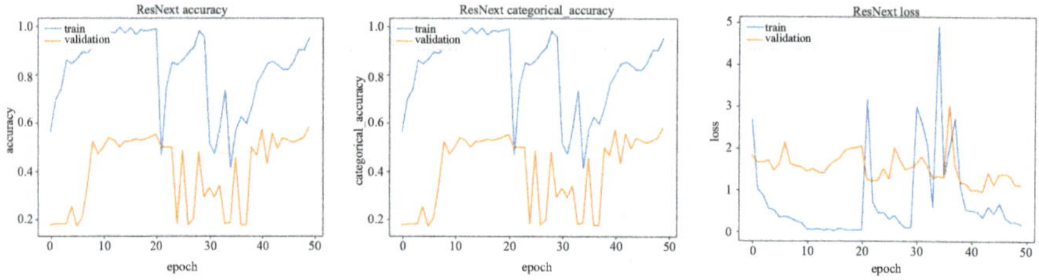

Figure 8.10 Graphical measurements of performance accuracy and losses for both training and validation dataset of the model ResNeXt

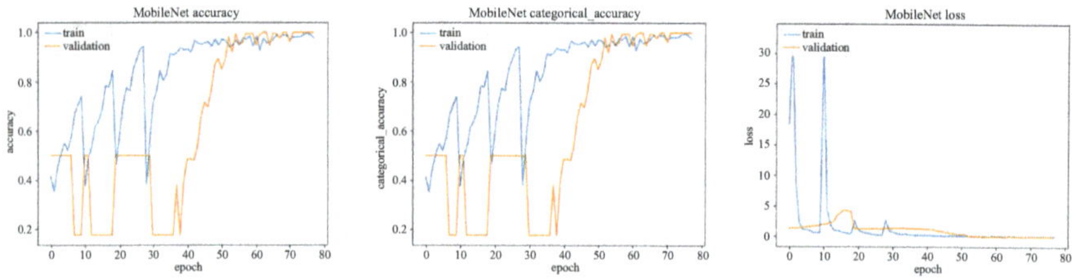

Figure 8.11 Graphical measurements of performance accuracy and losses for both training and validation dataset of the model MobileNet

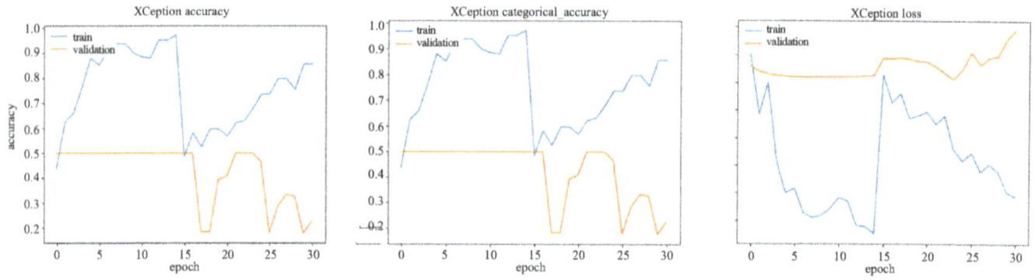

Figure 8.12 Graphical measurements of performance accuracy and losses for both training and validation dataset of the model XceptionNet

Figure 8.11 shows MobileNet having an average amount of validation accuracy 0.98–1.0 (approx) with an increasing epoch of 80. The model shows sudden surge in training at an epoch of 40, and at epoch 50, its performance accuracy measures to be 0.68–0.72. The categorical accuracy points to 0.68–0.72 at epoch 50 with a loss value of 0.2 (approx.) relative to 50 epochs. Considering total inference time and overall performance with loss percentage, MobileNet produced a much better result in this scenario.

Figure 8.12 shows XceptionNet having an average amount of validation accuracy 0.25 (approx.) with an increasing epoch of 30 which is considered as worst in this regard. The model shows consistency in training up to epoch of 15,

Table 8.6 Experimental test results obtained by applying MobileNet model on Plasmodium *microscopic slide image dataset mentioned in Section 8.4.1*

Model	Epoch	Accuracy	Cat Acc	Loss	Val Acc	Val Cat Acc	Val loss
MobileNet	33	0.9278	0.9278	0.1498	0.9903	0.990384	0.0573

Table 8.7 Performance-wise order of the considered CNN models

Model	XceptionNet	SE_ResNet	ResNeXt	MobileNet
Performance	Worst	<	<	Best

then changed to a sudden surge at an epoch of 19, and shows variable results afterwards. its performance accuracy measures to be 0.25 (approx.) in case of categorical classification too. The loss counts to be 1.6 (approx.) which is not considerable in this piece of research work.

For its better performance in training and validation data, MobileNet has been considered for test data results. It has obtained 92.78% accuracy in predicting test data with constrained parameters. Other parameter results obtained are shown in Table 8.6. The performance-wise result order is mentioned in Table 8.7.

8.5 Conclusion and future work

Cell microscopy has been a challenging task and identification of *Plasmodium* species is more complex due to their close approximation. These challenges have been considerably addressed to classify the species with an accurate output. Multiple models have been carried out while experimenting and four CNN models have been studied extensively based on the literature evidence as well as the experimental output for classifying the variants of *Plasmodium* species by optical analysis of conventional blood slides. For an experimental agility and for identifying the robustness of the algorithm, the input has been synthetically modified for different scales of analysis and bench marking the result for graphical and numerical comparison to identify the range of input to output. For the state-of-the-art comparison, the models have been analyzed on similar input set and the results have been bench marked comparing each other's performance and out of the models most efficient has been found to be MobileNet in understanding slide images. Automated analysis of various medical and healthcare pathological or physiological data can be analyzed in a short time along with human expertise as a decision support system. Such type of medical systems are found out to be significant in referential structure of generating medical diagnosis in a faster manner. State-of-the-art comparison of various model has given significant indication of efficiency as well as accuracy of identifying species in a sample. The experimental outcome can be cited for a significant reference in this field of research. These methodology can be improvised over time by large set of data and increased epoch for training.

References

[1] Gu J, Wang Z, Kuen J, *et al.* Recent advances in convolutional neural networks. *Pattern Recognition.* 2018;77:354–377.

[2] Chakradeo K, Delves M, and Titarenko S. Malaria parasite detection using deep learning methods. *International Journal of Computer and Information Engineering.* 2021;15(2):175–182.

[3] Fuhad K, Tuba JF, Sarker M, *et al.* Deep learning based automatic malaria parasite detection from blood smear and its smartphone based application. *Diagnostics.* 2020;10(5):329.

[4] Lumchanow W and Udomsiri S. Image classification of malaria using hybrid algorithms: convolutional neural network and method to find appropriate K for K-Nearest neighbor. *Indonesian Journal of Electrical Engineering and Computer Science.* 2019;16(1):382–388.

[5] Chaudhuri R and Deb S., "Machine learning approaches for microscopic image analysis and microbial object detection (MOD) as a decision support system," In *2022 First International Conference on Artificial Intelligence Trends and Pattern Recognition (ICAITPR)*, Hyderabad, India, 2022, pp. 1–6, doi: 10.1109/ICAITPR51569.2022.9844212.

[6] Shekar G, Revathy S and Goud EK, Malaria detection using deep learning, *2020 Fourth International Conference on Trends in Electronics and Informatics (ICOEI) (48184)*, Tirunelveli, India, 2020, pp. 746–750, doi: 10.1109/ICOEI48184.2020.9143023.

[7] Guo X, Khalid MA, Domingos I, *et al.* Smartphone-based DNA diagnostics for malaria detection using deep learning for local decision support and blockchain technology for security. *Nature Electronics.* 2021;4(8):615–624.

[8] Krishnadas P, Chadaga K, Sampathila N, Rao S, Swathi KS and Prabhu S. Classification of malaria using object detection models. *Informatics.* 2022; 9(4):76. https://doi.org/https://doi.org/10.3390/informatics9040076. Editor: Jiang Bian

[9] Rahman A, Zunair H, Rahman MS, *et al.* Improving malaria parasite detection from red blood cell using deep convolutional neural networks. arXiv preprint arXiv:190710418. 2019;.

[10] Kunwar S. Malaria detection using image processing and machine learning. arXiv preprint arXiv:180110031. 2018;.

[11] Kapoor R. *Malaria Detection Using Deep Convolution Neural Network.* University of Cincinnati; 2020.

[12] Pattanaik PA, Mittal M , Zubair Khan M, Panda SN. Malaria detection using deep residual networks with mobile microscopy, *Journal of King Saud*

University – Computer and Information Sciences, 2022; 34(5), 1700–1705. doi:10.1016/j.jksuci.2020.07.003.

[13] Vijayalakshmi A. Deep learning approach to detect malaria from microscopic images. *Multimedia Tools and Applications*, 2020;79:15297–15317.

[14] Hu J, Shen L and Sun G. Squeeze-and-excitation networks. In: *2018 IEEE/CVF Conference on Computer Vision and Pattern Recognition*, Salt Lake City, UT, USA, 2018, pp. 7132–7141, doi: 10.1109/CVPR.2018.00745.

[15] Szegedy C, Liu W, Jia Y, *et al.* Going deeper with convolutions. In: *2015 IEEE Conference on Computer Vision and Pattern Recognition (CVPR)*, Boston, MA, USA, 2015 pp. 1–9. doi: 10.1109/CVPR.2015.7298594.

[16] Xie S, Girshick R, Dollár P, Tu Z. and He K. Aggregated residual transformations for deep neural networks. In: *2017 IEEE Conference on Computer Vision and Pattern Recognition (CVPR)*, Honolulu, HI, USA, 2017, pp. 5987–5995, doi: 10.1109/CVPR.2017.634.

[17] Chen B, Ghiasi G, Liu H, *et al.*MnasFPN: learning latency-aware pyramid architecture for object detection on mobile devices. In: *2020 IEEE/CVF Conference on Computer Vision and Pattern Recognition (CVPR)*, Seattle, WA, USA, 2020, pp. 13604–13613. doi: 10.1109/CVPR42600.2020.01362.

[18] Singh KK, Siddhartha M and Singh A. Diagnosis of coronavirus disease (COVID-19) from chest X-ray images using modified XceptionNet. *Romanian Journal of Information Science and Technology*. 2020;23 (657):91–115.

[19] Yadav D, Jalal A, Garlapati D, *et al.* Deep learning-based ResNeXt model in phycological studies for future. *Algal Research*. 2020;50:102018.

[20] Chechliński Ł, Siemiatkowska B and Majewski M. A system for weeds and crops identification—reaching over 10 fps on raspberry pi with the usage of mobilenets, densenet and custom modifications. *Sensors*. 2019;19(17):3787.

[21] Francois C. Xception: deep learning with depthwise separable convolutions, *Proceedings of the IEEE Conference on Computer Vision and Pattern Recognition (CVPR)*, 2017, pp. 1251–1258.

Chapter 9

Medical image classification and retrieval using deep learning

Ajeet Singh[1], Satendra Kumar[1], Amit Kumar[1], Atul Pratap Singh[2], Hariom Sharan[3] and Dijana Capeska Bogatinoska[4]

Medical image classification and retrieval using deep learning is an innovative application of artificial intelligence in the field of medical imaging. This technique involves training deep neural networks to classify and retrieve medical images based on their content [1]. There are many techniques available nowadays which seems more genuine and useful, for example, artificial neural networks in the field of medical imaging. These techniques are simply having more ability to perform the variety of operations like classification data visualization image segmentations based on tumours active or inactive lesions.

In the medical image classification first, we trained our model according to the given information and data. When we trained our data, we were able to classify the new medical images based on this trained model and make predictions. This method is very useful in monitoring and planning the medical treatments and diagnosis of humans.

The prediction and report generation is a difficult task in medical imaging. The creation of a deep learning data model for predictions has many challenges like data availability, data effectiveness, and its integrity. The results generated from the analysis of medical images possess great risk with its implementation to the real world. However, the deep learning techniques improve a lot in the outcome and prediction model in the field of medical imaging to improve patient outcomes.

9.1 Medical images

In order to detect, regulate, control, and treat medical disorders, a variety of technologies are employed in medical imaging [2]. Each technology type offers distinct

[1]Department of Computer Science & Engineering, Moradabad Institute of Technology, India
[2]Department of Computer Science & Engineering, Subharti University, India
[3]Department of Computer Science & Engineering, Rama University, India
[4]Department of Computer Science, UIST Ohrid, Republic of North Macedonia

information regarding the bodily part being investigated or treated, related to potential illness, damage, or the efficacy of medical care.

9.1.1 Ultrasound images

In the field of healthcare imaging known as sonography, which is sometimes known as ultrasonic imaging, sound waves with high frequencies have been utilized to create images of the inside of human organs. Ultrasonic imaging does not expose people to ionizing radiation same as X-ray does.

The body's interior parts moving around and blood flowing via blood arteries can both be seen in real-time ultrasound images which are also captured. An ultrasonic exam involves placing a transducer (probe) on the skin or within a body orifice. A thin gel layer is then added to let the transducer's ultrasound waves travel through the body.

9.1.2 Magnetic resonance imaging

Another deep learning model used in medicine is called magnetic resonance imaging (MRI), which creates images of the inside organs of the body using electromagnetic radiation (electromagnetic fields energy) [3]. Unlike X-ray imaging, MRI does not involve ionizing radioactivity. Protons that exist in water and lipid cells are the main sources of signal in magnetic resonance (MR) pictures.

When doing an MRI test, an electric charge is run via wire coils to create a momentary magnetic field inside the patient's internal organs. A pair of transmitters and receivers uses radio waves to transmit and gather data, which is then utilized to create photographic files of the bodily portion being scanned. Gadolinium-based contrast agents (GBCAs), an intravenous (IV) drug, that are sometimes used during MRI examinations may be administered to alter the clarity of MR picture.

9.1.3 X-ray imaging for pediatric

Pediatric patients, defined as any human having age 21 years or less, may require medical X-ray imaging for the diagnosis and treatment of numerous medical conditions. Patient size is more important to consider than age in pediatric X-ray imaging. The size of patient defines the amount of radiation required to generate a good quality image for medical examination purpose. The use of devices and vulnerability settings intended for adults may lead to excessive radioactive contamination if used on younger individuals since paediatric patients are more sensitive to radiation than seniors.

It is advised that the participant's size and age be taken into account while employing the least amount of radiation necessary for medical X-ray imaging tests such as computed tomography (CT), fluoroscopy, and ordinary X-rays. Affected children are more radiosensitive than adults.

Due to their expected longer lifespans, paediatric patients are more at risk of developing cancer as a consequence of radioactive contamination. Exposure to radiation could be excess if applied to younger children at dose levels meant for grownups.

9.1.4 X-ray imaging for medical

For the detection and cure of medical disorders in newborn and adults, we have various medical visualizing and imaging methods or modalities. Ionizing radiation is used to create body's images in radiography (including mammography), computed tomography (CT), and fluoroscopy [4], a type of radiation with enough energy to potentially harmful DNA. Whereas fluoroscopy shows real-time monitoring of an operation using a standard X-ray image on a monitor or contrast agent passage, radiography saves a single image for subsequent analysis. Trans images of the person's interior organs and connective tissue are produced by combining the different X-ray images that are captured as the sensor traverses it around person's blood. CT calls for a higher radiography; a picture is taken and saved for further analysis. Mammography is a unique kind of imaging method used to capture the inside components of a specific design.

Fluoroscopy: In fluoroscopy, the migration of a bright agent by the body can be observed on screen in real time thanks to the projection of a continuous X-ray image.

CT: While the detector goes around the body of the patient, several X-ray pictures are taken. All individual images are combined by a computer to create cross-sectional views of inside organs and tissues.

9.2 Deep learning

Artificial neural networks are trained to perform tasks like natural language processing (NLP), speech recognition, and picture identification using a process called deep learning.

The development of graphics processing units (GPUs) in computers and the accessibility of enormous datasets for training neural networks have made deep learning possible. Convolutional neural networks (CNNs) that are used to recognize images, recurrent neural networks (RNNs) that are used to analyse natural language, and deep belief networks (DBNs) for unsupervised learning are some examples of popular deep learning methods.

9.2.1 Feed-forward neural networks

One sort of neural network made up of neurons having an input and output layer on a single layer is the feed-forward neural network. Even when applying nonlinear activation functions at the output layer, this network's linear approach restricts its capacity to handle complex data patterns.

Deep neural networks with more than two layers have been created to address this problem. Neural networks enable training on less datasets and can minimize the number of trainable parameters by approximating complex functions with fewer units.

The use a modified perceptron with numerous output units or a perceptron (Figure 9.1) in jobs involving complex data patterns is prohibited because they are considered to be linear models.

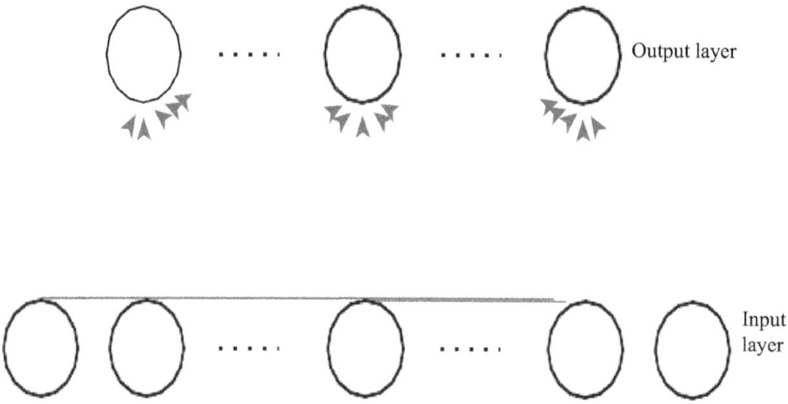

Figure 9.1 Single-layer neural network

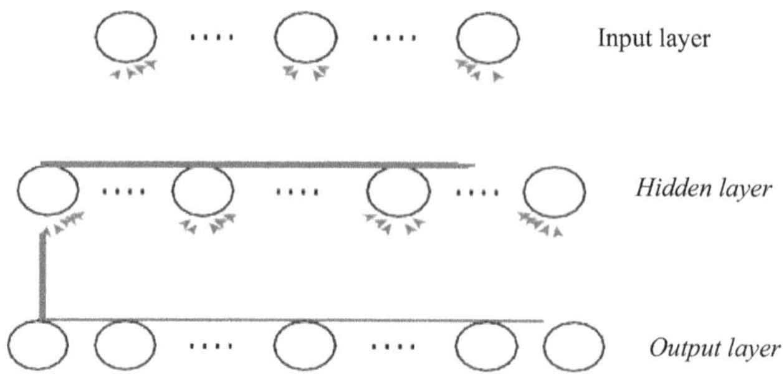

Figure 9.2 Multilayer neural network

9.2.2 Recurrent neural networks

For the connections that are used to create directed cycles, recurrent neural networks' present phase can take inputs from LSTM outputs. The output of the LSTM can remember all the past knowledge that the network is seen so far. Its inner storage allows it to access earlier inputs, which it uses as an input in the current phase. RNNs are frequently used for automated translation, handwriting recognition, time-series analysis, and NLP.

These neural connections with more than two layers have been designed to overcome this issue, with a loose assumption on the activation function. Neural networks enable training on less datasets and can minimize the number of trainable parameters by approximating complex functions with fewer units (Figure 9.2).

9.2.3 Convolutional neural networks

For (clinical) imaging, functional annotation among adjacent pixels or volume elements is essential; however, vectorization invariably removes this type of configural

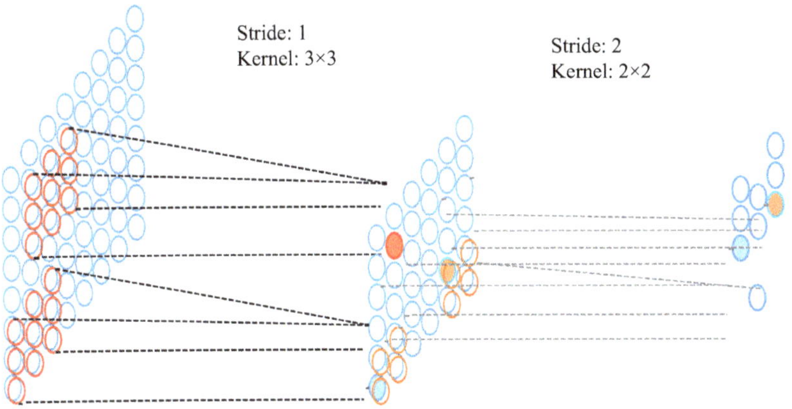

Figure 9.3 Three crucial processes (i.e., local receptive field, weight sharing, and subsampling) in CNNs [7]

and structural information [5]. CNNs are frequently employed for image-based tasks such as medical imaging. These networks use convolutional and pooling layers to best employ topological and configural knowledge, and use 2D or 3D photos as feed. Fully linked layers follow CNNs in a neural network with multiple layers similar in structure. Like a shallow neural network, [6] CNNs uses three techniques that significantly minimize the degrees of freedom in a model. Three factors, namely, subsampling, convolutional, and a convolutional networks area are considered (Figure 9.3).

9.3 Deep learning applications in medical images

Deep learning has many uses in diagnosis, treatment planning, and illness monitoring. Deep learning has developed into a potent tool for analysing and interpreting medical pictures. Examples of deep learning applications in medical imaging are shown below.

Deep learning algorithms are capable of accurately separating medical images into various structures such as organs or tumours, which can help in diagnosis and treatment formulation [8].

Computer-aided diagnosis (CAD): Deep learning models can be trained to find anomalies in medical images, including tumours or lesions, which helps radiologists make diagnosis.

9.3.1 Identification of anatomical structures

Deep learning can be used to identify and separate various retinal structures, including the macula or the optic disc, which can help in the detection and management of ophthalmological conditions. In order to recognize and segment various bone structures, including the spine and hip joint, deep learning algorithms have been applied. These techniques can help in the diagnosis and treatment of orthopaedic problems.

The number of anatomical signatures that may be recovered by computer procedures is primarily what determines if such methods are successful. Recent studies have shown that neural network-based computations have gained popularity for two reasons; in contrast to earlier studies that frequently used specific picture filters to extract anatomical signatures, deep learning technologies are now developed enough to address issues in the real world, and there are an increasing number of medical image datasets that are making it easier to explore large amounts of medical data.

9.3.2 Deep-learning-based organs and cell identification

The detection and segmentation of organs and body parts in medical imaging have showed considerable potential. Here we discuss few instances of deep learning applications for identifying organs and body parts. Deep learning algorithms can precisely identify organs in medical imaging such as liver or kidneys, which can help in the diagnosis and planning of treatments for disorders like cancer or liver disease.

For diagnosis and treatment planning of illnesses like tumour or fractures, deep learning models can be trained to recognize and segment various body parts such as the head and neck or pelvis. Detecting and segmenting blood vessels in medical pictures using deep learning algorithms can help in the detection and treatment of vascular disorders like atherosclerosis or aneurysms.

To identify and categories various types of breast tissue, such as fatty or glandular tissue, deep learning can be utilized. This can help in the early detection and diagnosis of breast cancer. In order to diagnose and treat lung disorders like pneumonia or emphysema, deep learning models can be trained to segment lung tissue in CT scans.

Anatomy-specific classification technique used for classifying medical photos anatomically according to specific organs or bodily parts using deep convolutional networks. They specifically used axial 2D CT scans to train their deep network. This research produced an average value of 0.98 and an anatomy-specific classification error of 5.89%. To overcome this constraint, real-world applications would need a finer level of separation than what is employed for just five body parts (for instance, they might need to differentiate between the aortic arch and cardiac regions) [4].

9.3.3 Deep learning for cell detection

To help in the detection and treatment of cancer, deep learning algorithms can be employed to find cancer cells in tissue samples. Red blood cells and white blood cells, for example, can be detected and classified by deep learning models, which can help in the detection and treatment of blood illnesses.

Cell features including size, shape, and texture can be analysed and quantified with the use of deep learning algorithms, which can precisely segment cells in photos. In order to identify and categorize various pathogens, such as viruses or bacteria, in medical images, deep learning can be utilized. This can help in the diagnosis and treatment of infectious diseases [9,10]. The identification and quantification of stem cells in tissue samples using deep learning algorithms can help in the development of regenerative treatments.

Much of the diversification associated with illness stages occurs at the cell layer. It is necessary to investigate each individual cell for correct diagnosis. Researchers have used deep CNNs to solve this issue; it is necessary to consistently and precisely detect and split lymphocytes from histopathological images that could significantly help tissue analytics for cancer scare [11]. In improvement of medical imaging, we were able to employ the deep CNN to identify mitosis in images of breast cancer histology. Each pixel in the photos was classified by their networks using a patch that was centred on the pixel [12].

9.4 Deep learning for segmentation

In medical imaging, deep learning has developed into a potent segmentation technique. Cancer detection and therapy may benefit from the correct segmentation of tumours in medical pictures using deep learning algorithms [13]. Deep learning models can be trained to separate out specific organs in medical images such as liver or kidneys, which can help in the diagnosis and planning of medical interventions for diseases like cancer or liver disease.

Deep learning can be utilized for vessel segmentation, which can help in the detection and management of vascular disorders like atherosclerosis and aneurysms.

The segmentation of several brain regions, including the corpus callosum and hippocampus, using deep learning algorithms can help in the detection and management of neurological illnesses. In order to segment various cardiac structures, such as the left ventricle or atrium, deep learning can be utilized. This can help in the diagnosis and treatment of cardiovascular problems [14,15].

For patients of all ages to have their brains quantitatively assessed, segmentation of brain pictures is necessary. Removing non-brain areas, such as the skull, is a crucial step in the preprocessing of brain images. Although modern approaches nevertheless struggle when paired with different modes and cells that have experienced pathogenic change, despite performing well on a period of time – computed tomography – to circumvent these limitations. Using a convolutional neural deep learning structure, which was not limited to T1-weighted MR data alone, we were able to restore the skull. In order to maximize system throughput, researchers trained their 3D CNN on minibatches of several crystals that have been greater than just the input data.

By constructing a deep connected system, the deep structure may specifically take an arbitrarily defined 3D patches as source; as an outcome, the outcome could consist of a chunk of projections with insight instead of just a solitary projection, as in a normal CNN. Although their product's sensibility was still around absolute, it produced the highest ordinary precision readings when compared to perfectly serviceable techniques (BET, ROBEX, and 3d Skull Strip) [16,17].

9.5 Conclusion

Deep learning for medical image classification has shown a lot of promise for increasing the precision and effectiveness of disease diagnosis. CNNs, in particular, have shown greater performance in pattern recognition in medical images, enabling more precise diagnosis and treatment.

One of the key benefits of employing deep learning for medical image classification is its capacity to automatically extract the most important elements from the photos. Particularly in large datasets, this has resulted in increased accuracy and decreased variability. Deep learning has shown great potential for improving the accuracy and efficiency of physiological structure detection in medical images, which can aid in the diagnosis and treatment of a wide range of diseases and disorders.

But in order for deep learning models to operate at their best, they need a lot of labelled data, which can be difficult to get in medical contexts due to privacy issues and the requirement for expert annotations. Also, the noise and artefacts in the photos may affect how well deep learning models function. Deep learning-based medical picture categorization has great potential for enhancing illness diagnosis and therapy, particularly in disciplines like radiology and pathology. We may anticipate even more developments in this area as technology advances and larger, more varied datasets become available.

References

[1] Brody H. Medical imaging. *Nature.* 2013;502:S81.

[2] Sharma, A K, Nandal, A, Dhaka, A and Dixit, R. "A survey on machine learning based brain retrieval algorithms in medical image analysis." *Health and Technology*. 2020;10:1359–1373. https://doi.org/10.1007/s12553-020-00471-0.

[3] Sharma, A K., Nandal, A, Dhaka, A, *et al.* "HOG transformation based feature extraction framework in modified Resnet50 model for brain tumor detection." *Biomedical Signal Processing and Control.* 2023;84:104737.

[4] Aljabri, M, AlAmir, M, AlGhamdi, M, *et al.* "Towards a better understanding of annotation tools for medical imaging: a survey." *Multimedia Tools and Applications*. 2022;81:25877–25911. https://doi.org/10.1007/s11042-022-12100-1.

[5] Ebrahimighahnavieh, M A, Luo, S and Chiong, R. (2020). "Deep learning to detect Alzheimer's disease from neuroimaging: a systematic literature review." *Computer Methods and Programs in Biomedicine*. 2020;187:105242.

[6] Sharma, A K., Nandal, A, Dhaka, A and Dixit, R. "Medical image classification techniques and analysis using deep learning networks: a review." In: Patgiri, R, Biswas, A and Roy, P. (eds.). *Health Informatics: A Computational Perspective in Healthcare. Studies in Computational Intelligence*, vol. 932; 2021. Singapore: Springer. https://doi.org/10.1007/978-981-15-9735-0_13.

[7] www.ncbi.nlm.nih.gov

[8] Sharma, A K, Nandal, A, Dhaka, A, Koundal, D, Bogatinoska, D C, and Alyami, H. "Enhanced watershed segmentation algorithm based modified ResNet50 model for brain tumor detection". *BioMed Research International, Hindawi*, 2022.

[9] Sharma, A K, Nandal, A and Dhaka, A. "Brain tumor detection and classification using machine learning techniques." *RAMSACT 2022*, Springer, Manipal University Jaipur, India, May 2022.

[10] Sharma, A K, Nandal, A and Dhaka, A. "Comparison between predictive model and transfer learning-based model for MRI classification." *RAMSACT 2022*, Springer, Manipal University Jaipur, India, May 2022.

[11] Sharma, A K, Nandal, A, Ganchev, T, *et al.* "Brain tumor classification using modified VGG modelbased transfer learning approach." In *20th International Conference on Intelligent Software Methodologies, Tools, and Techniques (SOMET-2021)*, Accepted in June 2021, Conference date 21–23 Sept. 2021, Cancun, Mexico.

[12] Dhaka, A, Nandal, A, Rosales, H G, *et al.* "Likelihood estimation and wavelet transformation based optimization for minimization of noisy pixels." *IEEE Access*. 2021;9:132168–132190.

[13] Wang L, Chen K C, Gao Y, *et al.* Automated bone segmentation from dental CBCT images using patch-based sparse representation and convex optimization. *Medical Physics*. 2014;41:043503.

[14] Suk H I, Lee S W and Shen D. Deep sparse multi-task learning for feature selection in Alzheimer's disease diagnosis. *Brain Structure and Function*. 2016;221:2569.

[15] Chen Y, Juttukonda M, Su Y, *et al.* Probabilistic air segmentation and sparse regression estimated pseudo CT for PET/MR attenuation correction. *Radiology*. 2015;275:562–569.

[16] Karpathy A and Li F. Deep visual–semantic alignments for generating image descriptions. In *Proceedings of the 2015 IEEE Conference on Computer Vision and Pattern Recognition*; Washington, DC: IEEE; 2015. pp. 3128–3137.

[17] Silver D, Huang A, Maddison CJ, *et al.* Mastering the game of Go with deep neural networks and tree search. *Nature*. 2016;529:484–489.

Chapter 10

Game theory, optimization algorithms and regularization techniques using deep learning in medical imaging

Ali Hamidoğlu[1]

Medical imaging (MI) is perhaps the most appropriate and promising domain in the field of biomedicine and healthcare systems, which is one of the most significant areas for artificial intelligence (AI) applications. In other words, MI is crucial for giving directions on disease detection and therapy as well as using and designing various AI perspectives. One of the main fundamental and crucial members of MI is medical image reconstruction, the main aim of which is to obtain excellent medical pictures for medical purpose at small expense and patient risk. Moreover, it becomes a highly difficult topic to construct picture correspondence through reliable image registration, which makes it essential to numerous therapeutic tasks like image fusion, creating organ atlases, and tracking tumor progression. The MI research group has created deep learning-based methods and reached the pioneering position in several implementations, consisting of image registration as well as image reconstruction recently. In this regard, mathematical models have proven crucial in the field of medical picture reconstruction and, more broadly, image restoration in computer vision. Early mathematical models were mostly created by using human knowledge or theories about the image that would be rebuilt. These hand-made models then gave way to data-driven modeling, which still largely relies on human design but now incorporates some model learning from observed data. Deep learning-based models have lately driven data-driven algorithms which are primarily depend on learning with limited human design. The main reason behind this achievement is that researchers and clinicians could be able to reach more data and computational resources which have nowadays become available. Both manually created models and data-driven models have benefits and drawbacks.

In this chapter, we investigate various mathematical algorithms and models which have used recently through deep learning perspectives in MI. In this regard, we mention about game theoretical platforms, several optimization as well as regularization techniques applied on MI. Many challenges of this field can be

[1]Department of Mathematics, Bahçeşehir University, Turkey

architected or overcome by means of game theory. Game theory is a powerful mathematical tool to support deep learning architectures. The majority of classification issues that frequently use a deep learning technique may be seen as competitive models such as Stackelberg games as well as cooperative models such as Nash games in MI. For having the best decision-making strategy in these games, many optimization algorithms have been developed and used to derive the best possible optimal solution. Moreover, we consider several meta-heuristic optimization algorithms that have been recently applied in the field of MI which become beneficial because of their adaptability, ease of use, and independence from particular problems. Then, we investigate several regularization techniques by means of deep learning used in MI. The chapter concludes with several remarks and future research directions in this field.

10.1 Introduction

Clinical data, picture analysis, and artificial intelligence (AI) models may all have the greatest potential to make a good, long-lasting impact on people's lives in a rather short amount of time. Computer vision and machine learning techniques have been more and more frequent in medical image procedure. This made medical imaging (MI) analysis possible to apply in different fields of medicine [1]. The term "machine learning" (ML) refers to a collection of methods for immediately detecting samples in data and using these samples to predict future information or support decision-making in the face of uncertainty [2]. Hence, it can be seen as a subset of AI. A special category of artificial neural network (ANN) called deep learning [3] is a component of ML which is inspired by the multilayered structure of human cognition. Due of its application to large-scale healthcare data, deep learning is currently receiving a lot of interest. Even though, ANNs were first developed in 1950, their ability to address real-world problems was severely constrained by vanishing gradient and overfitting issues, which made it difficult to train deep architectures and, more importantly, lack of training data makes it so impractical to implement those structures on computers. Luckily, because of the current possibility of massive resources, expended evaluation power from graphics processing units (GPUs), and advanced optimization algorithms and involvement of game theory in AI models, many constraints have, nevertheless, been overcome regarding to deep neural network (DNN). In numerous disciplines, including medical imaging, deep learning techniques have demonstrated amazing achievements in emulating humans. For example, the detection of structural abnormalities and the classification of such abnormalities into illness categories are frequent tasks in radiology practice [4]. Many ML algorithms with various mathematical foundations, and logical frameworks have been used to carry out such classification tasks since the 1980s. Hence, in the early 2000s, a number of computer-aided detection (CAD) processes were created as well as implemented in the clinical workflow. Yet, these systems have been linked to negative effects according to clinical investigations [5,6]. Recent deep learning technology has been used to

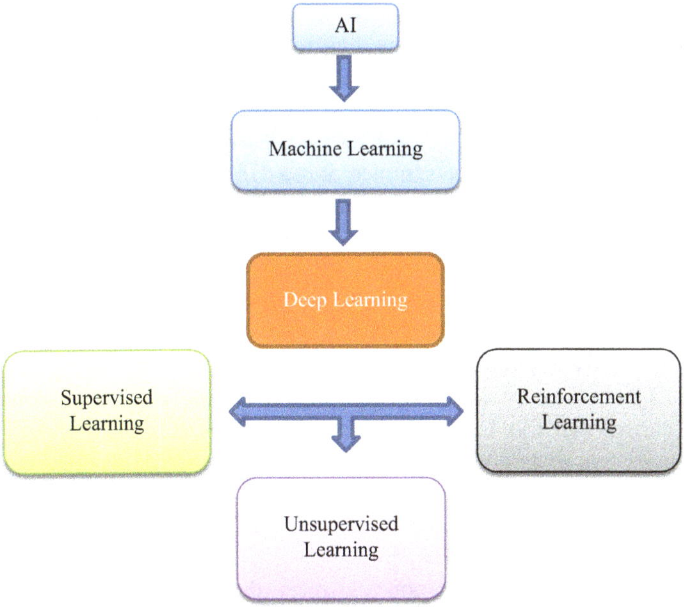

Figure 10.1 Simple diagram of deep learning in AI

overcome shortcomings of earlier CAD systems, which provides higher detection accuracy, and increase the productivity of human readers by allowing them to work on AI algorithms more efficiently instead of performing repeated tasks and classification processes. More precisely, deep learning may be employed to harvest medical big data for useful knowledge. Automatic lesion recognition, differential diagnosis recommendations, and the preparation of draft radiology reports are among practical benefits for this novel AI technology.

Deep learning applications in healthcare address a wide scope of concerns, infection monitoring from to cancer detection to individualized treatment options. Physicians now have access to a vast amount of data from many data sources, including radiological imaging, genetic patterns, and pathological imaging by means of X-ray, diffusion tensor imaging, computed tomography (CT), magnetic resonance imaging (MRI) and positron emission tomography [7,8]. Roughly speaking, deep learning uses neural networks made up of several convolutional nodes of artificial neurons to learn patterns in data structures. Here, a sort of cell known as an artificial neuron that functions similarly to a biological neuron in which it accepts several inputs, essentially does calculations, and in the end, outputs the result [9]. This straightforward calculation uses a nonlinear activation function before a linear source regular expression form [10]. There are several nonlinear activation functions of a network such as rectified linear unit [11], hyperbolic tangent transformation [12,13], and the sigmoid conversion [14].

As shown in Figure 10.1, ML may be classified as three kind of classes: unsupervised learning, supervised learning, and reinforcement learning (RL). Here, we shortly discus about these techniques and their interaction with deep learning in MI. Unsupervised and supervised are two major categories that can be used to group conventional ML approaches. The supervised learning paradigm's approaches use a set of established annotations, characteristics, and features to categorize objects in a pool. More precisely, supervised learning enables a model to learn a representation between input samples and a target variable. It demands for a significant quantity of well-labeled data. Such detailed data are difficult, labor-intensive, and time-consuming to obtain. However, unsupervised learning algorithms organize or cluster items in a pool based on how similar they are, and then use those clusters to categorize the unknowns. It uses unlabeled data as its foundation and learns from the statistics of the real data features that are essential and subtle. Additionally, the third type of learning, known as RL, enables a system to draw lessons from the experiences it has when interacting with its environment. For example, classifying and learning from inputs of both cancer and non-cancer images, feature learning could be implemented by supervised deep learning as well as by unsupervised learning of a single-layer network [15]. In addition, deep RL may be used to simulate the data augmentation task as a trial-and-error process for kidney tumor segmentation [10,16].

Learning from real data or non-real ones, deep learning is the research of different learning algorithms that employ several nonlinear processing layers. Each succeeding layer uses the output of the preceding layer as its input. In this process, higher-level features are put into practice by deep learning algorithms after being retrieved from lower-level features. The effectiveness of the model depends critically on the neural networks. Since the neural network framework can represent dynamic environments, those environments can be further represented as game platforms where these features can be regarded as strategies of players or simply players practicing decision making processes during algorithms. Recently, AI has modified various platforms appeared in MI by means of game theory to solve or mimic a variety of real-world issues, and it has been seen that performance improves when game theory is used [17–19].

Game theory is the branch of mathematics that studies how to deal with competitive or cooperative situations in which a participant's actions are likely to have a significant impact on how other individuals behaved. Game theory has been used in many fields of science and technology, including economics [20], social planning [21], medicine [22], management [23], military [24], and many other fields of life and natural science. Game models can be roughly characterized as competitive and cooperative games performed among players. To provide some game types in these categories, respectively, Stackelberg games [25] and Nash games [26] are seen as competitive and cooperative game mostly used in literature.

In Stackelberg games, there are a group of players competing each other under the rule of one leader and several followers. Here, the player with leader status performs the first move or chooses the first decision and then all decisions are followed by the rest of the team, namely followers. Becoming a leader in the game, most of the time, provide the best strategy for the leader, but this decision may not be the best for the rest of the team. Nash games, on the other hand, consider a common platform which is the best for all as a group of all players. In this regard, players work cooperatively to achieve the best strategy that no one wants to leave his/her own decision.

In this chapter, we investigate several cooperative and competitive games used as a mathematical tool in deep learning of MI. Moreover, several optimization techniques including meta-heuristics would be discussed as numerical techniques to minimize (or sometimes maximize) costs, pay-offs or errors caused by deep learning process of MI. In addition, regularizing those errors in order to get accurate MI, we discuss several regularizations perspectives as well. Finally, we conclude the chapter with several remarks and insights regarding to deep learning of MI.

10.2 Game theoretical aspects in MI

In this section, we investigate several game models applied in deep learning of MI. Before start, we provide a brief introduction in the theory of games. A game can be seen as any interaction among the players or the objects, in a cooperative or competitive playing ground. In general, games are characterized as cooperative, competitive, zero-sum, non-zero sum, perfect, imperfect, static and dynamic games and so on. In the following sections, we consider several of them and provide some brief definitions as well as their relations.

10.2.1 Cooperative games

Players act together cooperatively with a common goal either maximize or minimize the total profit. Nash game is an example for such games. As an application in MI, Nash game may be used for the evaluation of segmentation of heart ventricles from cardiac magnetic resonance cross-sections and lung fields from chest radiographs [27]. Moreover, Nash game was used in [28] for treating rheumatoid arthritis disorder disease (see e.g. Figure 10.2).

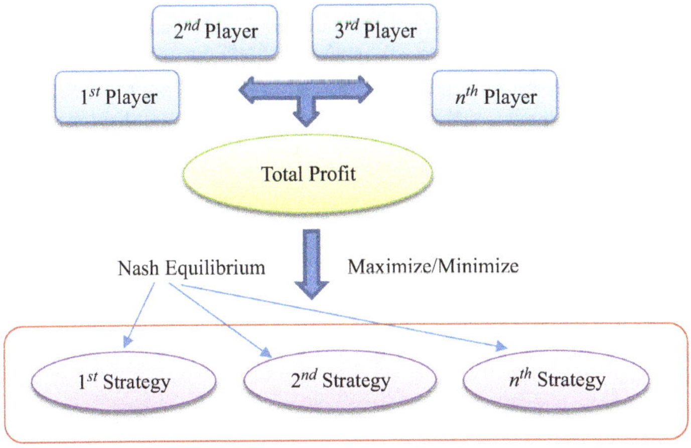

Figure 10.2 Nash equilibrium for n *number of players in a cooperative game*

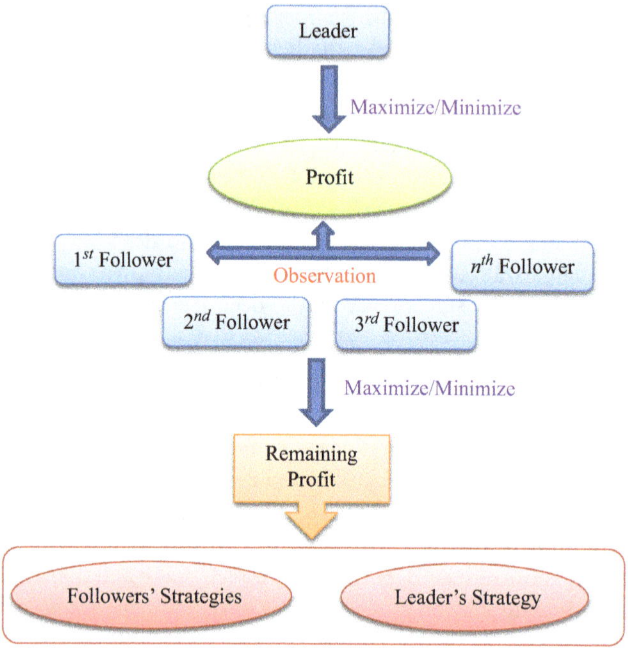

Figure 10.3 Stackelberg competition performed by n *number of followers against a leader in a competitive game*

10.2.2 Competitive games

Players are competing among each other to get the highest (or the lowest) revenue from the given profit. Stackelberg game is an example for such game models. In that game, a group of players competes against one another while being governed by one leader and a number of followers. Here, the person with leader status initiates the first action or makes the first decision, and the remainder of the team, or followers, subsequently follow all decisions. It was investigated in [25] that cancer treatments may be viewed as a Stackelberg game played between the doctor's therapy and the cancer cells' resistance tactics. The oncologist is the "leader" in this game, and the cancer cells are the "followers," responding to the treatment and changing as necessary (see e.g. Figure 10.3).

10.2.3 Zero-sum and non-zero-sum games

The value of a game becomes zero in zero-sum games. On the other side, zero-sum games can be characterized as the following manner: the sum of rewards for one team of players would be opposite to the other team of players and all players' total revenue equals to zero. However, if the value of a game is not equal to zero, then it is called as a non-zero sum game. Regarding to MI, It is crucial to understand that in a zero-sum game, radiologists and AI do not compete against one another. A bigger gain in the overall economic surplus that can be transferred and shared with

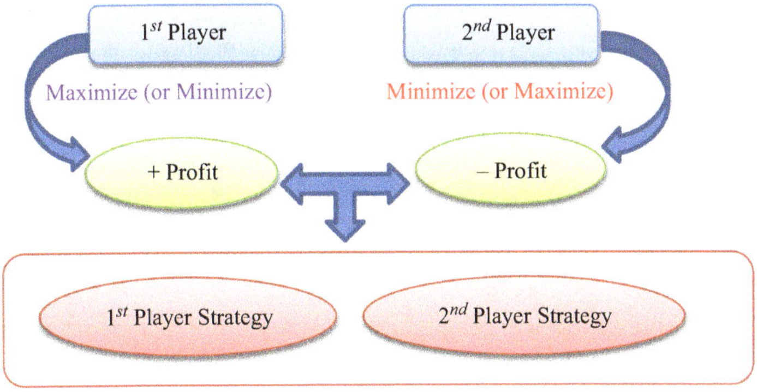

Figure 10.4 Simple diagram of two-player zero-sum game in a competitive environment

patients, providers, and payers to improve the overall value in healthcare will be realized as radiologists integrate AI technology to build a partnership [29]. Still, there have been innovative approaches used in zero-sum games to develop learning algorithms [30] (see e.g. Figure 10.4).

10.2.4 Deep learning in game theory

As an advance component of ML algorithms, deep learning is based on ANNs and other learning techniques, such as reinforcement, unsupervised, and supervised learning. For deep learning models, key properties are extracted from raw data using numerous layers of an ANN. In this section, we provide several works related to game theoretical applications in deep learning. For extended literature review, we provide the survey paper [31]. Moreover, Chivukula and Liu [32] develop an adversarial learning technique for convolutional neural networks (CNN) by creating a multiplayer stochastic game that targets the CNN's classification performance. In this regard, several two-player games are used to represent the multiplayer stochastic game where one player acts an intelligent adversary and the other player as the learner CNN interacting in each game in such a way that the payoff function of each player grows over time. Each two-player game is solved for the Nash equilibrium after a sequential non-cooperative Stackelberg game converges. In this model, the Nash equilibrium identifies a set of tactics (learner weights and evolutionary operations) from which neither the learner nor the adversary have any motivation to leave their decisions. In convex single layer problems, a simple two-player game is proposed in [33] as an equivalence between global minimizers of the training issue and Nash equilibrium.

In a simple supervised deep learning problem, one is given as a set of paired data $\{(x_k, y_k)\}_{k=1}^{N}$ such that $(x_k, y_k) \in \mathbb{R}^m \times \mathbb{R}^n$ and tries to learn from a predictor $f : \mathbb{R}^m \to \mathbb{R}^n$. In this regard, the following single layer learning problem is provided:

Problem 1 (Single-layer learning problem [33,34]): Let a loss function be given as $\ell : \mathbb{R}^n \times \mathbb{R}^n \to \mathbb{R}$ which is convex in the first component, let $\theta \in \mathbb{R}^{n\times m}$ be a transformation matrix. The training problem is to minimize $L(\theta) = \frac{1}{N}\sum_{k=1}^{N} \ell(\theta x_k, y_k)$ with respect to the parameters θ.

Now, we state its corresponding game theoretical problem version in the following way:

Problem 2 (Single layer learning game [33]): Consider a game environment between two players. The first player chooses a parameter matrix $\theta \in \mathbb{R}^{n\times m}$ whereas the second player chooses a set of N vectors $\{(a_k, b_k)\}_{k=1}^{N}$ with $a_k \in \mathbb{R}^n$ and $b_k \in \mathbb{R}$ such that $a_k^T\theta x_k + b_k \leq \ell(\theta x_k, y_k)$ for all $x_k \in \mathbb{R}^m$. Hence, one strategy of both player could be seen as a couple $(\theta, \{(a_k, b_k)\}_{k=1}^{N})$ and utility or pay-off function for the first player is given by

$$U_{first} = \frac{1}{N}\sum_{k=1}^{N} \left(a_k^T\theta x_k + b_k\right)$$

and pay-off function for the second player is given as $U_{second} = -\ U_{first}$. Here, the first player tries to maximize its utility function with its actions whereas the second player aims to minimize its pay-off function with its continuous actions. This makes the game to become a two-person zero-sum game with continuous actions.

To connect these two problems, we have the following theorem which states that there is always a Nash equilibrium in this game model and it is the global minimizer of Problem 1.

Theorem 1 (Existence of Nash equilibrium [33]): *The global minimizer for Problem 1 establishes a Nash equilibrium for Problem 2. More precisely, if θ^* is the global minimizer for Problem 1, then there exists a second player' strategy N; $\{(a_k, b_k)\}_{k=1}^{N}$ such that the couple (θ^*, $\{(a_k, b_k)\}_{k=1}^{N}$ gives the Nash equilibrium for Problem 2. On the other hand, if (θ^*, $\{(a_k, b_k)\}_{k=1}^{N}$ is the Nash equilibrium for the Problem 2, then θ^* becomes the global minimizer for Problem 1.*

In general, one can consider more layers in learning algorithms instead of one to build a deep learning perspective in game theory. In this case, a feedforward neural network is defined by a directed acyclic graph with vertices and edges [33]. Here, the matrix $\theta \in \mathbb{R}^{n\times m}$ would be more complex and players' strategies would be different, even the game mode may be changed according to this network architecture. Hence, characterization of deep learning problems as games enables one to connect methods for finding equilibrium to potential training algorithms (see e.g. Figure 10.5).

Game theory terminologies in deep learning are shown in Table 10.1 [35]:

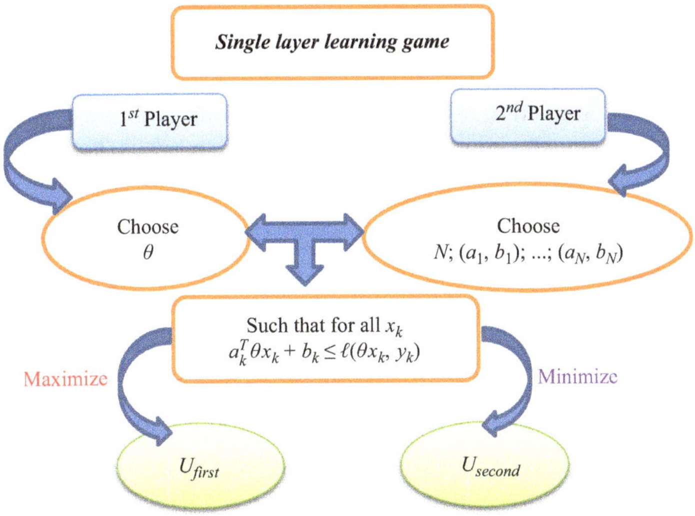

Figure 10.5 Single-layer learning game as a zero-sum two-player game

Table 10.1 Deep learning versus game theory [35]

Game theory	Deep learning
Player	Neuron unit
Action	Weight
Objective function	Objective function
Sub-goals/tasks	Features
StrategiesDesign of gameBody text1	OutputArchitecture design

10.3 Optimization techniques in MI

In this section, we investigate several optimization and meta-heuristics techniques applied in MI and the use in deep learning algorithms. Here, we start with commonly used optimization techniques used in MI.

10.3.1 Linear programming

Linear programming is an optimization technique that makes use of linear functions to determine the best possible output or outcome from a given set of parameters or list of requirements. The optimal method for distributing finite resources, including revenue, labor, people, equipment materials, time, space, and a range of other parameters, is typically determined by computer modeling or simulation. To reach the best outcome, either total profit is maximized or costs is minimized by means of this algorithm depending on the given problem. Sometimes, the discrete multi-labeling problem of medical picture registration with individual labels assigned to grid nodes could be used to formulate the problem. Markov random fields (MRFs) are a popular model for illustrating such issues [36]. MRFs were very popular in the field of computer vision in the late 1980s and early 1990s. Their biggest limitation at the time, nevertheless, was the lack of effective optimization methods for recovering their lowest potential. To meet this need, both deterministic and non-deterministic algorithms have been taken into consideration. Fast-PD [37], a recently proposed method, is used to accomplish this. This is an optimization method that effectively generates nearly optimal responses for a sizable class of NP-hard MRFs using concepts from linear programming duality concept. This technique offers a number of major improvements when used for the image registration task.

The multi-label task of dense image registration requires regularization parameters that are generally nonlinear functions like second- or first-order derivatives, and elastic equations. Additionally, it should be anticipated that the location of the control point within the image plane will affect the degree of resolution in the quantized search space. In other words, compared to smooth areas, the matching process will be more exact in regions with rich picture content such as edges and roughness. Finally, the technique ought to be computationally effective considering the high degree of freedom. The use of methods such as alpha-expansion is limited due to the requirements on the pairwise potentials. In this regard, Glocker *et al.* [38] considers effective linear programming by means of primal dual fundamentals to restore the cost function's lowest potential.

10.3.2 Nonlinear programming

The process of minimizing or maximizing a nonlinear objective function that is constrained by bound constraints, linear constraints, or nonlinear constraints, which can be inequalities or equalities, is the process that is referred to as nonlinear programming (NP), which is another commonly used optimization technique. In [39], NP approaches are proposed as a means of solving the non-rigid image registration issue in an effective manner while simultaneously retaining solution feasibility within the limits of rigidity. Non-rigid image registration is an essential technique utilized in the field of medical image analysis. By using a traditional non-rigid registration method, the entire image is distorted in a way that is not rigid. Although some parts of the image, including soft tissues, are able to bend in a way that is not rigid, some therapeutic procedures require a procedure called registration, which keeps specific parts of the image as rigid as possible (for example, bones). The deformation process causes objective functions and constraints to become nonlinear; as a result, we are motivated to use NP-optimization approaches in these kinds of situations.

In addition, parallel transmission is a powerful innovation that can reduce the RF field inhomogeneity that is inherent to ultra-high-field magnetic resonance imaging. This inhomogeneity cannot be eliminated entirely. In the beginning stages of development of this technology, the pulse construction was conceptualized as a least squares problem, and only the most fundamental power regularizations were implemented. This was done with the intention of controlling the specific absorption rate (SAR) and, as a result, improving the quality of medical care. Because this method was inefficient for many applications that were sensitive to the magnitude of the spin excitation rather than its phase, the magnitude least squares (MLS) issue was invented in 2007. This problem was developed in 2007 because the previous method was poor. The so-called variable exchange approach has been used to solve virtually all of the MLS problem, despite the fact that alternative, more powerful numerical optimization methods are readily available. This is because the problem was almost totally tackled by the pulse designer. In relation to this, the paper [40] makes use of a variety of NP approaches, each of which takes into consideration the stiff SAR and hardware restrictions in its solution.

In the past few years, technology has become widely available that can reconstruct PET images using a Maximum Likelihood estimation to improve the quality of the reconstructed images in MI. This has led to an increase in the number of medical institutions that are able to perform MI. Because of this, the degree of detail that is now visible in these photographs has been able to improve significantly. These methods, in general, call for the iterative minimization of a convex cost function, and they are accompanied by a whole host of issues that are connected to the computational complexity of their implementations. In this regard, the authors in [41] propose a fresh method for the reconstruction of pictures of the Row-Action type. This approach makes use of the duality that is present in NP in order to achieve rapid convergence while keeping the computational cost per iteration to a minimum. This innovative framework is investigated via the prism of simulated trials and the results of clinical PET data applications in [41].

10.3.3 Dynamical programming

Dynamic programming is an optimization technique that divides the whole problem into sub-problems, saves the results, and then optimizes the sub-problems to find the overall solution, which usually involves determining the maximum and minimum range of the algorithmic query. The maximum or minimum results obtained for sub-problems are gathered together which provides the maximum or minimum result for the whole problem. For dynamic time warping of voice signals, dynamic programming methods are extensively employed in speech processing. Here, this technique is used for contrast correction in image processing. Hence, the object picture becomes contrast-corrected by dynamic programming using one image as a reference and the grayscale value histograms of the photos which guarantees a nonlinear histogram warping and, as a result, an ideal nonlinear mapping between the images' grayscale values [42]. Assignments may be unclear if no path limitations are employed, meaning that more than one grayscale value of the reference picture may be assigned to the grayscale value of the object image. Hence, the most likely grayscale value is chosen in this situation through histogram-based analysis [42].

The authors in [43] propose a 2-D strain imaging technique in medical imaging which uses dynamic programming to minimize a cost function which consists of echo amplitude similarity and displacement continuity. Freehand palpation elastography produces strong and efficient strain images with up to 10% compression, proving that the method is more resistant to signal decorrelation than conventional correlation approaches. Tissue deformations are smooth, so incorporating smoothness into the cost function reduces decorrelation noise.

One of the most difficult things for the human eye to perform is to break up a picture into its component parts and then watch those parts change over time. The authors in [44] provide a method for discovering contour areas in complicated forms that is based on a user-selected initial contour and is relatively near the intended conclusion. This method was developed in order to locate contour regions. According to [45], doing further study on the data supplied by the user's chosen places and utilizing an optimal approach to identify outlines have a favorable effect on the picture segmentation process. This method makes use of dynamic programming, and it may be utilized in the creation of a broad variety of different forms. This is not an iterative process; rather, it is precise. In addition, the authors in [46] propose a novel method using dynamical programming for the purpose of segmenting objects in medical pictures that have closed curve bounds. A three-dimensional object in volumetric data can be segmented slice by slice if only one point inside the three-dimensional object of interest is specified as the pole for the polar transformation. This enables the segmentation of objects with very weak boundaries and makes it possible to segment three-dimensional objects in volumetric data.

In many biomedical imaging applications, it is important to be able to automatically find the edges of bone, organs, blood vessels, and cells. When making interactive apps, the goal is to either make it easier for experts to use their judgment or add user input to a processing pipeline that lets pictures be analyzed automatically. Most of the time, images used in biomedicine have a lot of noise and blurry edges. Consequently, trustworthy algorithms for contour and line detection are necessary. Dynamic programming is a well-known approach that satisfies these requirements in a variety of ways. In this regard, Ungru and Jiang [47] present a brief description of techniques and programs that make use of dynamic programming to address difficulties in the tough field of biomedical imaging. These strategies and programs aim to improve diagnostic accuracy and improve patient care.

10.3.4 Particle swarm optimization

In 1995, Kennedy and Eberhart [48] were the ones who initially presented the concept of particle swarm optimization (PSO). The idea behind this strategy stems from the observation that individuals in a flock of birds traveling together would learn from the experiences of all of the other members. To put it another way, while one bird may be flying around aimlessly looking for food, the other birds in the flock may be telling each other what they've seen so that the group as a whole can hunt better. When simulating a flock of birds, it is possible to make the assumption that each bird offers a way to find the optimal solution in a high-dimensional problem space, and that the flock's best solution is the best solution in the space.

This can be done by assuming that each bird provides a way to find the ideal solution in the space. Nonetheless, we can never verify whether or not the true global optimal solution can be discovered, and in most cases, it can't be. Because of this, the solution in question is referred to as a heuristic solution. Despite this, the PSO solution is typically quite close to the overall ideal.

Biomedical image registration, also known as the geometric alignment of 2D and/or 3D picture data, is utilized for computer-guided treatments, functional studies, diagnostics, and biomedical research. When registering images based on intensity values, it is usually necessary to optimize a similarity metric that is used to compare the images. Most of the time, the functions of these metrics in terms of the parameters of the transformation are not smooth and regular. Because of this, local approaches to optimization often fail, and it's common to need global solutions. PSO is a revolutionary evolutionary approach. In this context, [49] suggests that it is adjusted for single-slice 3D-to-3D biomedical image registration. In addition, a novel hybrid particle swarm strategy that incorporates initial user guidance is presented in [49]. Using three volumes of data from a variety of modalities, an investigation of multimodal registrations was carried out, utilizing initial orientations that were detached from reality. This was the case despite the fact that the hybrid technique also produced more registrations. These findings highlight the necessity for hybrid techniques for difficult registration issues and indicate the importance of particle swarm approaches in picture registration. Hybrid approaches are required for complex registration challenges.

Intensity-based registration algorithms have become increasingly popular as a solution to the fundamental problem of multimodal image co-registration in medical imaging. This is due to the fact that these algorithms make it possible to combine a number of images into a single representation, which in turn makes complementary information readily accessible and usable. These methods typically call for optimizing a similarity metric derived from the input photos in MI. Particularly when dealing with non-smooth fitness functions, local optimization approaches frequently produce low or bad convergence results to local optimum. This drawback motivates to use swarm intelligence approaches, like the PSO in optimizing fitness functions, which are incredibly powerful and providing efficient global optimization methods. Rundo *et al.* [50] focus on PSO for biological image

When it comes to the analysis of medical pictures, the term "segmentation" refers to the process of isolating the region of interest. Ahilan *et al.* [51] suggest using multilevel thresholding based on an optimization technique, which is more effective than traditional thresholding for picture analysis, in order to extract the region of interest from DICOM pictures and compress them for use in telemedicine applications. This can be done by reducing the file size of the images. PSO and various additional techniques targeted toward PSO optimization are utilized in order to arrive at an estimation of the threshold value in [51]. The classification and blending of prediction-based lossless compression techniques deliver effective results, particularly when contrasted with the JPEG loss and JPEG lossless methods of compression. The algorithms are tested at a variety of threshold levels; if the PSNR value comes out higher than expected, it indicates that the proposed segmentation approach is useful for the transfer of data in telemedicine. In addition, the segmentation of medical pictures provides clinicians with essential support in making decisions. In [52], the PSO approach is applied, and in order to minimize an energy function, hidden Markov random fields (HMRF) are utilized. The findings of the study indicate that the HMRF-PSO method is more effective in MI than the K-means and threshold-based approaches.

10.3.5 Simulated annealing algorithm

The simulated annealing (SA) procedure is quickly becoming one of the most widely used heuristic strategies for the resolution of optimization problems. Kirkpatrick and colleagues came up with SA [53] after being inspired by the annealing process that is used in metalworking. This theory considers the method of progressively cooling metals after they have been subjected to high heat, and it also specifies the appropriate molecular arrangements for metal particles that result in a decrease in the mass potential energy. In order to simulate the annealing process that metals go through, the SA algorithm will frequently proceed in an iterative manner in line with a temperature parameter that may be adjusted [53]. The SA method performs an iterative comparison of the results of the objective functions running with the current and adjacent points in the domain. If the result from the adjacent point provides a solution that is superior to the one that is being used, the SA method will use that result as the basis for the basis solution for the subsequent iteration. If this is not the case, the algorithm will finish the operation without attempting to find better results in a larger area. As a consequence of this, the algorithm may become stalled at either a local minimum or maximum, which results in the lack of a global maximum or minimum for the objective function.

In order to follow the progression of a disease or analyze medical data, registration techniques are employed to identify anatomic features from two or more images (CT, MRI, PET, etc.) collected at various times. The maximizing of a cost function that represents how comparable these photos are serves as the foundation for the registration procedure. The authors in [54] consider genetic algorithms (GA) and SA, to solve these types of optimization problems. A preliminary investigation of the use of SA for the reconstruction in microwave tomography is presented in [55]. Using simulated noise-free data, a simplified model of a human arm is rebuilt using three different techniques: SA, quenching, and the Newton-Kantorovich approach. These findings show that SA is still capable of accurately solving problems even when the two deterministic techniques are unsuccessful. This shows that heuristic approaches sometimes become powerful tools for obtaining global maximum or minimum of optimization problems derived from MI.

Watermarking is one method that may be used to protect patient information in medical pictures, such as those produced by magnetic resonance and ultrasound imaging. Frequency-domain analysis has been used in a lot of other ways that have been described in the research literature. Some examples of these approaches are digital wavelet transforms, cosine transforms, radon transforms, Fourier transforms, and so on. Most digital wavelet transform-based watermarking algorithms produce very good results even after being attacked many times. The method that performs digital wavelet transformations makes use of the scaling factor during the embedding and extraction processes. A user-selected number between 0 and 1 that falls outside of the effective range of a scaling factor is known as an ineffective scaling factor. The brute-force approach is one that may be utilized to solve an issue, despite the fact that it is a slow way. This method works by going over each and every situation that might potentially occur. By using SA as the basis for scaling factor calculation in frequency domain watermarking, the resulting watermarked pictures are more robust, resistant, and transparent. Elbasi [56] suggests a SA heuristic search algorithm as a means of determining the suitable scaling factor for an improved method of watermarking medical photographs that is also transparent, has a high data capacity, and is resistant to tampering.

10.3.6 Genetic algorithm

Natural selection and genetics are the theoretical foundations upon which a search-based optimization tool known as a GA is constructed [57]. When the GA method is used to find the best answer to a problem, the chromosomes are put to work to do a probabilistic search of the space of possible answers. In the context of these modeling applications, the number that is assigned to each chromosome is referred to as a gene, and each gene is assigned a decimal number that ranges from 0 to 9. This means that a wider range of possible values can be stored in fewer chromosomes, which is especially helpful for making models and designs work better. In order to establish the parameters that correspond to each set of chromosomes, a population of chromosomes is generated at random. After then, the framework is updated with these altered numerical values. The simulation of each parameter set in the population takes place, and the results are analyzed with the aid of a performance measure that is founded on the optimization of cost functions. When all of the cost values have been determined, they are put into ascending order using the chromosomes that are most appropriate for them. As was the case before, the best cost values may be inferred by beginning with their lowest values [58].

MI modalities provide a variety of medical images to aid physicians in the treatment of patients and the formulation of diagnoses. The intensity of each visual modality differs from one another. In order to produce fused medical images that can offer additional focus and image diagnostics motivated by the information for the medical examination, a number of researchers in the field of medical imaging have attempted to combine data that is both unnecessary and related that is taken from various types of medical images. Arif and Wang [59] propose a revolutionary strategy and method of fusion for multimodal medical images that is based on GA. This technique reduces uncertainties and scatter in the input image and may further enhance image fusion characteristics.

The authors in [60] describe a GA-based network evolution technique that uses a greedy exploration strategy based on experience and transfer learning to find the best genes and improve network architectures. This approach is based on the GA. Its evolutionary algorithm process is malleable, which enables it to seek out potentially fruitful neural networks by making use of the most cutting-edge modules currently available (e.g., residual blocks). The findings of the studies on computed tomography perfusion image denoising demonstrate that the GA is capable of choosing the most effective genes for the construction of high-performance networks.

10.4 Regularization techniques in MI

Inverse issues are those in which the goal is to reconstruct an object of interest from a series of measurements that are only implicit or partial. An example of a classic inverse issue in MI is the reconstruction of diagnostic data from a collection of values. This can be done in CT by reconstructing X-ray projections or in MRI by reconstructing spatial frequency information. The reconstruction of an event from its measurable data might be an example of an ill-posed inverse problem for a number of different reasons. For example, the reconstruction from noisy data in low-dose CT is ill-posed because the Radon transform inversion is ill-posed. This is the case because of the noise in the data. On the other hand, in the case of fast MRI, the primary problem that has to be addressed is a lack of clarity, and as a result, there is no particular solution that can be identified without incorporating prior information. It is common practice to impose certain attributes on the solution that have been predetermined by the issue by putting a regularization term in the problem in order to minimize the space for alternative solutions. This is done in order to make the problem easier to solve. The total variation minimization and sparse regularization procedures [61] as well as the finite-differences filter [62] are among the most used regularization methods. In contrast to the manual creation of priors utilized by the preceding methods, the regularization of the picture was learned by other operations as the image was being reconstructed. In these procedures, the regularization is implemented in a patch-wise fashion using the sparse approximation, which makes use of a dictionary that is learned in an unsupervised fashion [63]. Nevertheless, these learning-based techniques need a great deal of time due to the fact that the regularization is adaptive and is taught via an iterative reconstruction process. In addition, the unique dictionary learning methodology requires training on the dictionary in addition to an estimation of the current picture at each iteration for regularization, which creates issues in the use of the method in clinical practice.

In recent years, CNNs have been used to inverse issues of MI in a variety of different ways, including as direct full inversion techniques [64], post processing methods [65], learning iterative methods [66], and learned regularizers [67]. When used as post-processing approaches, the networks are trained to denoise pictures and remove artifacts from such images. The denoising and removal of artifacts occur as a result of the direct reconstruction of noisy or incomplete data. There is a significant possibility that the CNN's anticipated output would not be compatible with the data, despite the fact that several different network topologies have been proposed [68]. Methods have been devised in which the constructed networks describe unrolled iterative algorithms that employ adjoint operators in order to guarantee that the output image is an accurate representation of the collected raw data. These approaches may be regarded as learned iterative systems and have been applied effectively with a variety of imaging modalities.

The unique reconstruction method given in [69] clearly distinguishes between the employment of the NN, solution regularization, and coherence with the measured data. The regularization is provided in the form of an image prior that is formed by the output of a NN that has been trained in the past when using a Tikhonov regularization schema. This is because the Tikhonov regularization schema uses more complex and advanced network architectures to eliminate artifacts or noise.

Advanced models still need access to properly sizable and representative training sets, which is generally not attainable in clinically realistic contexts. Although the DNN is trained on data from a specific distribution, such as data recorded by a certain device manufacturer or patient group, it lacks generalization capacity when trained on small datasets. This makes it hard to generalize to data from a different distribution. In this context, the authors in [70] provide a simple but effective method for improving deep neural network generalization in the field of medical imaging classification. Since, the domain variability of medical images could be seen as compact, a novel linear-dependency regularization term is proposed [70] to identify the information that can be shared among medical data gathered from various sources, which makes it easier to generalize to "unseen" medical data.

When applying registration approaches in medical imaging, there are frequently seen sliding effects at tissue/organ borders. For instance, it is commonly known that during breathing, the lung and diaphragm rub up against the rib cage and the atria. This problem is not solved by conventional homogeneous smooth registration techniques. Recent research have used either robust regularization energy on the motion field or joint registration/segmentation to retain motion singularities. For physical deformations, nevertheless, tolerating all kinds of discontinuities is not stringent enough. In particular, the energy functional should prohibit flows that result in local vacuums or mass collisions. In this regard, the authors in [71] provide a regularization energy that represents a discriminative handling of various motion discontinuities by means of several regularization techniques such as the Helmholtz–Hodge decomposition and flow theory.

CNNs have demonstrated exceptional performance in a variety of medical imaging applications when they are provided with an abundant amount of labeled training data. However, obtaining expert annotations for medical data is often time- and money-consuming, posing a significant challenge for supervised learning systems. In this particular setting, the authors in [72] propose a one-of-a-kind semi-supervised deep learning technique for large-scale medical picture categorization. This method uses a regularization technique that uses the underlying knowledge to improve the ability of training models to tell different things apart. The goal of this method is to efficiently extract relevant information from unlabeled data.

10.5 Remarks and future directions

Medical research data sometimes contains a variety of data which is difficult or impossible to interpret using conventional statistical techniques. In order to deliver optimal or nearly optimal solutions to challenging data analysis and decision-making tasks in a fair period of time in MI, numerous optimization approaches and metaheuristic algorithms have been devised and deployed. Furthermore, designing game theoretical platforms and getting involve with different kinds of game models implemented in deep learning algorithms are motivating for the future development of MI. However, in general, the use of these algorithms efficiently in medicine is often unknown to physicians or medical experts. More collaboration with mathematicians and computer scientists would be crucial for those who could benefit state-of-art and practical use of deep learning applications from them to solve complex and sophisticated medical problems.

Apart from the enormous efforts and attention paid to MI, the future of deep learning in MI is not as close as it is in other imaging applications due to the complexity of this business. This is especially true in comparison to other imaging applications. The use of deep learning-based algorithms to interpret MI data is a fascinating and rapidly emerging area of research; nevertheless, progress in this field is being slowed down by a variety of difficulties. The lack of datasets, concerns over legality and privacy, the absence of specialized medical expertise, and the application of non-standard machine learning techniques are the primary factors contributing to these challenges and limitations associated with the use of deep learning techniques in medical imaging. A further significant shortcoming of management intelligence is that it does not have sufficient data integration and standards. Because of sensors and other components, there is now a large variation in the pictures that are produced since the nature of the data differs significantly from one piece of hardware to another. In addition, the range of applications required by the medical business mixes together a variety of datasets in order to enable learning algorithms that are more precise and effective. Interoperability lays the groundwork for big advances in the healthcare business, but it is still something that has to be accomplished. As was just shown, the mathematics involved in constructing a neural network are quite straightforward; however, the mechanism by which the output is generated is extremely convoluted. Despite the fact that huge amounts of data are input into machine-learning algorithms, which subsequently identify trends and develop predictive models, the primary challenge is in comprehending how a model works. The use of the deep learning model is continuous, and the vast majority of academics working in the field of medicine utilize it without having a clear grasp of how it functions or the reasons why it produces better results. Because of this, the application of deep learning algorithms is becoming more restricted in the field of MI.

As future directions, we can handle security problems in healthcare systems. Tackling malicious cyberattacks are very crucial for critical infrastructures in medical systems where personal health records are kept safely. Recently, several researchers propose RF Fingerprinting approach with deep learning to overcome malicious attacks especially spoofing attacks [73–75]. The work in [75–77] propose fine-grained data augmentation technique to improve the performance of deep learning-based RF fingerprinting technique. Besides, blockchain application in MI can be used for keeping security and privacy of medical images for interesting application such as driver safety and transparency [78]. Moreover, effective and reliable game theoretical platforms could be designed in deep learning algorithms to construct and to improve efficient and decision making processes regarding to data selecting methods and interoperability in MI.

10.6 Conclusion

The medical industry stands apart from every other industry in a major way. In this high priority area, consumers seek the finest level of care and services, regardless of cost. The health care industry has fallen short of society's expectations despite making up a major amount of national budgets. Usually, the analysis and interpretation of medical data is done by medical specialists. The sophistication of the images makes it difficult for a medical expert to understand them, and there are large variations among specialists due to the hard nature of their work. Deep learning is viewed as offering innovative and precise solutions for MI and is recognized as a significant technique for upcoming applications in the healthcare industry as a result of its success in other practical uses. In this chapter, we investigate game theoretical perspectives of deep learning regarding to MI as well as several optimization and regularization techniques for segmenting and categorizing medical images. Moreover, the limitations of deep learning methods encircled with game theory, optimization and regularization techniques regarding to MI are presented and open research issues and remarks are proposed at the end of chapter.

References

[1] Razzak, M. I., Naz, S. and Zaib, A. (2018). Deep learning for medical image processing: overview, challenges and the future. *Classification in BioApps: Automation of Decision Making*, *26*,323–350.

[2] Mitchell, T. M. (2007). *Machine Learning*. Vol. 1. New York: McGraw-hill.

[3] Sharma, A. K., Nandal, A., Dhaka, A., and Dixit, R., *et al*. "Medical image classification techniques and analysis using deep learning networks: a review," In *Health Informatics: A Computational Perspective in Healthcare*, Springer, pp. 233–258, 2021.

[4] Sharma, A. K., Nandal, A., Ganchev, T., and Dhaka, A. "Breast cancer classification using CNN extracted features: a comprehensive review," *Application of Deep Learning Methods in Healthcare and Medical Science (ADLMHMS-2020)*. Taylor and Francis: CRC Press; 2022.

[5] Fenton, J. J., Taplin, S. H., Carney, P. A., *et al.* (2007). Influence of computer-aided detection on performance of screening mammography. *New England Journal of Medicine*, *356*(14), 1399–1409.

[6] Sharma, A. K., Nandal, A., Dhaka, A., and Dixit, R. "A survey on machine learning based brain retrieval algorithms in medical image analysis," *Health and Technology*, Springer, vol. 10, pp. 1359–1373, 2020.

[7] Merjulah, R. and Chandra, J. (2019). Classification of myocardial ischemia in delayed contrast enhancement using machine learning. *In* D. Jude Hemanth, Deepak Gupta and Valentina Emilia Balas (eds.), *Intelligent Data Analysis for Biomedical Applications*, Academic Press, pp. 209–235.

[8] Amin, J., Sharif, M., Haldorai, A., Yasmin, M., and Nayak, R. S. "Brain tumor detection and classification using machine learning techniques," *RAMSACT 2022*, Springer, Accepted May 2022.

[9] Haskins, G., Kruger, U. and Yan, P. (2020). Deep learning in medical image registration: a survey. *Machine Vision and Applications*, *31*, 1–18.

[10] Zhou, S. K., Le, H. N., Luu, K., Nguyen, H. V. and Ayache, N. (2021). Deep reinforcement learning in medical imaging: a literature review. *Medical Image Analysis*, *73*, 102193.

[11] Nayak, D. R., Das, D., Dash, R., Majhi, S. and Majhi, B. (2020). Deep extreme learning machine with leaky rectified linear unit for multiclass classification of pathological brain images. *Multimedia Tools and Applications*, *79*, 15381–15396.

[12] Hamidoglu, A. (2016). On general form of the Tanh method and its application to nonlinear partial differential equations. *Numerical Algebra, Control and Optimization*, *6*(2), 175–181, 2016.

[13] Hamidoglu, A. (2017). On general form of Tanh method and its application to medical problems. In *Proceedings of the Tenth International Conference on Management Science and Engineering Management,* Springer Singapore, pp. 269–278.

[14] Kowsari, K., Sali, R., Ehsan, L., *et al.* (2020). HMIC: hierarchical medical image classification, a deep learning approach. *Information*, *11*(6), 318.

[15] Sharma, A. K., Nandal, A., Dhaka, A., and Polat, K., *et al*. "HOG transformation based feature extraction framework in modified Resnet50 model for brain tumor detection biomedical signal processing and control," *Biomedical Signal Processing & Control*, Elsevier, vol. 84, 2023.

[16] Sharma, A. K., Nandal, A., Dhaka, A., Koundal, D., Bogatinoska, D. C., and Alyami, H. (2022). Enhanced watershed segmentation algorithm based modified ResNet50 model for brain tumor detection. *BioMed Research International*, *Hindawi*.

[17] Li, L., Muneeswaran, V., Ramkumar, S., Emayavaramban, G. and Gonzalez, G. R. (2019). Metaheuristic FIR filter with game theory based compression technique-a reliable medical image compression technique for online applications. *Pattern Recognition Letters*, *125*, 7–12.

[18] Zhou, J., Li, J. and Di, X. (2020). A novel lossless medical image encryption scheme based on game theory with optimized ROI parameters and hidden ROI position. *IEEE Access*, *8*, 122210–122228.

[19] Liu, Y., Zhou, J., Liu, L., *et al.* (2022). FCP-net: a feature-compression-pyramid network guided by game-theoretic interactions for medical image segmentation. *IEEE Transactions on Medical Imaging*, *41*(6), 1482–1496.

[20] Hamidoğlu, A. (2021). A novel one target game model in the life insurance market. *International Journal of Management Science and Engineering Management, 16*(3), 221–228.

[21] Hamidoğlu, A. (2022). A game theoretical approach for finding near-optimal solutions of an optimization problem. *Optimization, 72*, 1–23.

[22] Hamidoğlu, A., Taghiyev, M. H. and Weber, G. W. (2021). On building two-player games with treatment schedules for the SIR model. *Azerbaijan Journal of Mathematics, 11*(2), 183–195.

[23] Hamidoğlu, A. and Taghiyev, M. H. (2021). On discrete game models with applications to management. In *Proceedings of the Fifteenth International Conference on Management Science and Engineering Management* vol. 1 15 Springer Nature Switzerland, pp. 195–207.

[24] Hamidoğlu, A., Taghiyev, M. and Weber, G. W. (2022). On construction of pursuit-evasion games in discrete control models. *Applied and Computational Mathematics, 21*(1), 52–60.

[25] Staňková, K., Brown, J. S., Dalton, W. S. and Gatenby, R. A. (2019). Optimizing cancer treatment using game theory: a review. *JAMA Oncology, 5*(1), 96–103.

[26] Kallel, M., Aboulaich, R., Habbal, A. and Moakher, M. (2014). A Nash-game approach to joint image restoration and segmentation. *Applied Mathematical Modelling, 38*(11–12), 3038–3053.

[27] Ibragimov, B., Likar, B. and Pernus, F. (2012). A game-theoretic framework for landmark-based image segmentation. *IEEE Transactions on Medical Imaging, 31*(9), 1761–1776.

[28] Zhang, F., Shan, G. S., Gao, H. and Jia, T. Z. (2019). Rheumatoid arthritis analysis by Nash equilibrium game analysis. *Journal of Medical Imaging and Health Informatics, 9*(7), 1382–1385.

[29] Mandal, S., Greenblatt, A. B. and An, J. (2018). Imaging intelligence: AI is transforming medical imaging across the imaging spectrum. *IEEE Pulse, 9* (5), 16–24.

[30] Wu, D. and Lisser, A. (2022). Using CNN for solving two-player zero-sum games. *Expert Systems with Applications, 204*, 117545.

[31] Hazra, T. and Anjaria, K. (2022). Applications of game theory in deep learning: a survey. *Multimedia Tools and Applications, 81*(6), 8963–8994.

[32] Chivukula, A. S. and Liu, W. (2018). Adversarial deep learning models with multiple adversaries. *IEEE Transactions on Knowledge and Data Engineering, 31*(6), 1066–1079.

[33] Schuurmans, D. and Zinkevich, M. A. (2016). Deep learning games. *Advances in Neural Information Processing Systems*, 29, 1–9.

[34] Nisan, N., Roughgarden, T., Tardos, E. and Vazirani, V. V. (eds.) (2007). *Algorithmic Game Theory*. Cambridge University Press.

[35] Tembine, H. (2019). Deep learning meets game theory: Bregman-based algorithms for interactive deep generative adversarial networks. *IEEE Transactions on Cybernetics, 50*(3), 1132–1145.

[36] Geman, S. and Geman, D. (1984). Stochastic relaxation, Gibbs distributions, and the Bayesian restoration of images. *IEEE Transactions on Pattern Analysis and Machine Intelligence*, *6*, 721–741.

[37] Komodakis, N. and Tziritas, G. (2007). Approximate labeling via graph cuts based on linear programming. *IEEE Transactions on Pattern Analysis and Machine Intelligence*, *29*(8), 1436–1453.

[38] Glocker, B., Komodakis, N., Tziritas, G., Navab, N. and Paragios, N. (2008). Dense image registration through MRFs and efficient linear programming. *Medical Image Analysis*, *12*(6), 731–741.

[39] Luong, D. V., Rueckert, D. and Rustem, B. (2011). Incorporating hard constraints into non-rigid registration via nonlinear programming. In Benoit M. Dawant, David R. Haynor (eds.), *Medical Imaging 2011: Image Processing,* SPIE, Lake Buena Vista (Orlando), FL, USA, vol. 7962, pp. 298–308.

[40] Hoyos-Idrobo, A., Weiss, P., Massire, A., Amadon, A. and Boulant, N. (2013). On variant strategies to solve the magnitude least squares optimization problem in parallel transmission pulse design and under strict SAR and power constraints. *IEEE Transactions on Medical Imaging*, *33*(3), 739–748.

[41] Li, M. and Kudo, H. (2007). Statistical PET image reconstruction using duality of nonlinear programming. *Electronics and Communications in Japan (Part II: Electronics)*, *90*(11), 122–131.

[42] Wagenknecht, G., Kaiser, H. J., Sabri, O. and Buell, U. (2000). Dynamic programming algorithm for contrast correction in medical images. In Edward R. Dougherty, Jaakko T. Astola (eds.), *Nonlinear Image Processing XI,* SPIE, San Jose, CA, USA, vol. 3961, pp. 216–226.

[43] Rivaz, H., Boctor, E., Foroughi, P., Zellars, R., Fichtinger, G. and Hager, G. (2008). Ultrasound elastography: a dynamic programming approach. *IEEE Transactions on Medical Imaging*, *27*(10), 1373–1377.

[44] Terzopoulos, D., Platt, J., Barr, A. and Fleischer, K. (1987). Elastically deformable models. In *Proceedings of the 14th Annual Conference on Computer Graphics and Interactive Techniques,* Association for Computing Machinery, New York, NY, USA, pp. 205–214.

[45] Geiger, D., Gupta, A., Costa, L. A. and Vlontzos, J. (1995). Dynamic programming for detecting, tracking, and matching deformable contours. *IEEE Transactions on Pattern Analysis and Machine Intelligence*, *17*(3), 294–302.

[46] Zhang, Y., Matuszewski, B. J., Shark, L. K. and Moore, C. J. (2007). A novel medical image segmentation method using dynamic programming. In *International Conference on Medical Information Visualisation-BioMedical Visualisation (MediVis 2007),* IEEE, Zurich, Switzerland, pp. 69–74.

[47] Ungru, K. and Jiang, X. (2017). Dynamic programming based segmentation in biomedical imaging. *Computational and Structural Biotechnology Journal*, *15*, 255–264.

[48] Kennedy, J. and Eberhart, R. (1995). Particle swarm optimization. In *Proceedings of ICNN'95-International Conference on Neural Networks,* IEEE, Perth, WA, Australia, vol. 4, pp. 1942–1948.

[49] Wachowiak, M. P., Smolíková, R., Zheng, Y., Zurada, J. M. and Elmaghraby, A. S. (2004). An approach to multimodal biomedical image registration utilizing particle swarm optimization. *IEEE Transactions on Evolutionary Computation*, *8*(3), 289–301.

[50] Rundo, L., Tangherloni, A., Militello, C., Gilardi, M. C. and Mauri, G. (2016). Multimodal medical image registration using particle swarm optimization: a review. In *2016 IEEE Symposium Series on Computational Intelligence (SSCI),* IEEE, Athens, Greece, pp. 1–8.

[51] Ahilan, A., Manogaran, G., Raja, C., *et al.* (2019). Segmentation by fractional order darwinian particle swarm optimization based multilevel thresholding and improved lossless prediction based compression algorithm for medical images. *IEEE Access*, *7*, 89570–89580.

[52] Ait-Aoudia, S., Guerrout, E. H. and Mahiou, R. (2014). Medical image segmentation using particle swarm optimization. In *2014 18th International Conference on Information Visualisation,* IEEE, Paris, France, pp. 287–291.

[53] Kirkpatrick, S., Gelatt Jr, C. D. and Vecchi, M. P. (1983). Optimization by simulated annealing. *Science*, *220*(4598), 671–680.

[54] Ait-Aoudia, S. and Mahiou, R. (2007). Medical image registration by simulated annealing and genetic algorithms. *In Geometric Modeling and Imaging (GMAI'07),* IEEE, Zurich, Switzerland, pp. 145–148.

[55] Garnero, L., Franchois, A., Hugonin, J. P., Pichot, C. and Joachimowicz, N. (1991). Microwave imaging-complex permittivity reconstruction-by simulated annealing. *IEEE Transactions on Microwave Theory and Techniques*, *39*(11), 1801–1807.

[56] Elbasi, E. (2020). Enhancing the robustness of watermarked medical images using heuristic search algorithm. In *2020 7th International Conference on Internet of Things: Systems, Management and Security (IOTSMS)*, IEEE, Paris, France, pp. 1–5.

[57] Grefenstette, J. J. (1993). Genetic algorithms and machine learning. *In Proceedings of the Sixth Annual Conference on Computational Learning Theory,* CA, USA, pp. 3–4.

[58] Murray-Smith, D. J. (2012). 6-experimental modelling: system identification, parameter estimation and model optimisation techniques. *Modelling and Simulation of Integrated Systems in Engineering*, 165–214.

[59] Arif, M. and Wang, G. (2020). Fast curvelet transform through genetic algorithm for multimodal medical image fusion. *Soft Computing*, *24*(3), 1815–1836.

[60] Liu, P., El Basha, M. D., Li, Y., Xiao, Y., Sanelli, P. C. and Fang, R. (2019). Deep evolutionary networks with expedited genetic algorithms for medical image denoising. *Medical Image Analysis*, *54*, 306–315.

[61] Lustig, M., Donoho, D. L., Santos, J. M. and Pauly, J. M. (2008). Compressed sensing MRI. *IEEE Signal Processing Magazine*, *25*(2), 72–82.

[62] Block, K. T., Uecker, M. and Frahm, J. (2007). Undersampled radial MRI with multiple coils. Iterative image reconstruction using a total variation constraint. *Magnetic Resonance in Medicine: An Official Journal of the*

International Society for Magnetic Resonance in Medicine, *57*(6), 1086–1098.

[63] Wang, Y. and Ying, L. (2013). Compressed sensing dynamic cardiac cine MRI using learned spatiotemporal dictionary. *IEEE Transactions on Biomedical Engineering*, *61*(4), 1109–1120.

[64] Zhu, B., Liu, J. Z., Cauley, S. F., Rosen, B. R. and Rosen, M. S. (2018). Image reconstruction by domain-transform manifold learning. *Nature*, *555* (7697), 487–492.

[65] Schwab, J., Antholzer, S. and Haltmeier, M. (2019). Deep null space learning for inverse problems: convergence analysis and rates. *Inverse Problems*, *35*(2), 025008.

[66] Gupta, H., Jin, K. H., Nguyen, H. Q., McCann, M. T. and Unser, M. (2018). CNN-based projected gradient descent for consistent CT image reconstruction. *IEEE Transactions on Medical Imaging*, *37*(6), 1440–1453.

[67] Schlemper, J., Caballero, J., Hajnal, J. V., Price, A. and Rueckert, D. (2017). A deep cascade of convolutional neural networks for MR image reconstruction. In *Information Processing in Medical Imaging: 25th International Conference, IPMI 2017, Boone, NC, USA*, June 25–30, 2017, *Proceedings 25*, Springer International Publishing, Boone, NC, USA, pp. 647–658.

[68] Han, Y. and Ye, J. C. (2018). Framing U-Net via deep convolutional framelets: application to sparse-view CT. *IEEE Transactions on Medical Imaging*, *37*(6), 1418–1429.

[69] Kofler, A., Haltmeier, M., Schaeffter, T., *et al.* (2020). Neural networks-based regularization for large-scale medical image reconstruction. *Physics in Medicine & Biology*, *65*(13), 135003.

[70] Li, H., Wang, Y., Wan, R., Wang, S., Li, T. Q. and Kot, A. (2020). Domain generalization for medical imaging classification with linear-dependency regularization. *Advances in Neural Information Processing Systems*, *33*, 3118–3129.

[71] Ruan, D., Esedoglu, S. and Fessler, J. A. (2009). Discriminative sliding preserving regularization in medical image registration. In *2009 IEEE International Symposium on Biomedical Imaging: From Nano to Macro*, IEEE, Boston, MA, USA, pp. 430–433.

[72] Wang, X., Chen, H., Xiang, H., Lin, H., Lin, X. and Heng, P. A. (2021). Deep virtual adversarial self-training with consistency regularization for semi-supervised medical image classification. *Medical Image Analysis*, *70*, 102010.

[73] Comert, C., Gul, O. M., Kulhandjian, M., *et al.* (2023). Secure design of cyber-physical systems at the radio frequency level: machine and deep learning-driven approaches, challenges and opportunities. In: Traore, I., Woungang, I., and Saad, S. (eds). *Artificial Intelligence for Cyber-Physical Systems Hardening. Engineering Cyber-Physical Systems and Critical Infrastructures*, vol 2. Springer, Cham.

[74] Comert, C., Kulhandjian, M., Gul, O. M., *et al.* (2022). Analysis of augmentation methods for RF fingerprinting under impaired channels. In

Murtuza Jadliwala (ed.), *Proceedings of the 2022 ACM Workshop on Wireless Security and Machine Learning*, New York, NY, USA, pp. 3–8.

[75] Gul, O. M., Kulhandijan, M., Kantarci, B., Touazi, A., Ellement, C. and D'Amours, C. On the impact of CDL and TDL augmentation for RF fingerprinting under impaired channels. *48th Wireless World Research Forum (WWRF 2022)*, 7–9 November 2022, UAE, pp. 1–6.

[76] Gul, O. M., Kulhandjian, M., Kantarci, B., Touazi, A., Ellement, C. and D'Amours, C. (2022). Fine-grained augmentation for RF fingerprinting under impaired channels. In *2022 IEEE 27th International Workshop on Computer Aided Modeling and Design of Communication Links and Networks (CAMAD),* IEEE, Paris, France, pp. 115–120.

[77] Gul, O. M., Kulhandijan, M., Kantarci, B., Touazi, A., Ellement, C. and D'Amours, C. (2023). Secure industrial IoT systems via RF fingerprinting under impaired channels with interference and noise. *IEEE Access*, *11*, 26289–26307.

[78] Gul, O. M. (2022). Blockchain-enabled internet of things (IoTs) platforms for vehicle sensing and transportation monitoring. *Machine Learning, Blockchain Technologies and Big Data Analytics for IoTs: Methods, Technologies and Applications*, 351–373.

Chapter 11

Data preparation for artificial intelligence in federated learning: the influence of artifacts on the composition of the mammography database

Heráclio A. da Costa[1], Vitor T.R. Ribeiro[2] and Edmar C. Gurjão[2]

Artificial intelligence algorithms use databases to train a model to posterior utilization, so the algorithm's performance is highly dependent on the database quality to represent the behavior of the modeled system. In this chapter, we evaluate the influence of image artifacts on mammograms in training an artificial intelligence algorithm using a federated learning technique. Using a simulation of motion and dust artifacts, we evaluated different scenarios of a database divided into three clients for the application of the federated learning technique. Applying the transfer learning technique, we adapted an open-source convolutional neural network to classify images into three categories: images without microcalcifications, benign microcalcifications, and suspicious microcalcifications, considering both cases before and after the addition of simulated artifacts. The artifact's addition to databases for federated learning training has influenced the performance of the artificial intelligence model. Depending on the size of the database and the type of artifact, it may have a beneficial or less harmful effect than their exclusion. Using images with artifacts can benefit the training of algorithms instead of excluding images with artifacts in the composition of the databases.

11.1 Introduction

In recent years, benefiting from the significant increase in the medical data volume in diagnostic imaging, the development of tools based on machine learning (ML) for diagnostic purposes and data distribution by cloud computing technology has configured a fertile field of research [1–3]. However, despite the enormous potential of computational tools based on artificial intelligence and the growing

[1]Postgraduate Program in Health Science and Technology of State University of Paraíba, Brazil
[2]Department of Electrical Engineering of Federal University of Campina Grande, Brazil.

volume of data, there are still limitations to developing the models and their use in clinical practice.

The medical image segmentation and automatic classification demonstrated promising results, especially with the greater computational processing power, which allowed the development of powerful neural network models structured in multiple layers, called deep learning models, with convolutional neural networks being the most utilized models for medical image analysis [4–6]. In these networks, filters are applied to the images for attribute extraction, allowing the progression of relevant information in the network and resizing of matrices, resulting in efficient computational cost without loss of essential characteristics of the images.

The creation of databases for properly training these networks usually requires sharing data from different medical centers, seeking an increase in volume and a more significant data diversity, reducing potential biases in the trained models [7]. However, data transmission implicates legal and ethical questions involving the protection of sensitive data [8], generating the need for protection measures ranging from simple anonymization to more advanced encryption resources and based on neural networks [1,9].

A recent proposal related to information sharing privacy issues between multiple centers and reducing the size of transmitting data in ML training algorithms on medical images is federated learning [10]. This technique allows the local training of a model and the transmission of training results with the information embedded without sharing sensitive data from the used databases. The local information is transmitted to a central server, generating a central model that can be readily used or redistributed to new local iterations. However, by having access only to the results of local training, aggregating the information on the central is made without knowledge of the data set used in each service [10], leading to potential suspicions about the quality of the images used in training, as in the case of the presence of artifacts that may compromise the final quality of the image, with lost or distorted information and consequent harm to the results of training.

Face the potential harm artifacts can generate errors in interpreting images and possible interference in the ML process. This chapter proposes to evaluate the influence of images with artifacts in the training databases on the performance of a neural network based on federated learning technique to guide recommendations regarding the treatment of these data in the composition of databases. The chosen approach evaluated a model in the ability to analyze microcalcifications in mammograms, a question of great relevance for the diagnosis of breast cancer in screening exams and already present in the clinical practice of using artificial intelligence.

11.2 Federate learning

Federated learning is a machine learning technique that allows the merging of information and distribution of models across different devices, such as computers, smartphones, and wearable devices, in a collaborative and remote manner without sharing sensitive data. It was first proposed in [10]. The idea is to share information from the local training of a machine learning model to a central server, which is responsible for combining this information with results from various devices. A new model with the same structure but modified weights is generated and redistributed to the devices for new local training in a pre-determined or continuous cycle. This technique enables users to benefit from a model trained with rich data without the need for central storage of this data, saving storage resources and massive data transmission, and preserving the raw data privacy between the users themselves and the central server [10].

The technique was designed with some key points for optimization, with a focus on local training with personal data and massive distribution, as well as imbalances in data volume between clients, initially assuming a synchronous update mode with communication rounds and redistribution of the updated central model for new local training. The proposed FederatedAveraging algorithm considers the volume of databank from each client and proceeds with a weighted average of the gradient results.

However, new workflow designs are being developed, including flows where the data combination process also takes place in a decentralized and dynamic manner and without the need to share all information of the model every training round, as well as variations of the technique for more efficient information combination and in highly heterogeneous data environments [11].

Federated learning already has a wide range of uses, with applications in areas such as health, recommendation mechanisms, the Internet of Things, and edge computing [12]. The use of this technique is already used in health with a focus on image diagnosis, but there are still some challenges to overcome [6]. Roth *et al.* [13] conducted a multicenter international study for the development and practical application of a classification model for mammary density using mammography images. In this study, a previously designed model for image classification was used and the FederatedAveraging algorithm originally proposed by Brendan McMahan *et al.* [10] was implemented for aggregating the results of local trainings, starting from databases with heterogeneous characteristics in terms of data volume, image resolution, and acquisition techniques. Although high accuracy was not achieved in the models, it was possible to demonstrate that federated learning improved performance and better generalization ability. More recently, due to the COVID-19 pandemic, there has been an increase in interest and acceleration in the implementation of the technique [14], with various collaborative efforts in the diagnosis of COVID-19 through federated learning [15–18]. Other areas of interest in the medical field have embraced the federated learning technique such as mortality risk models for various clinical conditions [15].

Despite being highly promising, the technique has some limitations to be explored, such as concerns related to malicious agents regarding the upload of suspicious information, the potential high number of clients generating high communication cost, the heterogeneity of data and the heterogeneity of systems and devices [19,20]. In the use for image diagnosis, when preparing medical image data from databases, the reliability of labeling data, the quality of images, the standardization of information and the suitability of the training performer are information of great importance and are unknown by the central server and other participants when applying federated learning [10]. Local data with bias, intentional or not, the use of data with parameters outside the proposed standard and security attacks, such as interception of transmissions, invasions of local databases, and model inversion attacks, where it is possible to recreate images, are potential problems that have opened the field for the study [12,15,21].

Tolpegin *et al.* [22] showed that simply reversing the labels of training images in a convolutional neural network on one client was able to negatively impact the overall model, demonstrating the need for protection methods against malicious agents. Intentional poisoning of a model does not seem reasonable when it comes to collaboration between health centers for training a network for medical diagnosis. However, concerns about the quality of the locally used images are relevant, especially in relation to image artifacts and unintentional label failures in the data.

Another challenge relates to the long-term storage of the databases initially used for the training. In the case of medical images, the costs of long-term storage can become major problems given the volume and size of the files, and there are already studies on how to optimize federated learning over time without necessarily training with the initial data, with new data being used in place of the initial data [23].

11.3 Methodology

Using a public database with mammography images in DICOM format, without loss of quality and previous notes on suspicious microcalcifications [24], we proceeded to select fields with standardized dimensions in each mammography image (area of interest of 228 $\times$ 228 pixels), categorizing them into three groups: the presence of suspicious microcalcifications; the presence of benign microcalcifications; and without microcalcifications, containing only breast tissue (Figure 11.1). The area selection focusing on the microcalcifications analysis allowed an increase in the database size, without loss of conformity for the classification of microcalcifications by the BI-RADS system [25].

Part of the images in each category was randomly selected and allocated for exclusive use on models' accuracy evaluation. The remaining images were divided into three clients simulating three different medical centers, in a ratio of 1:2:3, named clients A, B, and C. Next, data augmentation techniques that did not change the quality of the images were implemented (for example, rotation and flipping)—the final number of images for each customer is in Table 11.1.

A convolutional neural network with partially pre-established architecture and open source was selected [26], implemented in Python using the Keras and tensor flow libraries. The output layers of the model were adapted for three possible results: parenchyma without microcalcifications, the presence of benign microcalcifications, and the presence of suspicious microcalcifications. Federated learning was performed by training a model for each client database. The central

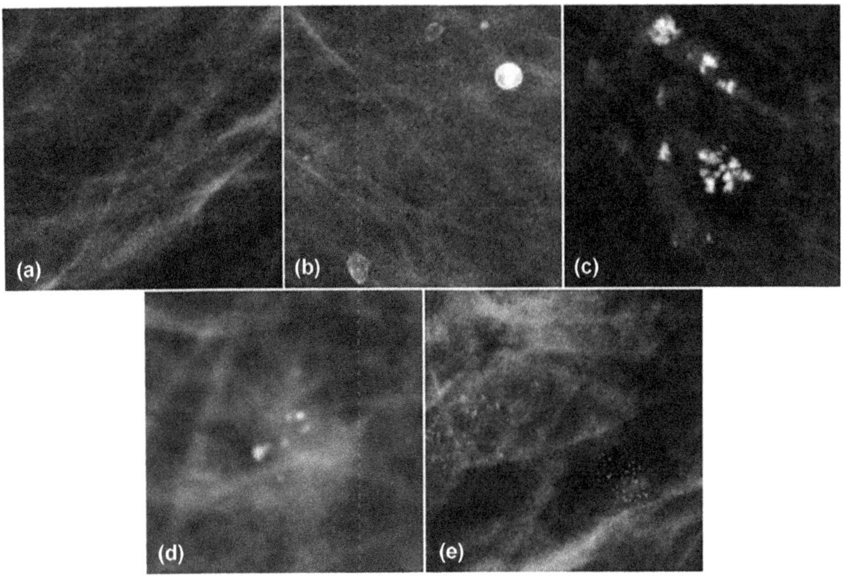

Figure 11.1 *Examples of images selected from areas of interest obtained from the original database: (a) normal breast parenchyma; (b) benign microcalcifications; (c) microcalcifications suspicious for malignancy. Example of images with artifact applications: (d) image with motion artifact due to application of blur; (e) addition of simulated dust artifact in the lower right quadrant of the image.*

Table 11.1 *Number of images in clients databases A, B, and C after data augmentation techniques*

Class	**Client A**	**Client B**	**Client C**
Normal	602	399	204
Benign	594	397	204
Suspect	248	174	84
Total	1444	970	492

combination of data follows the FederatedAveraging algorithm initially proposed by Brendan McMahan *et al.* [10]. This algorithm uses a weighted average, maintaining the information proportionality according to the number of images in each database. The model evaluation uses ROC curves and the areas under the curve (AUC) for each category studied using python programming and calculation of overall accuracy of each federated model on testing data.

The image artifacts addition in the databases was done through simulation on the python platform, using blur resources to simulate motion artifacts and the addition of tiny foci of close to white tone simulating dust artifacts that can lead to confusion with microcalcifications (Figure 11.1). The artifacts insertion was in four different scenarios, forming new databases: (1) addition of motion artifact in 10% of client A's images; (2) addition of motion artifact in 10% of client C's images; (3) added dust artifact to 10% of client A's images; (4) added dust artifact to 10% of customer images C.

The training uses a federated learning technique for each artifact addition scenario to generate new ROC curve graphs, allowing the comparison of AUC values from the resulting modes. Thus, the insertion of artifacts influence was evaluated by directly comparing the variation of these values.

Finally, to evaluate the performance of the models in a scenario where possible images exclusions were carried out for the composition of the databases for training, we reproduced the federated learning in two other situations: (1) with a random reduction of 10% of the images from client A; (2) 10% random reduction of client C images.

11.4 Results

Starting from the data without adding artifacts, after local training with 10 iterations and subsequent federation with FederatedAveraging, the resulting model obtained an AUC of 0.96 for the group of normal images, 0.90 for the benign microcalcifications group, and 0.88 for the suspicious microcalcifications group, with an overall accuracy of 78.45%. Results for the model AUC modification in each scenario with the addition of artifacts to the databases are in Table 11.2.

In the motion artifacts scenario, the overall accuracy after client A database modification increased to 81.63%. However, applying the process in the client C database, the accuracy remained stable at 78.45%. Changing for dust artifacts, we obtained a drop in accuracy to 73.7% of the central model when we changed the images in client A's database and a drop to 67.54% when changing client C's database. Facing such an accuracy drop we proceeded with a new evaluation considering image exclusion with the artifacts in each scenario to evaluate the impact of reducing the number of images rather than the exclusion of images that led to a worsening in the classification capacity of the model. After image deletion on client A, the accuracy dropped to 65.19%, reaching 76.10% when the deletion was performed on client C.

Table 11.2 AUC values for each category in the different scenarios evaluated

Category	Without artifact	Movement artifact in A	Movement artifact in C	Dust artifact in A	Dust artifact in C
Normal	0,96	0,94	0,93	0,93	0,94
Benign	0,90	0,90	0,91	0,90	0,88
Suspect	0,88	0,92	0,91	0,91	0,87

11.4.1 Discussion

Brendan McMahan *et al.* [10] proposed that federated learning consists of new ML algorithms training that minimizes privacy and large database transmission. Such an algorithm provides several local pieces of training, with a record of the stochastic gradient descent that is then sent to a central server responsible for combining them into one single model without sharing the user database. This technique allows users to benefit from a trained model with rich data without the need for central storage of all the training data and to maintain the privacy of the raw data between the users themselves and the central server. The COVID-19 pandemic increased the interest in the technique since many affected people led to several collaborative efforts in diagnosing COVID-19 through federated learning [14,16–18,27,28] allowing to build knowledge from exams proceeded in a significant number of different centers without sharing sensitive data.

The labeling data reliability, the image quality, the standardization of information, and the trust in the local model training are topics of concern and usually unknown by the central server and by the other participants on applying federated learning Biased local data, intentional or not, as well as the use of data with parameters outside the standard pattern are potential problems, leading to a field for the study of tools to mitigate possible damage to the central model and look to identify a lack of equity in the composition of the databases [21,29,30,31]. To understand how to standardize the composition of databases, we need to understand the potential interference that may contain artifacts in medical images can cause in the training of ML models.

Results of motion artifacts demonstrate that using these images for model training can increase the accuracy, obtaining the most expressive improvement precisely when the interference was in the client with the highest number of images in the database. Furthermore, even being an artifact that may difficult image interpretation by radiologists, the blur technique used to simulate the artifact is one of the resources used as data augmentation technics to improve databases for ML [11].

However, adding dust artifacts led to model degradation with AUC reduction, especially when interference occurs on the client with the lowest number of images. This result may have occurred due to the decrease in absolute numbers of interference-free images in an already small database, making the proper training of the local model unfeasible and leading to the poisoning of the central model. The values for client C's local training accuracy before adding the artifacts was 56.97%, with an AUC of 0.82, 072, and 0.65 for the groups of normal images, with benign microcalcifications and with suspicious microcalcifications. Such values were initially low, demonstrating the inadequate training capacity of the model with such a quantity of images, claiming attention for the possibility of entry barriers to clients with such performance.

When utilizing the manipulated images in the client with largest database, the drop of performance was smaller than in the scenario when we opted to exclude those images, suggesting that in such cases the maintenance of images with artifacts in the database may be less deleterious than downsizing the database.

11.5 Conclusions

The study's results demonstrate that, when faced with local clients' database with minimally adequate dimensions for ML model training, the presence of image artifacts can benefit model performance. We also observed that in certain situations, the presence of images with artifacts could be less harmful than their exclusion from the database of ML training. These findings encourage medical institutions to work together in adopting a standardized protocol to maintain the images with artifacts on the composition of databases when proceeding with a federated learning initiative. However, our study was limited because it did not use real artifacts from mammography images and made use of a limited original number of images, as our objective was not a creation of an algorithm for clinical use.

Other diagnostic modalities, as computed tomography and magnetic resonance, has their own types of image artifacts [32,33]. The use of those images and of different machine training techniques can generate divergent results regarding the use of images with artifacts in the composition of databases, with the necessity of new studies to approach such situations.

References

[1] Sahi A., Lai D. and Li Y. Security and privacy preserving approaches in the eHealth clouds with disaster recovery plan. *Comput Biol Med.* 2016;78:1–8.

[2] Sharma A.K., Nandal A., Dhaka A. and Dixit R. A survey on machine learning based brain retrieval algorithms in medical image analysis. In *Health Technology*, Springer, Berlin. 2020, vol. 10, pp. 1359–1373.

[3] Panayides A.S., Pattichis M.S., Leandrou S., Pitris C., Constantinidou A. and Pattichis C.S. Radiogenomics for precision medicine with a big data analytics perspective. *IEEE J Biomed Health Inform.* 2019;23(5):2063–2079.

[4] Panayides A.S., Amini A., Filipovic N.D., *et al.* AI in medical imaging informatics: current challenges and future directions. *IEEE J Biomed Health Inform.* 2020;24(7):1837–1857.

[5] Sharma A.K., Nandal A., Dhaka A. and Dixit R. Medical image classification techniques and analysis using deep learning networks: a review. In *Health Informatics: A Computational Perspective in Healthcare*, Springer, Singapore. 2021, pp. 233–258.

[6] Chartrand G., Cheng P.M., Vorontsov E., *et al.* Deep learning: a primer for radiologists. *Radiographics.* 2017;37(7):2113–2131.

[7] Rouzrokh P, Khosravi B, Faghani S, *et al.* Mitigating bias in radiology machine learning: 1. Data handling. *Radiol Artif Intell.* 2022;4(5):e210290.

[8] van Panhuis W.G., Paul P., Emerson C., *et al.* A systematic review of barriers to data sharing in public health. *BMC Public Health.* 2014;14:1144.

[9] Shah A. and Gulati R. Privacy preserving data mining: techniques, classification and implications – a survey. *Int J Comput Appl.* 2016;137:40–46. 10.5120/ijca2016909006.

[10] Brendan McMahan H., Moore E., Ramage D., Hampson S. and Agüera y Arcas B. Communication-efficient learning of deep networks from decentralized data. In *Proceedings of the 20th International Conference on Artificial Intelligence and Statistics*, AISTATS 2017.

[11] Sharma A.K., Nandal A., Zhou L., Dhaka A. and Wu T. Brain tumor classification using modified VGG model-based transfer learning approach. In *Proceedings of the 20th International Conference on Intelligent Software Methodologies, Tools, and Techniques* (SOMET-2021), June 2021, Conference date Sept. 21–23, 2021, Cancun, Mexico.

[12] Wen J., Zhang Z., Lan Y., Cui Z., Cai J. and Zhang W. A survey on federated learning: challenges and applications. *Int J Mach Learn Cybern.* 2023;14(2);513–535.

[13] Roth H.R., Chang K., Singh P. *et al.* Federated learning for breast density classification: a real-world implementation. *Lecture Notes in Computer Science* 2020;12444LNCS:181–191.

[14] Naz S., Phan K.T. and Chen Y-P. A comprehensive review of federated learning for COVID-19 detection. *Int J Intell Syst.* 2022;37(3):2371–2392.

[15] Oh W. and Nadkarni G.N. Federated learning in health care using structured medical data. *Adv Kidney Dis Health.* 2023;30(1):4–16.

[16] Sharma M.K., Dhiman N., Mishra V.N., Mishra L.N., Dhaka A. and Koundal D. Post-symptomatic detection of COVID-2019 grade based mediative fuzzy projection. *Comput Electr Eng.* 2022;101:108028.

[17] Liang H., Guo Y., Chen X., *et al.* Artificial intelligence for stepwise diagnosis and monitoring of COVID-19. *Eur Radiol.* 2022;32(4):2235–2245.

[18] Zhou L., Nandal A., Dhaka A., *et al.* Artificial neural network dual hesitant fermatean fuzzy implementation in transportation of COVID-19 vaccine. *J Organ End User Comput.* 2023;35(2):1–23.

[19] Zhang K., Song X., Zhang C. and Yu S. Challenges and future directions of secure federated learning: a survey. *Front Comput Sci.* 2022;16(5):165817.

[20] Darzidehkalani E., Ghasemi-Rad M. and van Ooijen P.M.A. Federated learning in medical imaging: Part II: methods, challenges, and considerations. *J Am Coll Radiol.* 2022;19(8):975–982.

[21] Panda A., Mahloujifar S., Bhagoji A.N., Chakraborty S. and Mittal P. SparseFed: mitigating model poisoning attacks in federated learning with sparsification. ArXiv. December 2021. Available at: https://arvix.org/abs/2112.06274. Accessed on: 12 Jan. 2022.

[22] Li B., Chen S. and Peng Z. New generation federated learning. *Sensors.* 2022;22(21):8475.

[23] Sharma A.K., Nandal A., Ganchev T. and Dhaka A. Breast cancer classification using CNN extracted features: a comprehensive review. (2022). In R. Tanwar, P. Kumar, M. Kumar and N. Nandal (eds.). *Application of Deep Learning Methods in Healthcare and Medical Science*, Taylor and Francis, CRC Press, New York. 2022, pp. 147–164.

[24] Loizidou K., Skouroumouni G., Pitris C. and Nikolaou, C. Digital subtraction of temporally sequential mammograms for improved detection and classification of microcalcifications. *Eur Radiol Exp*. 2021;5(1):40.

[25] Sickles E.A., D'Orsi C.J., Bassett L.W. *et al.* ACR BI-RADS® mammography. In: C.J. D'Orsi, E.A. Sickles, E.B. Mendelson *et al.* (Eds.) *ACR BI-RADS® Atlas, Breast Imaging Reporting and Data System.* American College of Radiology, Reston, VA. 2013, pp. 121–140.

[26] Tan M. and Le Q.V. Efficientnet: rethinking model scaling for convolutional neural networks. *ArXiv*. 2019. Available at: https://arxiv.org/abs/1905.11946. Accessed on: Jan 12, 2023.

[27] Zhang W., Zhou T., Lu Q. Dynamic-fusion-based federated learning for COVID-19 detection. *IEEE Internet Things J.* 2021;8(21):15884–15891.

[28] Samuel O., Omojo A.B., Onuja A.M., *et al.* IoMT: a COVID-19 healthcare system driven by federated learning and blockchain. *IEEE J Biomed Health Inform*. 2023;27(2):823–834.

[29] Kairouz P., McMahan H.B., Avent B., *et al.* Advances and open problems in federated learning. *Found Trends Mach Learn*. 2021;14(1–2):1–210.

[30] Rieke, N., Hancox, J., Li, W. *et al.* The future of digital health with federated learning. *npj Digit Med*. 2020;3(1):119.

[31] Tolpegin V., Truex S., Gursoy M.E. and Liu L. Data poisoning attacks against federated learning systems. *Lecture Notes in Computer Science* 2020;12308LNCS:480–501.

[32] Barrett J. and Keat N. Artifacts in CT: recognition and avoidance. *Radiographics*. 2004;24(6):1679–1691+1796.

[33] Noda C., Ambale Venkatesh B., Wagner J.D., Kato Y., Ortman J.M., Lima J. A.C. Primer on commonly occurring MRI artifacts and how to overcome them. *Radiographics*. 2022;42(3):E102–E103.

Chapter 12

Spatial cognition by the visually impaired: image processing with SIFT/BRISK-like detector and two-keypoint descriptor on Android CameraX

Dmytro Zubov[1], *Ayman Aljarbouh*[1], *Andrey Kupin*[2] *and Nurlan Shaidullaev*[1]

Over 250 million people worldwide are visually impaired or blind today. Guide dogs and white canes are still the most preferred methods of aid because existing smart assistive devices are often unaffordable and/or not accurate/fast enough. In this work, Android phones are employed to provide an affordable and reliable solution to support the spatial cognition of the visually impaired. A new image processing approach has been proposed. In preprocessing, a two-dimensional Gaussian function downscales the image and removes noise. A pair of keypoints is selected from a maximum of 300 keypoints localized by the SIFT method. Their coordinates allow the calculation of the orientation and scale of the image descriptor, employing locations of these two keypoints and the distance between them. Then, the BRISK algorithm designs a two-keypoint binary descriptor of specific shape with 291 points located near two keypoints of interest (for example, two centers of the human eye pupils) in an image with a maximum dimension size of 180 pixels. The Hamming distance threshold value and the Hamming distance value after five AdaBoost weak classifiers were chosen to be 19,317 and 18,558, which shows a 99.5% recognition accuracy for the validation sample and 100% in real life experiments. To implement the proposed approach, a multi-threaded Java Android application was developed to improve spatial cognition by the visually impaired and blind using the CameraX library. The execution time is approximately 0.8 s on the smartphone Doogee S96 Pro for a knowledge base with six template objects.

12.1 Introduction

Up-to-date smartphones with high-quality cameras, fast processors, and different mobile OSs, along with restricted access to assistive hardware, such as Raspberry

[1]Department of Computer Science, University of Central Asia, Kyrgyzstan
[2]Department of Computer Systems and Networks, Kryvyi Rih National University, Ukraine

Pi wearable devices [1–3], increase the need for the development of new and improvement of existing spatial cognition technologies and soft-/hardware for VIBs, of which there are over 250 million worldwide today. [4] The new COVID-19 pandemic reality provides an additional stimulus to Java Android applications [5–7], because of the availability of low-cost Android smartphones and their market share is above 70% for over four years.

Different assistive mobile applications for spatial cognition, such as Google Maps for GPS navigation and Google Lookout [8] for identification of banknotes, packaged foods, documents, and other objects, have become freely accessible by VIBs in the last decade. However, the detection and recognition of images remain challenging because of the unknown category, scale, and orientation of the image, optical distortion, noise, different light intensities, smartphone camera parameters, and many other factors. In image processing, feature descriptors are a popular research topic and a challenging task because different approaches are employed to detect and recognize heterogeneous objects in images and video streams. A detector locates keypoints in an image that meets predefined requirements. Image matching uses feature descriptors; for example, a part of a human face can uniquely describe the template image. The Lucas–Kanade algorithm [9] has high computational complexity because it minimizes the sum of squared errors between the template and the current image scaled to the size of the template.

In general, software applications implement algorithms with either strict requirements to maximize the recognition accuracy or to minimize execution times. The first case can be represented by the SIFT approach [10–12] that precisely identifies keypoints despite the unknown scale and orientation of the image with random variations in brightness or color information. However, existing real-life applications of SIFT require high-performance computing [13]. The second case can be represented by the combination of FAST [14] and BRISK [15] approaches, which have much less computational complexity but much less robust to changes in image rotation and scale. Some real-time applications such as SLAM [16] must employ probabilistic methods [17] for data association in image matching.

After detecting the keypoints, a descriptor is computed for each one of them. Subsequently, image classification algorithms are applied. Sometimes, VIBs provide only one photo of the template image that they want to recognize. When training data are lacking, machine learning algorithms, such as convolutional neural networks [18,19], SVM [20–22], and k-nearest neighbors [21–23], are unfeasible. Syntactic methods [24] in machine learning show a recognition accuracy higher than 90% for the images of buildings and shop fronts. However, only a few papers (19%) presented methods to learn the structure of grammar from data [24]. Hence, machine learning algorithms are not discussed in this work.

The weighted majority voting scheme [25] is a promising method with a parallel structure that integrates multilayer perceptron classifiers and multinomial logistic regression employing three different accident datasets. However, other approaches, such as AdaBoost weak classifiers [26,27], have less computational complexity in terms of multistage analysis of different segments of the image.

AdaBoost classifiers can improve classification accuracy and reduce the computational complexity of image classification.

Computer vision libraries, e.g., open-source and cross-platform OpenCV [28,29], advance vision-based applications by providing optimized codes for basic vision infrastructure. However, the recognition accuracy is insufficient [30]. For instance, the ORB [31–33] and boosted efficient binary local image descriptor algorithms showed 63.20% and 77.57% of inliers, respectively.

12.1.1 Contribution

To develop a new mobile application that has a recognition accuracy rate of about 99% and an affordable price (zero in the ideal case), two problems have been solved in this work to support VIBs:

1. A new image processing approach has been proposed. In preprocessing, a two-dimensional Gaussian function downscales the image and removes noise. A pair of keypoints is selected from a maximum of 300 keypoints localized by the SIFT method. The BRISK algorithm designs a two-keypoint binary descriptor of specific shape with 291 points located near two keypoints of interest in an image with a maximum dimension size of 180 pixels. The Hamming distance threshold value and the Hamming distance value after five AdaBoost weak classifiers were chosen to be 19,317 and 18,558, which shows a 99.5% recognition accuracy for the validation sample and 100% in real-life experiments.
2. A multi-threaded Java Android application was developed to improve spatial cognition by the VIB using the CameraX library [7]. The execution time is approximately 0.8 s on the smartphone Doogee S96 Pro for a knowledge base with six template objects.

12.2 Related work

VIBs have primarily relied on haptic or auditory cues [34] because assistive devices are not generally accepted and standardized and are sometimes unreliable and unaffordable [35,36]. Internet of Things/Everything [37–39] implements edge computing, including image processing. In [40], 25 assistive devices were compared using range (>5 m, 1 m–5 m, <1 m), coverage (outdoor/indoor/both), time (night/day/both), real-time/non-real-time mode, and object type (static/dynamic). Seventy two percent of the presented products have a minimum of three characteristics that do not meet the customer requirements, e.g., the obstacle detection distance is less than 5 m because of the ultrasonic limits; device Eye Subs detects only static objects outdoors. Prices are still high for commercially released assistive devices with computer vision. The OrCam MyEye wearable assistive device [41] is an example that costs USD 3,500. It recognizes the text and speaks it aloud to the VIB. Many other papers, for example [42–46], were mainly related to navigation aids employing computer vision.

SIFT [10–12] is a pioneering feature description method invented by David Lowe [10,12]. In SIFT, the DoG is first used to determine the gradient changes, and then a feature description is conducted. Despite the precise identification of keypoints, some reports have shown an insufficient recognition accuracy, for example 53.0% for automatic Chinese food identification [47].

SURF [48–50] was proposed by Herbert Bay. In SURF, a Hessian matrix locates keypoints and employs an integral image to be scale invariant. SURF is faster than SIFT because it only requires a Hessian matrix of different sizes to represent different scales (after precomputing the integral image). Then, keypoints around each keypoint are identified, and the direction of most keypoints is considered as the direction of the keypoint to achieve rotational invariance. The feature descriptor is based on the sum of Haar-wavelet responses around the keypoint.

Rosten and Drummond proposed a FAST corner detector [14]. In FAST, the segment test criterion operates with a circle of 16 pixels around the corner candidate. Mair *et al.* also put forward the AGAST corner detector [51]. AGAST switches between two or more binary decision trees after the pixel neighborhood changes. Both algorithms quickly search for initial keypoints and use binary descriptors; however, they are sensitive to blur and noise.

Tola *et al.* [52] proposed the descriptor DAISY, originally designed for dense utilization [13,52,53]. In [54], the local invariant feature is based on the phase congruency model that is invariant to contrast variation and illumination; its main limitation is the high computational complexity. In the HOG method [55–58], measurements are made around keypoints with different orientations and radii. Then, different orientation gradients are estimated, and positive directions are used to design the descriptor as a flower with overlapping areas. In [53], the SVM algorithm was applied to the object classification after all DAISY local descriptors were transformed into fixed-length feature vectors using the bag-of-features model, and it was shown that the recognition accuracy varied from 44.17% to 48.61%.

The ORB algorithm proposed by Rublee *et al.* [31] improves the rotational invariance of BRIEF, the calculation of different keypoint directions, and the initial keypoint filtering mechanism. First, the FAST method locates keypoints in the image. Second, keypoints are filtered to identify the most reliable keypoints. Finally, the orientations of keypoints regions are calculated according to the intensity centroid of each keypoint to establish the binary descriptors.

The BRIEF algorithm [32,33] was proposed by Michael Calonder. In many cases, it outperforms SURF and SIFT in terms of recognition accuracy and speed. BRIEF computes pairwise pixel intensity comparisons to design an image patch as a binary string. In the comparisons, the spatial locations of pixels are random. The BRIEF algorithm is fast, and the descriptor length is adjustable. However, image patches must be smoothed to remove noise, the spatial sampling pattern needs to be rotated to satisfy rotation invariance, and the descriptor does not scale well because of discrete pixel sampling.

Leutenegger *et al.* proposed BRISK [15], which was inspired by BRIEF. It also computes pairwise comparisons to design a descriptor for the image patch. BRISK employs a fixed sampling pattern with circular regions and Gaussian smoothing instead of randomly selected sampling points.

Cronje presented the BFROST algorithm [59]. BRFOST is robust to noise and scales well because it employs the intensity of square regions for comparison instead of discrete pixels, but the sampling pattern is not rotation invariant.

Alahi *et al.* proposed the FREAK method [60] using a concept of the human retina to design descriptors and claimed this approach could show better results than BRISK, SURF, and SIFT. In FREAK [60–63], a cascade of binary strings is calculated by comparing image intensities in the retinal sampling pattern.

Gradient-based feature descriptors, e.g., SIFT, SURF, FAST, and DAISY, are mainly differentiated by small modifications in algorithms on how smoothed-gradient histograms are estimated. Binary feature descriptors, such as ORB, BRIEF, BRISK, BFROST, and FREAK, are mainly differentiated by small modifications in the noise filters applied and pairwise comparisons performed.

Local tetra pattern [64,65] represents the spatial structure of the local texture employing the direction of center grey pixel. However, this method needs additional research to identify its computational complexity.

Many other papers are related to keypoints finding and feature descriptors. Examples are as follows: MALCOM [66]; local descriptors based on principal components analysis with SIFT scale space peak selection, keypoint localization and orientation assignment [67]; HOG feature vectors with a linear SVM for object/non-object classification [68]; GLOH, an extension of the SIFT descriptor [69]; and HOOFR [70].

12.3 Methodology

The proposed approach (see Figure 12.1) starts with image capture on an Android CameraX. Then, the image is converted from RGB to greyscale using the luminosity function [71] and downscaled with noise removal by employing the two-dimensional Gaussian function [11]. The conjunction of downscaling and noise removal reduces the execution time. Keypoints are calculated using the SIFT-like algorithm with the subsequent selection of the best pair of keypoints suitable for the BRISK-like feature descriptor design. Image matching is implemented employing the Hamming distance [62]. A diagram of the proposed approach is shown in Figure 12.2.

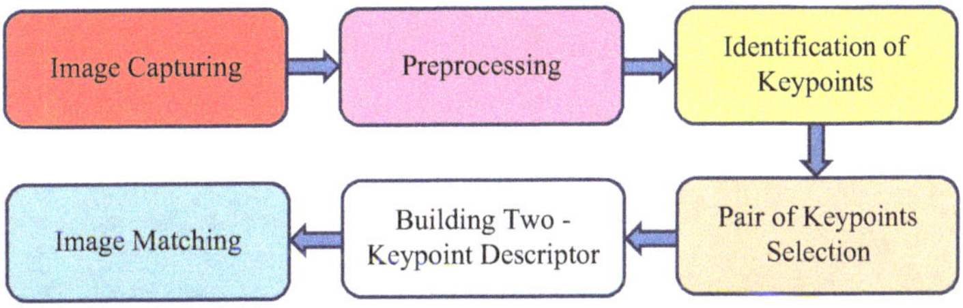

Figure 12.1 The architecture of the proposed approach

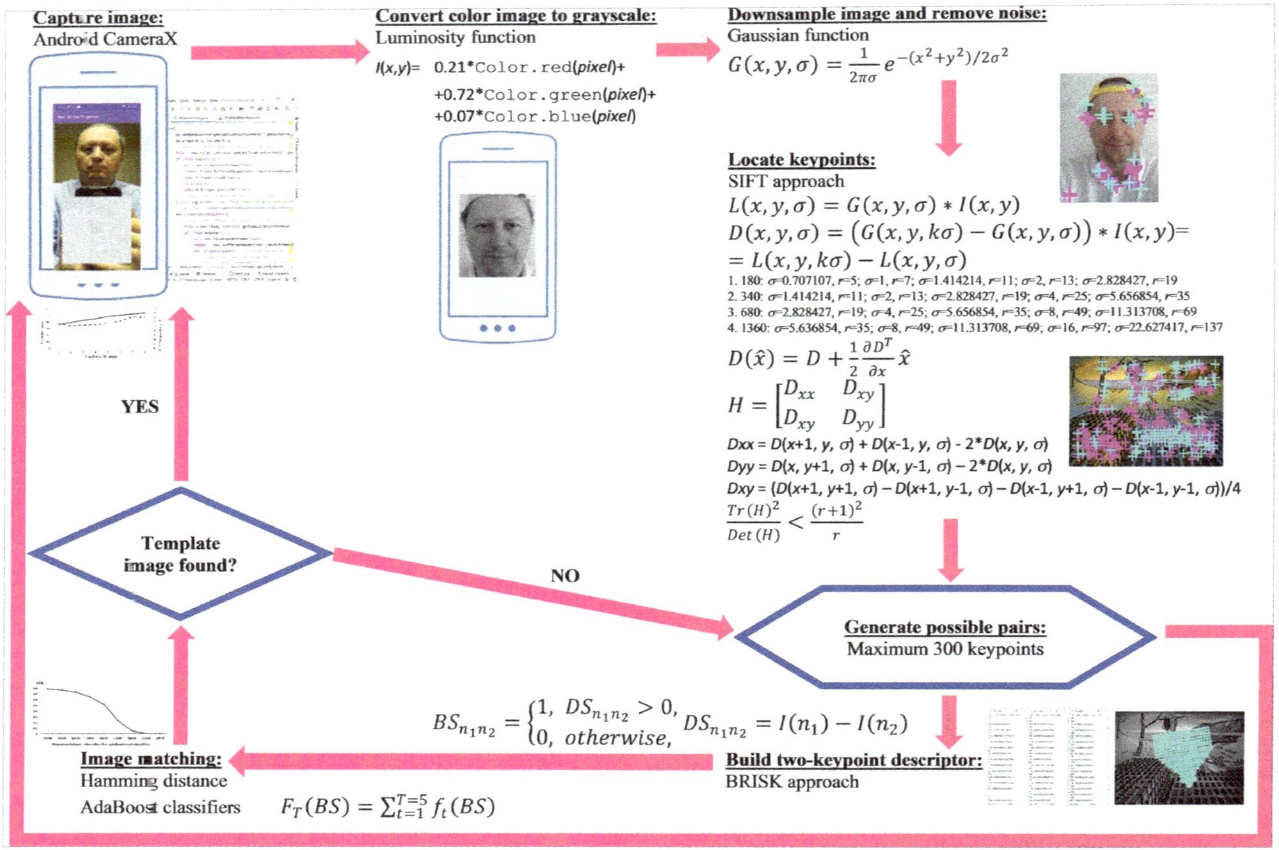

Figure 12.2 A diagram of the proposed approach

12.3.1 Problem formulation

In image matching, the feature descriptor is designed to be invariant to rotation, shear, translation, and scale, which are the four basic operators in affine transformation. In this work, the BRISK-like descriptor with two keypoints identified by the SIFT-like algorithm has specific shape with a set of points. Scale, rotation, and translation invariance are achieved by analyzing the region around these two keypoints, considering their locations and the distance between them. Shear invariance is partially achieved using the Hamming distance threshold value [71]. The classical problems of template image detection and recognition are solved using the cascade of AdaBoost weak classifiers [26].

12.3.2 Identification of all keypoints on the template image: SIFT-like approach

The two major steps in the identification of all keypoints in the template image using the SIFT-like approach are as follows:

1. Scale space construction: design of DoG and Gaussian pyramids.
2. Keypoint localization: keypoint candidates are selected from the extrema in scale space, and then keypoints are filtered according to their stability.

12.3.2.1 Scale space construction

To find the locations of keypoints, the cascade filter approach is employed. Scale-invariant locations are discovered by the scale function that searches for stable features in different scales. The filtered image $L(x, y, \sigma)$ is the product of convolution (i.e., the scale space function):

$$L(x,y,\sigma) = G(x,y,\sigma) * I(x,y), \tag{12.1}$$

where $G(x, y, \sigma)$ is the two-dimensional Gaussian function (i.e., the scale space convolution kernel), $I(x, y)$ is the input greyscale image, σ is the population standard deviation, "*" is the convolution operation, x and y are coordinates of kernel points.

The two-dimensional Gaussian function $G(x, y, \sigma)$ is as follows:

$$G(x,y,\sigma) = \frac{1}{2\pi\sigma} e^{-\left(x^2+y^2\right)/2\sigma^2}. \tag{12.2}$$

In this work, the RGB image is converted to a greyscale image $I(x, y)$ using the luminosity function [70] and Java methods `red`, `green`, and `blue` of the Android class `Color`:

$$I(x,y) = 0.21 * \texttt{Color.red}(\mathit{pixel}) + 0.72 * \texttt{Color.green}(\mathit{pixel}) + \\ + 0.07 * \texttt{Color.blue}(\mathit{pixel}), \tag{12.3}$$

where *pixel* is the smallest addressable element in a raster image represented by the RGB model [70] with red, green, and blue values.

To detect stable locations of keypoints in scale space, the DoG function $D(x, y, \sigma)$ is calculated from the difference between two neighbor scales separated by a constant multiplicative factor k convolved with the image $I(x, y)$:

$$D(x, y, \sigma) = (G(x, y, k\sigma) - G(x, y, \sigma)) * I(x, y) = \tag{12.4}$$

$$= L(x, y, k\sigma) - L(x, y, \sigma). \tag{12.5}$$

The DoG function approximates the scale-normalized Laplacian of Gaussian $\sigma^2\nabla^2 G$. Maxima and minima of $\sigma^2\nabla^2 G$ find the most stable image features compared to the gradient, Harris, or Hessian corner detectors.

To construct $D(x, y, \sigma)$, the input image is incrementally convolved with two-dimensional Gaussian functions using the population standard deviation σ to produce different images. The bottom image is initially convolved with a two-dimensional Gaussian function $G(x, y, \sigma)$ using σ, and then the convolution operation is reiterated with a further incremental smoothing of σ to obtain the second bottom image. The bottom DoG function is acquired by subtracting images—the second bottom from the bottom. These steps are repeated until $(s + 3)$ images (s is the number of intervals; $s = 2$ in this work) are generated and thus $(s + 2)$ images are in the stack of DoG images on each octave or pyramid level.

To get the next pyramid level, the initial image is downscaled by employing the Java method `createScaledBitmap` of the Android class `Bitmap`. Four pyramid levels are used in this work.

Four octaves (i.e., images with a maximum dimension size of 180, 340, 680, and 1360 pixels) are blurred five times to obtain five scales with the following population standard deviations σ and orders r of the square matrix in the Gaussian blur operator:

1. 180: $\sigma = 0.707107$, $r = 5$; $\sigma = 1$, $r=7$; $\sigma=1.414214$, $r = 11$; $\sigma = 2$, $r = 13$; $\sigma = 2.828427$, $r = 19$.
2. 340: $\sigma = 1.414214$, $r = 11$; $\sigma = 2$, $r = 13$; $\sigma = 2.828427$, $r = 19$; $\sigma = 4$, $r = 25$; $\sigma = 5.656854$, $r = 35$.
3. 680: $\sigma = 2.828427$, $r = 19$; $\sigma = 4$, $r = 25$; $\sigma = 5.656854$, $r = 35$; $\sigma = 8$, $r = 49$; $\sigma = 11.313708$, $r = 69$.
4. 1360: $\sigma = 5.636854$, $r = 35$; $\sigma = 8$, $r = 49$; $\sigma = 11.313708$, $r = 69$; $\sigma = 16$, $r = 97$; $\sigma = 22.627417$, $r = 137$.

12.3.2.2 Keypoint localization

After the scale space extrema are detected, low-contrast keypoints are suppressed and some are deleted which are located on the edges.

Detecting local extrema

To detect the local minima and maxima of the DoG function $D(x, y, \sigma)$ as keypoint candidates, each sample point is compared to its nine neighbors in scales above/below and the eight nearest neighbors in the current image. This point is selected as the keypoint candidate if the light intensity is greater or less than all neighbors under comparison.

To reliably identify extrema, the right frequency of sampling must be selected in the scale- and image-domains. Lowe [10,12] showed that three intervals at each pyramid level are the best, but two are also reliable. Three intervals are selected in this work. Lowe pointed out that the total number of keypoints increases if the total number of pyramid levels grows, but the percentage of correctness (i.e., repeatability) decreases. In this work, four levels are applied, as suggested by Lowe.

Rejecting low-contrast keypoints

Points are rejected (i.e., they are poorly localized along the edge or sensitive to noise) if they have low contrast. To reject low-contrast keypoints, a three-dimensional quadratic function fits the local sample points to find the location of the maximum. This approach uses the following Taylor expansion of the DoG function $D(x, y, \sigma)$:

$$D(x) = D + \frac{\partial D^T}{\partial x} + \frac{1}{2} x^T \frac{\partial^2 D^T}{\partial x^2} x, \tag{12.6}$$

where D and its derivatives are evaluated at the sample point, $\mathbf{x}=(x, y, \sigma)^{\mathrm{T}}$ is the offset from a particular point. The location of the extremum $\widehat{x}$ is determined by taking the derivative of D with respect to $\mathbf{x}$ and setting it to zero:

$$\widehat{x} = \frac{\partial^2 D^{-1}}{\partial x^2} \frac{\partial D}{\partial x}. \tag{12.7}$$

In practice, the Hessian matrix and derivative of D are approximated by the differences between neighbor sample points, solving a 3×3 linear system. The extremum lies closer to a different sample point if the offset $\widehat{x}$ is larger than 0.5 in any dimension. In this case, sample points are changed, and interpolation is performed near another point. The final offset $\widehat{x}$ is added to the location of its sample point to obtain an interpolated estimate for the extremum location.

By taking the function value at the extremum $D(\widehat{x})$, unstable extrema with low contrast are rejected. Lowe suggested deleting extrema with a value $|D(\widehat{x})|<0.03$ that was employed in this work. The extremum $D(\widehat{x})$ can be found by substituting (1.7) into (1.6):

$$D(\widehat{x}) = D + \frac{1}{2} \frac{\partial D^T}{\partial x} \widehat{x}. \tag{12.8}$$

Eliminating edge responses

Using only the low-contrast rejecting criterion is not sufficient to eliminate edge responses because the DoG function has a strong response at the edges. Lowe showed that edges can be detected using a 2×2 Hessian matrix [10].

A poorly defined extremum in the DoG function has a large principal curvature across the edge but a small curvature in the perpendicular direction. The Hessian matrix computes principal curvatures:

$$H = \begin{bmatrix} D_{xx} & D_{xy} \\ D_{xy} & D_{yy} \end{bmatrix}, \tag{12.9}$$

where derivatives are determined by taking differences between neighbor points [11]:

$$Dxx = D(x+1, y, \sigma) + D(x-1, y, \sigma) - 2 * D(x, y, \sigma), \tag{12.10}$$

$$Dyy = D(x, y+1, \sigma) + D(x, y-1, \sigma) - 2 * D(x, y, \sigma), \tag{12.11}$$

$$\begin{aligned} Dxy = {} & (D(x+1, y+1, \sigma) - D(x+1, y-1, \sigma) - D(x-1, y+1, \sigma) \\ & - D(x-1, y-1, \sigma))/4. \end{aligned} \tag{12.12}$$

To ensure that the ratio of principal curvatures is less than a certain threshold r, the following inequality must be true:

$$\frac{Tr(H)^2}{Det(H)} < \frac{(r+1)^2}{r}. \tag{12.13}$$

Lowe suggested r=10 (i.e., the left part of inequality (1.13) greater than 12.1) to eliminate keypoints.

Two examples of images with their octaves and keypoints are shown in Figures 12.3 and 12.4. In Figure 12.3, a photo of the corresponding author Dmytro Zubov is used (the initial image is of size 2560×1920 pixels):

(A) Scaled image of size 180×135 pixels with 48 keypoints (sign "+" designates the keypoint employing two lines of ten-pixel length and two-pixel width; two different colors represent two intervals: fuchsia—first, second, and third scales, turquoise—second, third, and fourth scales).
(B) Scaled image of size 340×255 pixels with 77 keypoints.
(C) Scaled image of size 680×510 pixels with 101 keypoints.
(D) Scaled image of size 1360×1020 pixels with 118 keypoints.

The execution time varies significantly at different program starts. It was approximately 5 min on the smartphone Doogee S96 Pro for the image and four octaves presented in Figure 12.3.

Figure 12.4 shows a sample image without a human face—oil on canvas painting "The Disintegration of the Persistence of Memory" by Spanish surrealist Salvador Dalí (the initial image is of size 2048×1536 pixels):

(A) Scaled image of size 180×135 pixels with 439 keypoints.
(B) Scaled image of size 340×255 pixels with 581 keypoints.
(C) Scaled image of size 680×510 pixels with 950 keypoints.
(D) Scaled image of size 1360×1020 pixels with 1228 keypoints.

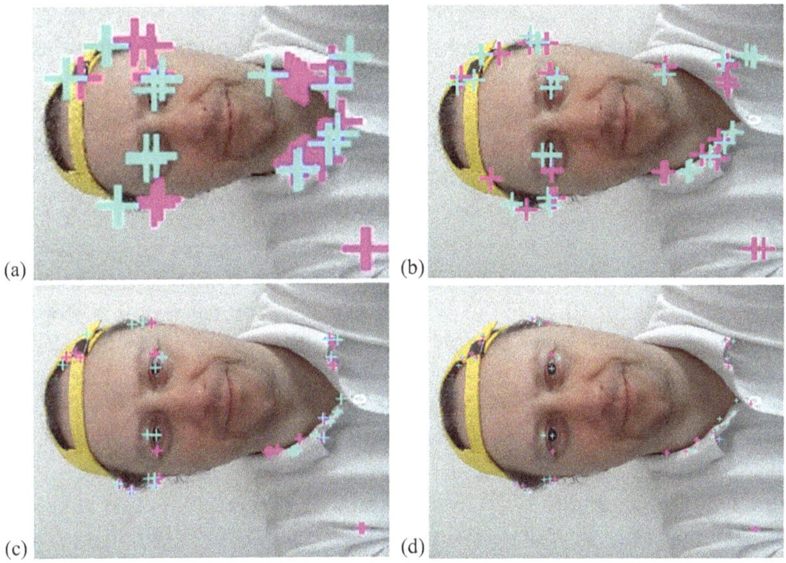

Figure 12.3 Four octaves with keypoints: an example of the image with a human face

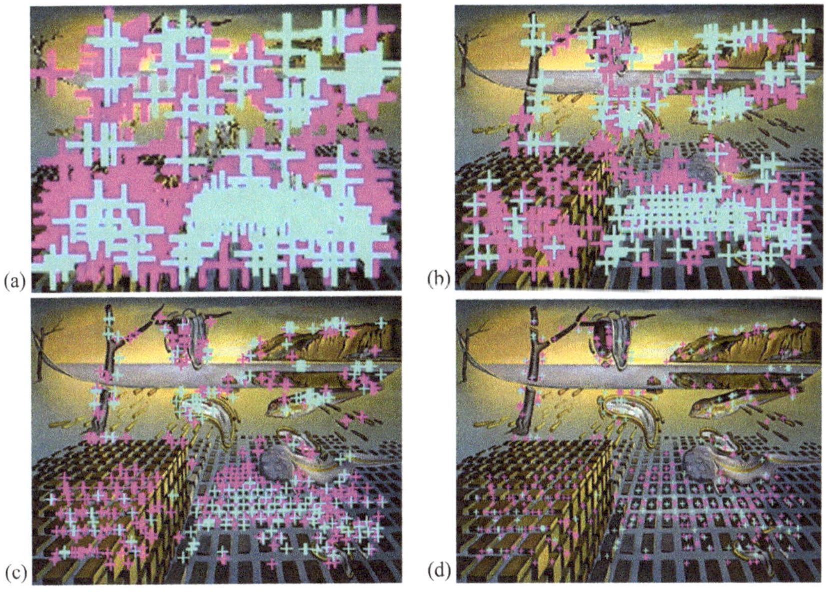

Figure 12.4 Four octaves with keypoints: an example of the image without a human face

The execution time varies significantly at different program starts. It was approximately 6 min on the smartphone Doogee S96 Pro for the image and four octaves presented in Figure 12.4.

12.3.3 Identification of two keypoints to design the template image feature descriptor: BRISK-like approach

To recognize specific object in an image, the traditional approach is to find the local image gradients and intensities around only one keypoint using a normalized correlation measure. Two classical problems, finding the scale and orientation, should be solved before that. In this work, a two-keypoint descriptor is proposed to model objects as a collection of parts [73] employing a 291-point feature descriptor from the template image, followed by the BRISK-like [15] method to recognize specific object in the image. The scale and orientation are calculated using the locations of these two keypoints and the distance between them.

Two keypoints of interest can be located manually by a human expert and/or a machine learning algorithm [74]. In this work, an expert selects two keypoints of interest by analyzing keypoints with a stable location in different octaves; for example, two centers of the eye pupils are the natural choice for the human face. Figure 12.5 shows two greyscale images with a maximum dimension size of 1360 pixels with (keypoints: $x = 470$, $y = 561$; $x = 469$, $y = 312$) and without (keypoints: $x = 535$, $y = 367$; $x = 881$, $y = 312$) human faces. The grey values are calculated pixel-by-pixel using (12.3) after the image is downscaled by employing the Java method `createScaledBitmap` of the class `Bitmap`.

Currently, most binary descriptors do not consider the specificity of the objects in an image. This work assumes that human faces will be detected and recognized at least. Hence, the sampling pattern has specific shape that includes the eyes, eyebrows, nose, and center of the lips. As shown below, this sampling pattern can be applied to detect and recognize other objects, such as road signs and paintings. It includes 291 points that are divided into three groups: 0–24, 25–82 and 83–290 (see Figure 12.6). Figure 12.6 (D) shows the feature descriptor with all points. In Figure 12.6 with an example of a human face, two keypoints of interest are located

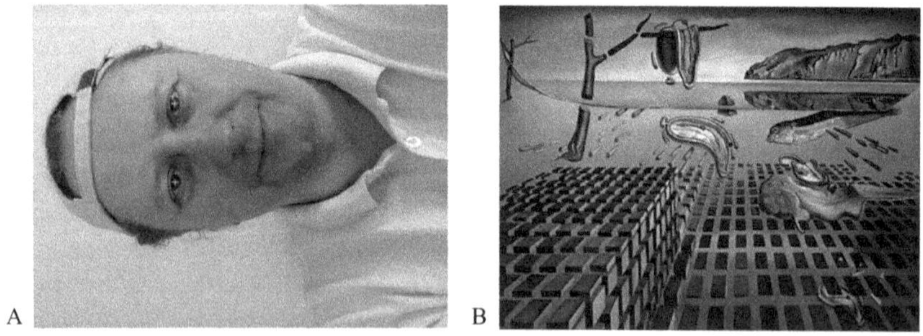

Figure 12.5 Examples of greyscale images of size 1360×1020 pixels with two keypoints of interest, with (A) and without (B) a human face

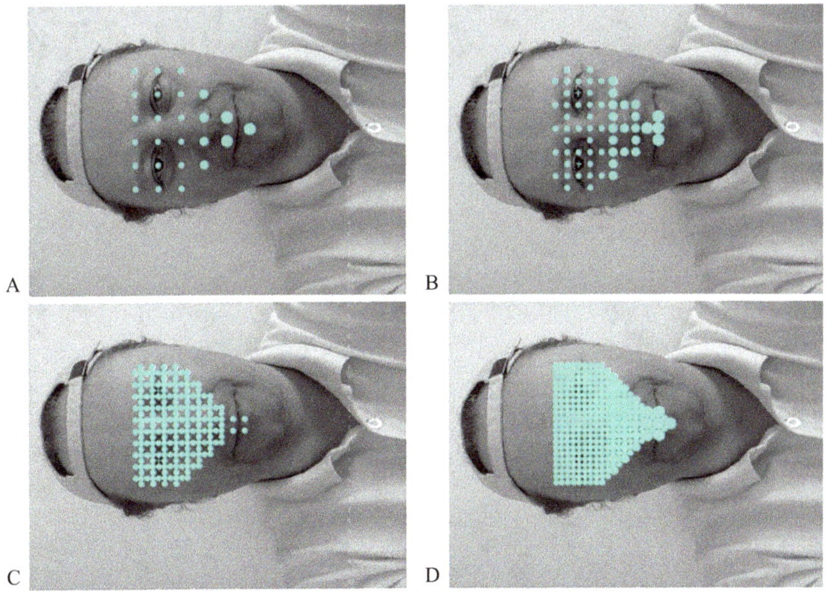

Figure 12.6 Different groups of points in a feature descriptor in the example with a human face: 0–24 (A), 25–82 (B), 83–290 (C), 0–290 (D)

in the centers of eye pupils. The distance between two points is calculated using the distance between two keypoints of interest. For instance, two points are on the line between the centers of eye pupils, and their distance to the closest point in the first group is 1/3 of the distance between two keypoints of interest. Each point n is characterized by the average pixel intensity $I(n)$ in a circle with specific radius. In this work, three different radii are employed: 1/24, 1/16, and 1/12 of the Euclidean distance $E_{n_1n_2}$ between two keypoints of interest n_1 and n_2.

Figure 12.7 shows the feature descriptor for the oil on canvas painting "The Disintegration of the Persistence of Memory" by Spanish surrealist Salvador Dalí.

Object detection and recognition are based on the binary string BS formed by comparing the average pixel intensity $I(n_1)$ at point n_1 with another average pixel intensity $I(n_2)$ at point n_2:

$$BS_{n_1n_2} = \begin{cases} 1, & DS_{n_1n_2} > 0, \\ 0, & otherwise, \end{cases} \tag{12.14}$$

where the string DS is calculated as follows:

$$DS_{n_1n_2} = I(n_1) - I(n_2). \tag{12.15}$$

For specific point n_1, absolute differences

$$abs(DS_{n_1n_2})$$

are sorted in descending order within the appropriate group:

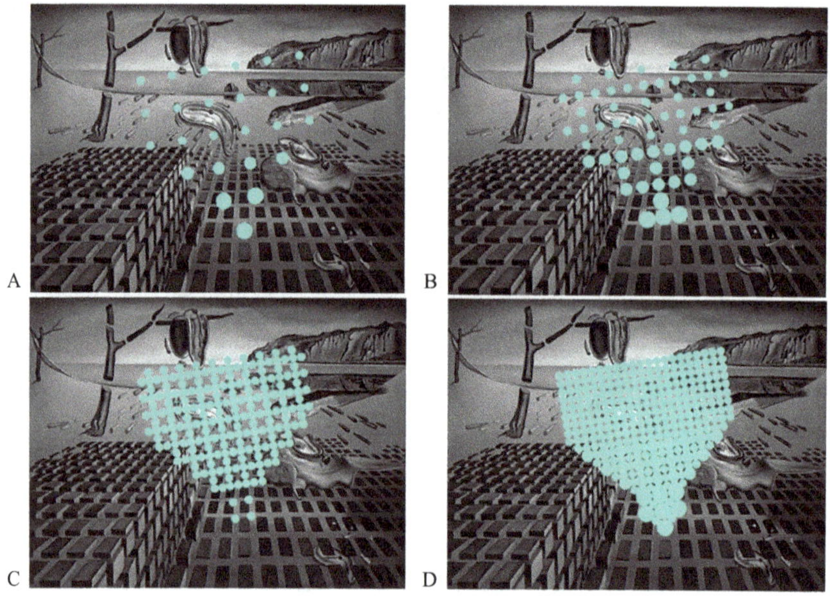

Figure 12.7 Different groups of points in a feature descriptor in the example without a human face: 0–24 (A), 25–82 (B), 83–290 (C), 0–290 (D)

1. Points 0–24: the first 12 absolute differences are considered in the feature descriptor for each point.
2. Points 25–82: the first 32 absolute differences are considered in the feature descriptor for each point.
3. Points 83–290: the first 100 absolute differences are considered in the feature descriptor for each point.

Hence, 22,956 = 300 + 1856 + 20,800 = 25*12 + 58*32 + 208*100 absolute differences

$$abs(DS_{n_1 n_2})$$

are considered in the feature descriptor. Figure 12.8 presents examples of differences for the first (A), second (B), and third (C) groups of feature descriptor points.

This approach is inspired by the BRIEF [33] and BRISK [15] methods, where image patches can be classified by employing a small number of pairwise intensity comparisons.

In the prototype, the knowledge base includes the following files placed on the internal storage of the smartphone:

1. File "root.txt" includes only the number N which is the number of objects in the knowledge base.

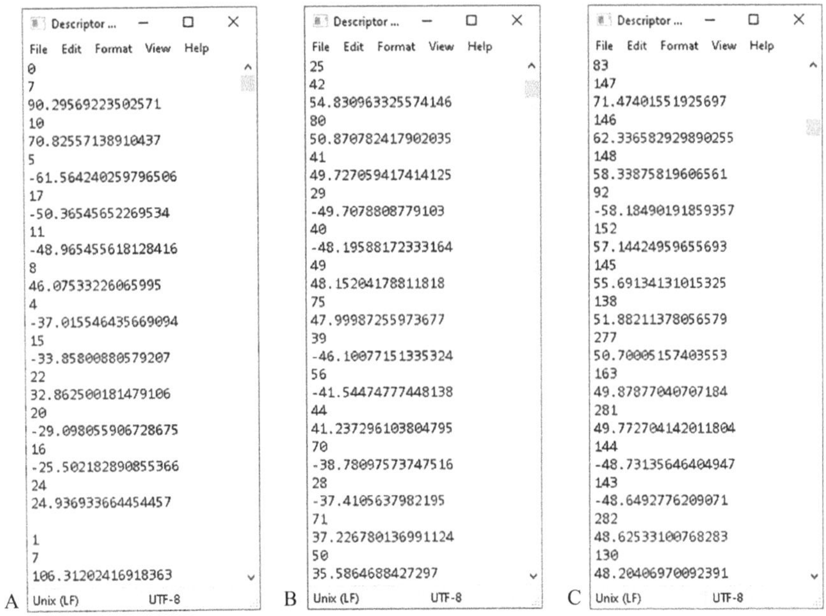

Figure 12.8 Examples of differences for the first (A), second (B), and third (C) groups of feature descriptor points

2. *N* files "d0.txt", "d1.txt", …, "dN.txt" include information on *N* image feature descriptors.
3. *N* files "Name0.mp3", "Name1.mp3", …, "NameN.mp3" are audio files with information regarding *N* template images.
4. File "Description.txt" includes a brief textual description of *N* template images.

In this work, the knowledge base consists of six template images (N=6): two male and one female faces, the oil on canvas painting "The Disintegration of the Persistence of Memory" by Spanish surrealist Salvador Dalí and two Ukrainian road signs—"Pedestrians only" and "No pedestrians." Images are available for download at https://drive.google.com/drive/folders/1GTMWxbv2Xn_JYujE-7ZCxG_0Uu5e_OU0?usp=sharing. The execution time was approximately 5 s on the smartphone Doogee S96 Pro for two images presented in Figures 12.6 and 12.7.

12.3.4 Fast binary feature matching

The Hamming distance is the similarity metric that is employed to calculate the distance between two binary strings *BS* calculated by (12.14) in this work. Five AdaBoost weak classifiers [26] identify objects using binary decision trees [51].

To speed up object detection and recognition, the maximum number of key-points is set to be 300 and the maximum dimension size of the image is set to be 180 pixels. This image is formed from the original image with 25% top/left/right/

Table 12.1 An example of the pseudocode for the parallel calculation of the scales

Algorithm: Parallel calculation of scales
Main process: Start the first thread to calculate the first scale
Main process: Start the second thread to calculate the second scale
Main process: Start the third thread to calculate the third scale
Main process: Start the fourth thread to calculate the fourth scale
Main process: Calculate the fifth scale
Main process: Wait for the first thread to be terminated
Main process: Wait for the second thread to be terminated
Main process: Wait for the third thread to be terminated
Main process: Wait for the fourth thread to be terminated

bottom margins to avoid the optical distortion effect [75,76]. In this work, a Gaussian kernel is used to downscale the image [77,78]: the order of the Gaussian square matrix equals to integer number calculated as the double Euclidean distance between two points in the original image with 180-point grid plus 1, the population standard deviation σ equals to the order of Gaussian matrix divided by 8. Four parallel computational threads form an image with a maximum dimension size of 180 pixels to reduce the execution time.

The image with a maximum dimension size 180 pixels is blurred five times to obtain five scales with the following population standard deviation σ and an order r of the square matrix in the Gaussian blur operator: $\sigma = 0.353553$, $r = 5$; $\sigma = 0.5$, $r = 5$; $\sigma = 0.707107$, $r = 7$; $\sigma = 1$, $r = 9$; $\sigma = 1.414214$, $r = 11$. Five parallel computational threads calculate the scales to reduce the execution time. An example of the pseudocode for the parallel calculation of the scales is shown in Table 12.1.

The four DoG images are calculated using (12.5) in four parallel computational threads to reduce the execution time.

Two parallel computational threads identify keypoints of interest in first/second/third and second/third/fourth scales. If the number of identified keypoints is greater than 300, only 300 keypoints with Euclidean distances closer to the center of the image are considered.

Initially, the feature matching algorithm excludes keypoint pairs with feature descriptor points that lie outside the image borders. Then, the AdaBoost classifier is applied:

$$F_T(BS) = \sum_{t=1}^{T=5} f_t(BS), \tag{12.16}$$

where T=5 and each $f_t(BS)$ is a weak classifier that takes a binary string BS as input and returns the Hamming distance between this sequence and the template, indicating the class of the object. If the Hamming distance is greater than the Hamming distance threshold value [72], the analyzed image belongs to the template class.

The first weak classifier, $t = 1$, in the first group of points 0-24 with 12 differences is the Hamming distance $f_1(BS)$ between two sequences designed by two keypoints of interest 7 and 10, which represent two centers of the eye pupils on the human face. If the Hamming distance is greater than 20, the analyzed image can be considered as of the template class.

The second weak classifier, $t = 2$, in the first group of points 0-24 with 12 differences is the Hamming distance $f_2(BS)$ between two sequences designed using 23 points from 0 to 24 except 7 and 10. If the Hamming distance is greater than 240, the analyzed image can be considered as of the template class.

The third weak classifier, $t = 3$, in the second group of points 25–82 (the first two differences are considered) is the Hamming distance $f_3(BS)$ between two sequences designed using 58 points from 25 to 82. If the Hamming distance is greater than 93, the analyzed image can be considered as of the template class.

The fourth weak classifier, $t = 4$, in the second group of points 25–82 (the last 30 differences are considered) is the Hamming distance $f_4(BS)$ between two sequences designed using 58 points from 25 to 82. If the Hamming distance is greater than 1485, the analyzed image can be considered as of the template class.

The fifth weak classifier, $t = 5$, in the third group of points 83–290 with 100 differences is the Hamming distance $f_5(BS)$ between two sequences designed using 208 points from 83 to 290. If the Hamming distance is greater than 16720, the analyzed image can be considered as of the template class.

If the Hamming distance value after five AdaBoost weak classifiers is larger than the Hamming distance threshold value, the analyzed image belongs to the template class.

12.4 Implementation, results, and discussion

12.4.1 Implementation

Java Android application [5,6] titled "WeCanSeeTogether" implements the proposed approach to recognize different objects in front of VIBs. Three Java Android applications of API SDK level 29 have been developed in Android Studio 4.0:

1. Identification of all keypoints in the template image: SIFT-like approach. A screenshot of the Java Android application titled "InitialImageKeypoints" is shown in Figure 12.9 (A). Image processing is invoked in the event `onCreate`.
2. Identification of two keypoints of interest to design the template image feature descriptor: BRISK-like approach. A screenshot of the Java Android application titled "DisplayTwoKeypoints" is shown in Figure 12.9 (B). Image processing is invoked in the event `onCreate`.
3. Detection and recognition of the template object in the current image: a combination of SIFT and BRISK approaches. A screenshot of the multi-threaded Java Android application titled "WeCanSeeTogether" is shown in Figure 12.9 (C). Image processing begins when the end-user clicks a button on the mobile application activity. CameraX API [7] is employed to accept a surface for

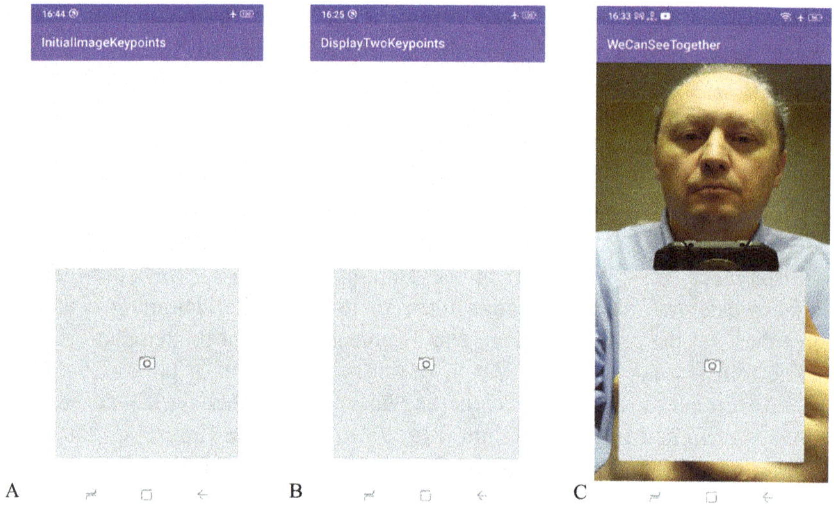

Figure 12.9 Screenshots of Java Android applications titled "InitialImageKeypoints" (A), "DisplayTwoKeypoints" (B), and "WeCanSeeTogether" (C)

displaying a preview (the use case "Preview") and to capture/save photos (the use case "Image capture") in this work. The recognized object (if found) is then pronounced to the VIB via the smartphone headphone.

Smartphone Doogee S96 Pro was mainly used in experiments: OS—Android 10; processor—Helio G90; battery—6350 mAh; memory—8 GB + 128 GB; display—6.22 inch; rear camera—48 MP; rugged standards—IP68/IP69K/MIL-STD-810G. The presented approach was also successfully tested on smartphones Samsung M31 SM-M315F/DSN with Android 12 OS and Samsung Galaxy A52 SM-A52F/DS with Android 11 OS.

12.4.2 Results and discussion

The image matching algorithm with the ensemble of five above-described AdaBoost weak classifiers was tested on the validation sample with 200 freely usable images downloaded from the website https://unsplash.com: 100 images were found with the keyword "face"; 100 images were found with the keyword "nature." Images are available for download at https://drive.google.com/drive/folders/1GTMWxbv2Xn_JYujE-7ZCxG_0Uu5e_OU0?usp=sharing. Figure 12.10 shows the FPR [79] for the above-described knowledge base with six template images at different Hamming distance values after five AdaBoost weak classifiers. The maximum value of $F_5(BS)$ is 20,217 for all false positives and 19,317 is the preceding value. Hence, the Hamming distance threshold value is chosen to be 19,317, which shows a 99.5% recognition accuracy on the validation sample. Real-life experiments showed a 100% recognition accuracy.

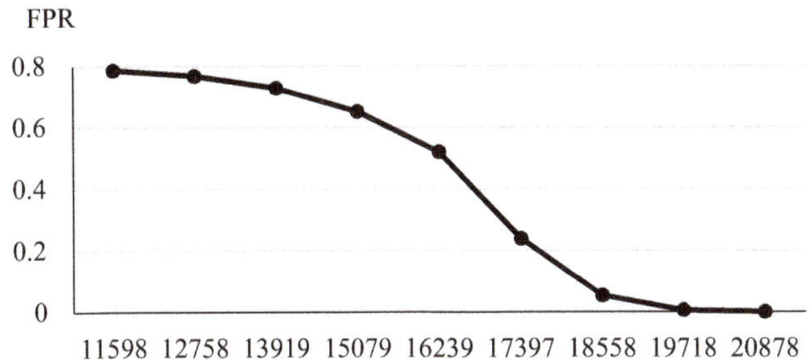

Figure 12.10 Caption FPR for a knowledge base with six template images at different Hamming distance values after five AdaBoost weak classifiers

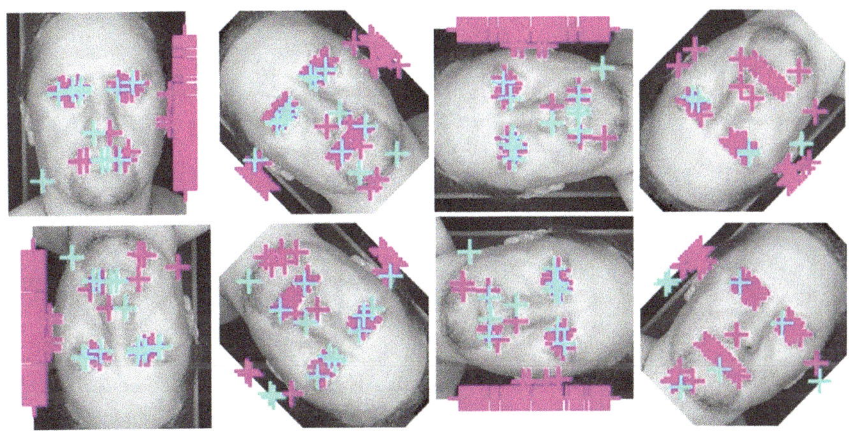

Figure 12.11 Experiment with successful identification of the human face with rotation and scaling: Greyscale images with keypoints

The above-stated Hamming distance value 18,558 is employed in five AdaBoost weak classifiers (1.16) because it is close to the Hamming distance threshold value 19,317. The maximum range that the algorithm can tolerate for image scale difference is determined by the Hamming distance threshold value 19,317 in this work. If the image descriptor is not big enough and the Hamming distance cannot reach the threshold value 19,317, the template image is not detected.

An example of the experiment with the successful identification of the human face with rotation and scaling in the image with dimensions 5312 and 5056 pixels is shown with keypoints in Figure 12.11. An example of successful identification of a human face in a real-life experiment is presented in Figure 12.12.

Android application "WeCanSeeTogether" receives images of different dimensions because of the various cameras installed on smartphones. Examples are as follows: 3264×2448 pixels on Doogee S96Pro, 4624×3468 pixels on Samsung M31 SM-M315F/DSN, 4032×3024 pixels on Samsung Galaxy A52 SM-A52F/DS.

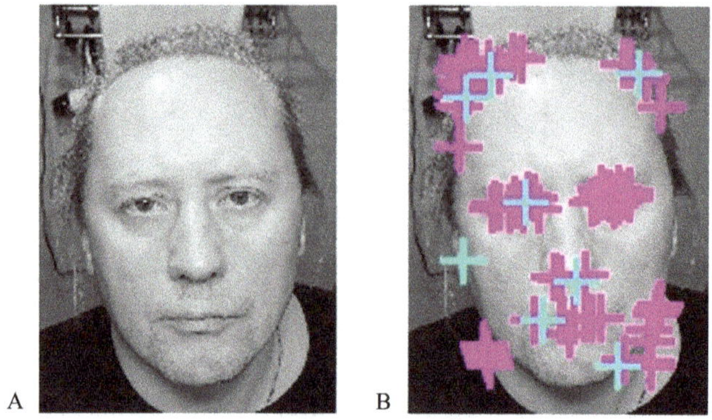

Figure 12.12 An example of successful identification of a human face: Images without (A) and with (B) keypoints

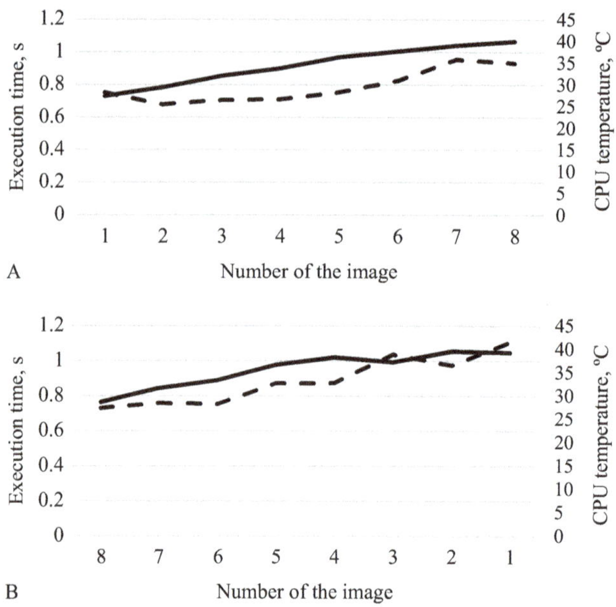

Figure 12.13 Dependencies between the execution time (solid line), the CPU temperature (dashed line) and the number of the image in Figure 12.11 for the smartphone Doogee S96Pro: straight (A) and reverse (B) orders

Hence, the execution time depends on the smartphone model since the larger image needs more computational operations to be converted to the target size of 180×135 pixels. Figure 12.13 depicts the dependencies between the execution time (starts after the image downsampling and ends with the image matching), the CPU temperature, and the number of the image in Figure 12.11 (the top left image is 1; the bottom right image is 8) for the smartphone Doogee S96Pro. Figure 12.13 (A) uses straight order, and Figure 12.13 (B) employs reverse order of images. The Android application "CPU Monitor" [80] measures CPU temperature that is continuously increased because experiments start one after another. The execution time is approximately 0.8 s in the presented charts. The values are averages from three consecutive experiments.

12.5 Conclusions

In this work, a feature-based detection and recognition algorithm that employs a combination of SIFT and BRISK ideas is proposed. The Java Android application with CameraX library was prototyped with an emphasis on spatial cognition by VIBs: half of the mobile application activity area is occupied by the button that invokes the image processing programming code.

Three Java Android applications were developed:

1. Java Android application titled "InitialImageKeypoints": Identification of all keypoints in the template image using the SIFT-like approach.
2. Java Android application titled "DisplayTwoKeypoints": Identification of two keypoints of interest to design the template image feature descriptor using the BRISK-like approach.
3. Multi-threaded Java Android application titled "WeCanSeeTogether": Detection and recognition of the template object in the current image using SIFT and BRISK approaches.

The feature descriptor is designed from 291 points that are divided into three groups (0–24, 25–82, and 83–290) and located near two keypoints of interest (e.g., two centers of the human eye pupils) in the image with a maximum dimension size of 180 pixels. The coordinates of two keypoints of interest allow the calculation of the orientation and scale of the image feature descriptor.

Initially, the feature matching algorithm excludes the keypoint pairs with the feature descriptor points outside the image borders. Then, five AdaBoost weak classifiers are applied on groups of points 0–24, 25–82, and 83–290 employing the Hamming distance that must be greater than 18,558. The Hamming distance threshold value was chosen to be 19,317 for all groups of points, which shows a 99.5% recognition accuracy on the validation sample and 100% in real-life experiments. The execution time was approximately 0.8 s on the smartphone Doogee S96 Pro for a knowledge base with six template objects.

The most likely prospects for further development of this work are the automatic identification of two keypoints of interest on the template image and parallelized multi-target detection and recognition with prioritized analysis of the most frequent objects in the AdaBoost classifier.

Abbreviations

AGAST:	Adaptive and generic accelerated segment test
API:	Application programming interface
BFROST:	Binary features from robust orientation segment tests
BRIEF:	Binary robust independent elementary features
BRISK:	Binary robust invariant scalable keypoints
CPU:	Central processing unit
DoG:	Difference of Gaussians
FAST:	Features from accelerated segment test
FPR:	False positive rate
FREAK:	FAST retina keypoint
GLOH:	Gradient location and orientation histogram
HOG:	Histogram of oriented gradients
HOOFR:	Hessian ORB – overlapped FREAK
MALCOM:	Machine-learned compact descriptor
ORB:	Oriented FAST and Rotated BRIEF
OS:	Operating system
RGB:	Red-Green-Blue model
SDK:	Software development kit
SIFT:	Scale-invariant feature transform
SLAM:	Simultaneous localization and mapping
SURF:	Speeded up robust features
SVM:	Support vector machine
VIB:	Visually impaired and blind

Data availability

The data supporting the conclusions of this work are available from the corresponding author upon request.

Conflicts of interest

The authors declare no conflicts of interest for publication.

Acknowledgments

This work and the research behind it have the support of the universities where the authors have been conducting the presented project. The authors sincerely appreciate the management and colleagues of the University of Central Asia and Kryvyi Rih National University for their patience and kind assistance in the completion of this work.

References

[1] Shilpashree K.S., Lokesha H., and Shivkumar H. 'Implementation of image processing on Raspberry Pi'. *International Journal of Advanced Research in Computer and Communication Engineering*. 2015; **4**(5): 199–202.

[2] Pajankar A. *Raspberry Pi Image Processing Programming: Develop Real-Life Examples with Python, Pillow, and SciPy*. California: Apress; 2017. p. 145.

[3] Kirpan O.R., Baviskar P.I., Khawase S.D., Mankar A.S., and Ramteke K.A. Object detection on Raspberry Pi. *International Journal of Engineering Science and Computing*. 2017; 7(3): 5360–5362.

[4] World Health Organization. Blindness and Visual Impairment [online]. 2022. Available from http://www.who.int/news-room/fact-sheets/detail/blindness-and-visual-impairment [Accessed 28 Jan 2023].

[5] Martinez-Alpiste I., Golcarenarenji G., Wang Q., and Alcaraz-Calero J.M. 'Smartphone-based real-time object recognition architecture for portable and constrained systems'. *Journal of Real-Time Image Processing*. 2022; 19: 103–115.

[6] Bhandari S., Maid S., Malve S., Thakkar H., and Sonawane N. 'Object and scene recognition in an Android based mobile application'. *International Journal of Advanced Research in Computer and Communication Engineering*. 2015; 4(4): 185–188.

[7] Google Developers. CameraX Architecture [online]. 2022. Available from https://developer.android.com/training/camerax/architecture [Accessed 28 Jan 2023].

[8] Al-Heeti A. Google Expands Lookout App for People who are Blind or Vision-impaired [online]. 2020. Available from https://www.cnet.com/tech/mobile/google-expands-lookout-accessibility-app-across-android-adds-new-features/. [Accessed 28 Jan 2023].

[9] Baker S. and Matthews I. 'Lucas-Kanade 20 years on: a unifying framework'. *International Journal of Computer Vision*. 2004; 56: 221–255.

[10] Lowe D.G. 'Distinctive image features from scale-invariant keypoints'. *International Journal of Computer Vision*. 2004; 60(2): 91–110.

[11] Rey-Otero I. and Delbracio M. 'Anatomy of the SIFT method'. *Image Processing On Line*. 2014; 4: 370–396.

[12] Lowe D.G. *Method and Apparatus for Identifying Scale Invariant Features in an Image and Use of Same for Locating an Object in an Image. U.S. Patent No. US006711293B1*, 2004.

[13] Chatoux H., Lecellier F., and Fernandez-Maloigne C. 'Comparative study of descriptors with dense key points'. *Proceedings of the 23rd International Conference on Pattern Recognition*; Cancún, Mexico, Dec 2016. New York: IEEE; 2016. pp. 1988–1993.

[14] Rosten E. and Drummond T. 'Machine learning for high-speed corner detection'. In Leonardis A., Bischof H., and Pinz A. (eds.). *Proceedings of the 9th European Conference on Computer Vision*; Graz, Austria, May 2006. New York: Springer; 2006. pp. 430–443.

[15] Leutenegger S., Chli M., and Siegwart R.Y. 'BRISK: binary robust invariant scalable keypoints'. *Proceedings of the IEEE International Conference on Computer Vision*; Barcelona, Spain, Nov 2011. New York: IEEE; 2011. pp. 2548–2555.

[16] Davison A.J., Molton N.D., Reid I., and Stasse O. 'MonoSLAM: real-time single camera SLAM'. *IEEE Transactions on Pattern Analysis and Machine Intelligence*. 2007; 29(6): 1052–1067.

[17] Chli M. and Davison A.J. 'Active matching'. *Proceedings of the 10th European Conference on Computer Vision*; Marseille, France, Oct 2008. New York: Springer; 2008. pp. 72–85.

[18] Ponzio F., Urgese G., Ficarra E., and Di Cataldo S. 'Dealing with lack of training data for convolutional neural networks: the case of digital pathology'. *Electronics*. 2019, 8(3): 256.

[19] Shorten C. and Khoshgoftaar T.M. 'A survey on image data augmentation for deep learning'. *Journal of Big Data*. 2019; 6: 60.

[20] Hsu C-W. and Lin C-J. 'A comparison of methods for multiclass support vector machines'. *IEEE Transactions on Neural Networks*. 2002; 13(2): 415–425.

[21] Vani V. and Geetha M.K. 'Automatic tumor classification of brain MRI images'. *International Journal of Computer Sciences and Engineering*. 2016; 4(10): 144–151.

[22] Kambi Beli I.L. and Guo C. 'Enhancing face identification using local binary patterns and K-nearest neighbors'. *Journal of Imaging*. 2017, 3(3): 37.

[23] Song Y., Kong X., and Zhang C. 'A larges-scale K-nearest neighbor classification algorithm based on neighbor relationship preservation'. *Wireless Communications and Mobile Computing*. 2022, 2022: 7409171.

[24] Dias Pedro R.W., Nunes F.L.S., and Machado-Lima A. 'Using grammars for pattern recognition in images: a systematic review'. *ACM Computing Surveys*. 2013; 46(2): 26.

[25] Awan S., Mehmood Z., Nazeer Chaudhry H., et al. 'Profiling casualty severity levels of road accident using weighted majority voting'. *Computers, Materials & Continua*. 2022; 71(3): 4609–4626.

[26] Viola P. and Jones M. 'Rapid object detection using a boosted cascade of simple features'. *Proceedings of the IEEE Computer Society Conference on Computer Vision and Pattern Recognition*; Kauai, HI, USA, Dec 2001. New York: IEEE; 2001. pp. 511–518.

[27] Periyasamy N., Thamilselvan P., and Sathiaseelan J.G.R. 'A comparative study of ANN, k-means and AdaBoost algorithms for image classification'. *Proceedings of the International Conference on Innovations in Information, Embedded and Communication Systems*; Coimbatore, India, March 2015. Coimbatore, India: Karpagam College of Engineering; 2015. pp. 1–5.

[28] Howse J. *Android Application Programming with OpenCV 3*. Birmingham: Packt Publishing; 2015. p. 190.

[29] Mishra S., Verma V., Akhtar N., Chaturvedi S., and Perwej Y. 'An intelligent motion detection using OpenCV'. *International Journal of Scientific Research in Science, Engineering and Technology*. 2022; **9**(2): 51–63.

[30] Suárez I. *Improving* Your Image Matching Results by 14 % with One Line of Code: Towards Data Science [online]. 2021. Available from https://towardsdatascience.com/improving-your-image-matching-results-by-14-with-one-line-of-code-b72ae9ca2b73 [Accessed 28 Jan 2023].

[31] Rublee E., Rabaud V., Konolige K., and Bradski G. 'ORB: an efficient alternative to SIFT or SURF'. In Leonardis A., Bischof H., and Pinz A. (eds.). *Proceedings of the Ninth European Conference on Computer Vision*; Graz, Austria, May 2006. New York: Springer; 2006. pp. 404–417.

[32] Wang C-M. 'Real time non-rigid surface detection based on binary robust independent elementary features'. *Journal of Applied Research and Technology*. 2015; 13(2): 297–304.

[33] Calonder M., Lepetit V., Strecha C., and Fua P. 'BRIEF: binary robust independent elementary features'. *Proceedings of the 11th European Conference on Computer Vision*; Heraklion, Crete, Greece, Sep 2010. New York: Springer; 2010. pp. 778–792.

[34] Bălan O., Moldoveanu A., Moldoveanu F., and Butean A. 'Auditory and haptic spatial cognitive representation in the case of the visually impaired people'. *Proceedings of the 22nd International Congress on Sound and Vibration*; Florence, Italy, Jul 2015. New York: Curran Associates, Inc.; 2015. pp. 4396–4403.

[35] Zubov D. 'Mesh network of eHealth intelligent agents for visually impaired and blind people: a review study on Arduino and Raspberry Pi wearable devices'. In Taser P. Y. (ed.). *Emerging Trends in IoT and Integration with Data Science, Cloud Computing, and Big Data Analytics*. Hershey, PA: IGI Global; 2022. pp. 240–271.

[36] Lee J., Kim D., Ryoo H.-Y., and Shin B.-S. 'Sustainable wearables: wearable technology for enhancing the quality of human life'. *Sustainability*. 2016; 8(5): 466.

[37] Domingo M. C. 'An overview of the Internet of Things for people with disabilities'. *Journal of Network and Computer Applications*. 2012; 35: 584–596.

[38] Islam S.M.R., Kwak D., Kabir M.H., Hossain M., and Kwak K.-S. 'The Internet of Things for health care: a comprehensive survey'. *IEEE Access*. 2015; 3: 678–708.

[39] Farahani B., Firouzi F., Chang V., Badaroglu M., Constant N., and Mankodiya K. 'Towards fog-driven IoT eHealth: promises and challenges of IoT in medicine and healthcare'. *Future Generation Computer Systems*. 2018; 78(2): 659–676.

[40] Elmannai W. and Elleithy K. 'Sensor-based assistive devices for visually-impaired people: current status, challenges, and future directions'. *IEEE Sensors Journal*. 2017; 17(3): 565.

[41] Holton B. MyReader and MyEye from OrCam: Text and Item Recognition at the Touch of a Finger: AccessWorld 18(2) [online]. 2017. Available from https://www.afb.org/aw/18/2/15244, last accessed 2022/03/08 [Accessed 28 Jan 2023].

[42] Varde S. and Panse M.S. 'Computer vision based travel aid for blind'. *International Journal of Latest Technology in Engineering, Management & Applied Science*. 2015; 6(8): 58–63.

[43] Mali S., Padade S., Mote S., and Omkar R. 'An eye for a blind: assistive technology'. *International Research Journal of Engineering and Technology*. 2017; **4**(4): 2681–2685.

[44] Patil P.R. and Patil B.V. 'A review on a visually impaired person: challenging in future'. *International Journal of Scientific Research in Science, Engineering and Technology*. 2016; **2**(2): 1313–1316.

[45] Pooja S., Shimi S.L., and Chatterji S. 'A review on obstacle detection and vision'. *International Journal of Engineering Sciences & Research Technology*. 2015; **4**(1): 1–11.

[46] Poggi M. and Mattoccia S. 'A wearable mobility aid for the visually impaired based on embedded 3D vision and deep learning'. *Proceedings of the IEEE Symposium on Computers and Communication*; Messina, Italy, Jul 2016. New York: IEEE; 2016. pp. 208–213.

[47] Häggqvist V. and Lundberg P. Image Comparing and Recognition: Food Classification (Dissertation) [online]. 2015. Available from http://urn.kb.se/resolve?urn=urn:nbn:se:kth:diva-177474 [Accessed 28 Jan 2023].

[48] Oyallon E. and Rabin J. 'An analysis of the SURF method'. *Image Processing On Line*. 2015; 5: 176–218.

[49] Bay H., Tuytelaars T., and Gool L.V. 'SURF: speeded up robust features'. In Leonardis A., Bischof H., and Pinz A. (eds.). *Proceedings of the Ninth European Conference on Computer Vision*; Graz, Austria, May 2006. New York: Springer; 2006. pp. 404–417.

[50] Bilal M., Habib H.A., Mehmood Z., Yousaf R.M., Saba T., and Rehman A. 'A robust technique for copy-move forgery detection from small and extremely smooth tampered regions based on the DHE-SURF features and mDBSCAN clustering'. *Australian Journal of Forensic Sciences*. 2021; 53(4): 459–482.

[51] Mair E., Hager G.D., Burschka D., Suppa M., and Hirzinger G. 'Adaptive and generic corner detection based on the accelerated segment test'. *Proceedings of the 11th European Conference on Computer Vision*; Heraklion, Crete, Greece, Sep 2010. New York: Springer; 2010. pp. 183–196.

[52] Tola E., Lepetit V., and Fua P. 'DAISY: an efficient dense descriptor applied to wide-baseline stereo'. *IEEE Transactions on Pattern Analysis and Machine Intelligence*. 2010; 32(5): 815–830.

[53] Zhu C., Bichot C.-E., and Chen L. 'Visual object recognition using DAISY descriptor'. *Proceedings of the IEEE International Conference on Multimedia and Expo*; Barcelona, Spain, Jul 2011. New York: IEEE, 2011. pp. 1–6.

[54] Ye Y., Shan J., Hao S., Bruzzone L., and Qin Y. 'A local phase based invariant feature for remote sensing image matching'. *ISPRS Journal of Photogrammetry and Remote Sensing*. 2018; 142: 205–221.

[55] Sharma M. and Ghosh H. 'Histogram of gradient magnitudes: a rotation invariant texture-descriptor'. *Proceedings of the IEEE International Conference on Image Processing*; Québec City, Canada, Sept 2015. New York: IEEE, 2015. pp. 4614–4618.

[56] Chandrashekar T.R. and Gautam, A.K. 'Face recognition based on histogram of oriented gradients, local binary pattern and SVM/HMM classifiers'. *International Journal of Engineering Sciences & Research Technology*. 2014; 3(8): 344–352.

[57] Ye Y., Bruzzone L., Shan J., Bovolo F., and Zhu Q. 'Fast and robust matching for multimodal remote sensing image registration'. *IEEE Transactions on Geoscience and Remote Sensing*. 2019; 57(11): 9059–9070.

[58] Cheon M., Lee W., Hyun C.-H., and Park M. 'Rotation invariant histogram of oriented gradients'. *International Journal of Fuzzy Logic and Intelligent Systems*. 2011; 11(4): 293–298.

[59] Cronje J. BFROST: Binary Features from Robust Orientation Segment Tests Accelerated on the GPU [online]. 2011. Available from https://www.researchgate.net/publication/258888239_BFROST_Binary_Features_from_Robust_Orientation_Segment_Tests_accelerated_on_the_GPU [Accessed 29 Jan 2023].

[60] Alahi A., Ortiz R., and Vandergheynst P. 'FREAK: fast retina keypoint'. *Proceedings of the IEEE Conference on Computer Vision and Pattern Recognition*; Providence, RI, USA, Jun 2012. New York: IEEE, 2012. pp. 510–517.

[61] Tan S. Y., Arshad H., and Abdullah A. 'A new illumination invariant feature based on FREAK descriptor in RGB color space'. *Journal of Theoretical and Applied Information Technology*. 2016; 93(1): 207–214.

[62] Chen Y., Xu W., and Piao Y. 'Target matching recognition for satellite image based on the improved FREAK algorithm'. *Mathematical Problems in Engineering*. 2016; 1848471: 1–9.

[63] Gomez C.H., Medathati K., Kornprobst P., Murino V., and Sona D. 'Improving FREAK descriptor for image classification'. *Proceedings of the 10th International Conference on Computer Vision Systems*; Copenhagen, Denmark, Jul 2015. Berlin: Springer; 2015. pp. 14–23.

[64] Nawaz M., Mehmood Z., Nazir T., et al. 'Image authenticity detection using DWT and circular block-based LTRP features'. *Computers, Materials & Continua*. 2021; 69(2): 1927–1944.

[65] Nawaz Khayam K., Mehmood Z., Nazeer Chaudhry H., et al. 'Local-tetra-patterns for face recognition encoded on spatial pyramid matching'. *Computers, Materials & Continua*. 2022; 70(3): 5039–5058.

[66] Philip L. and Timmaraju A.S. Learning Binary Descriptors from Images [online]. 2014. Available from https://cvgl.stanford.edu/teaching/cs231a_-winter1415/prev/projects/LeeTimmaraju_report.pdf [Accessed 29 Jan 2023].

[67] Hamida S., Azizi R. and Maaloul A. 'Intelligent system of M-vision based on optimized SIFT'. *Journal of Computer and Communication*. 2016; 4: 52–62.

[68] Dalal N. and Triggs B. 'Histograms of oriented gradients for human detection'. *Journal of Computer Vision and Pattern Recognition*. 2005; 1: 886–893.

[69] Mikolajczyk K. and Schmid C. 'A performance evaluation of local descriptors'. *IEEE Transactions on Pattern Analysis and Machine Intelligence*. 2005; 27(10): 1615–1630.

[70] Nguyen D.-D., El Ouardi A., Aldea E., and Bouaziz S. 'HOOFR: an enhanced bio-inspired feature extractor'. *Proceedings of the 23rd International Conference on Pattern Recognition*; Cancún, Mexico, Dec 2016. New York: IEEE; 2016. pp. 2977–2982.

[71] Memon M.M., Hashmani M.A., Junejo A.Z., Rizvi S.S., and Arain A.A. 'A novel luminance-based algorithm for classification of semi-dark images'. *Journal of Applied Sciences*. 2021; 11(18): 8694.

[72] Guo J.-M., Chang L.-Y., and Lee J.-D. 'An efficient and geometric-distortion-free binary robust local feature'. *IEEE Sensors Journal*. 2019; 19(10): 2315.

[73] Helmer S. and Lowe D.G. 'Object class recognition with many local features'. *Proceedings of the 2004 Conference on Computer Vision and Pattern Recognition Workshop*; Washington, DC, USA, June–July 2004. New York: IEEE; 2004. pp. 187–187.

[74] Altwaijry H., Veit A., and Belongie S.J. 'Learning to detect and match keypoints with deep architectures'. In Wilson R.C., Hancock E.R., and Smith W.A.P. (eds.). *Proceedings of the British Machine Vision Conference*; York, UK, Sep 2016. York: BMVA Press; 2016. pp. 49.1–41.12.

[75] Liu Y., Cheng D., Wang Q., *et al.* 'Optical distortion correction considering radial and tangential distortion rates defined by optical design'. *Results in Optics*. 2021; 3: 100072.

[76] Xiong P., Wang S., Wang W., Ye Q., and Ye S. 'Model-independent lens distortion correction based on sub-pixel phase encoding'. *IEEE Sensors Journal*. 2021; 21(22): 7465.

[77] Chau S., Bouabid S., and Sejdinovic D. 'Deconditional downscaling with Gaussian processes'. *Proceedings of the 35th Conference on Neural Information Processing Systems*; Virtual-only Conference, UK, Dec 2021. NeurIPS Online Proceedings: NeurIPS; 2021. pp. 1–13.

[78] Cengiz Oeztireli A. and Gross M. 'Perceptually based downscaling of images'. *ACM Transactions on Graphics*. 2015, 34(4): 1–10.

[79] Tharwat A. 'Classification assessment methods'. *Applied Computing and Informatics*. 2021; 17(1): 168–192.

[80] Chakraborty A. 10 Best Apps to Monitor CPU Temperature on Android in 2022 [online]. 2022. Available from https://techviral.net/apps-to-monitor-cpu-temperature-on-android/ [Accessed 29 Jan 2023].

Chapter 13

Feature extraction process through hypergraph learning with the concept of rough set classification

K. Anitha[1] *and Ajantha Devi*[2]

Hypergraph learning is the process of representing data in the form of edge-connected graph. This learning process enables data scientists to extract optimal number of consistent features. Hypergraph clustering is an extensive property in the process of feature extraction. Here, clusters are framed with vertices of hypergraph in which vertices in each cluster are strongly connected. In this process, features are represented by vertices of hypergraph and edges are generated with high level of connectivity between features. After forming hypergraph clusters, *n* number of possible hypergraph subclusters have to be generated based on standard algorithms and finally clusters that contain feasible features are collected. This process can be used in social networking, bioinformatics, and computational intelligent system. In this chapter, we incorporate the principles of rough set dimensionality reduction process with hypergraph. Reduct is the unique term used in Rough set which defines possible subsets from knowledge base. In this chapter, we have proposed the reduct generation process using Rough hypergraph which is implemented in an autistic spectrum disorder dataset.

13.1 Introduction

Hypergraph is the generalization of conventional graph in which edges are connected with *n* number of vertices and these edges are coined as hyper-edges. In contrast, in a conventional graph, an edge may be represented as a pair of vertices and links just two vertices. However, in a hypergraph, the hyper-edges can have any number of vertices, allowing for the representation of more complicated interactions. Applications for hypergraphs generate major impact in social networking, bioinformatics, machine learning, and deep learning. Hypergraphs have

[1]Department of Mathematics, SRM Institute of Science and Technology, India
[2]Research Head, AP3 Solutions, India

the ability to record high-order interactions between variables or entities, which is often difficult to represent using traditional graphs.

More formally, a hypergraph is a pair (V, E), where V is a set of vertices, and E is a set of hyper-edges. A hyper-edge is a subset of V. So, instead of having edges that connect two vertices, a hyper-edge is a set of vertices that are connected.

For example, consider a hypergraph with vertices $\{A, B, C, D, E\}$ and hyper-edges $\{AB, BCD, ACDE\}$. In this hypergraph, the hyper-edge AB connects vertices A and B, the hyper-edge BCD connects vertices B, C, and D, and the hyper-edge $ACDE$ connects vertices A, C, D, and E.

Hypergraphs defined through an incidence matrix, in which rows and columns are demonstrated with vertices and hyper-edges, where the rows correspond to the vertices and the columns correspond to the hyper-edges. An entry in the matrix is 1 if the vertex is part of the hyper-edge, and 0 otherwise.

Hypergraphs have several important properties, including:

Degree: The number of hyper-edges incident on a vertex is degree of the vertex $\deg(v_i)$.

Transversals: The non-empty intersection of set of vertices and hyper-edges are called transversals. The minimum cardinality is noted as transversal number of the graph.

The vertices of a hypergraph represent the basic elements or entities of the system being modeled, while the hyper-edges represent relationships or interactions between these entities. Unlike in a standard graph, where edges connect only two vertices, hyper-edges can connect any number of vertices. This makes hypergraphs a more flexible and expressive way to represent complex relationships between entities. Hypergraphs can be represented graphically by drawing vertices as points and hyper-edges as curves that connect these points. In this representation, a hyper-edge is drawn as a polygon whose vertices are the vertices of the hyper-edge. Alternatively, hypergraphs can be represented by an incidence matrix, where each row represents a vertex and each column represents a hyper-edge, and the entries indicate whether a vertex is part of a hyper-edge. Hypergraphs can have various structures, depending on the relationships between vertices and hyper-edges. For example, a hypergraph can be connected, meaning that every vertex is connected to at least one other vertex through a hyper-edge. Alternatively, a hypergraph can be disconnected, meaning that there are vertices that are not connected to any other vertices. Hypergraphs can also have cycles, which occur when a hyper-edge contains a subset of vertices that are already part of another hyper-edge. Cycles can make it difficult to compute certain properties of hypergraphs such as transversals and colorings. Consider a group of people who have different interests. Let's say we have five people: Alice, Bob, Carol, Dave, and Eve. We can represent their interests as vertices and create a hypergraph connecting the people with the interests they have:

Alice: Cooking, Hiking, Reading
Bob: Hiking, Painting
Carol: Cooking, Gardening, Painting
Dave: Reading, Writing
Eve: Gardening, Writing

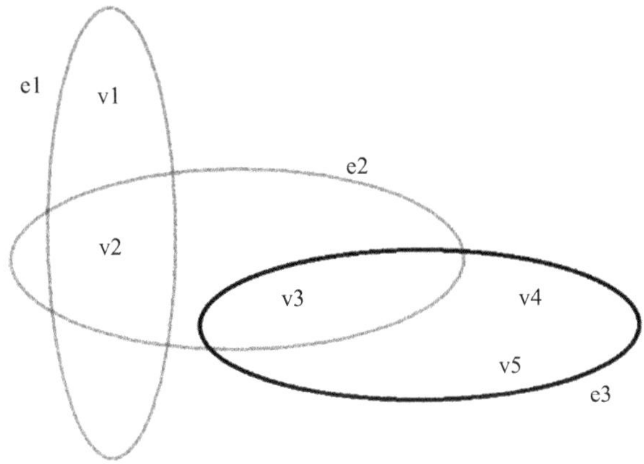

Figure 13.1 Structure of hypergraph

The hypergraph for this example would have eight hyper-edges, each connecting the people who share a common interest. For example, one of the hyper-edges would be {Alice, Carol}, since they both like cooking. Another hyper-edge would be {Bob, Carol}, since they both enjoy painting.

This example is just one possible hypergraph. In general, a hypergraph can have any number of vertices and hyper-edges, and each hyper-edge can connect any number of vertices (Figure 13.1).

13.2 Rough set theory

13.2.1 Preliminaries

Rough set is a computational framework used to handle the uncertainty and incomplete information. It was put forth by the Polish mathematician Zdzisław Pawlak with his most acclaimed publication [1]. It addresses the imprecision in data by dividing the data into sets of indiscernibility. The basic idea behind rough sets is to classify objects based on the attributes that they possess. Attributes are characteristics or features that can be used to distinguish one object from another. In rough sets, these attributes are represented as sets of values that an object can take. The rough set theory defines set approximations of a set named upper and lower approximation, which represents the boundaries of a set of indiscernibility. The lower approximation collects the objects definitely belonging to the target concept, whereas other one collects possibility related to target. The non-empty variation between set approximation is termed as boundary regions, which represents the objects that are uncertain.

13.2.1.1 Information system

An information system is a formal description of universe of discourse/objects that can be used to apply rough set theory to extract patterns hidden from the data. It

Table 13.1 Information system

Objects	Conditional attributes				
Make	**Model**	**Year**	**Price**	**Fuel efficiency**	**Transmission**
Honda	Civic	2015	12000	35	Manual
Toyota	Corolla	2018	15000	32	Automatic
Ford	Mustang	2014	22000	25	Manual
Honda	Accord	2019	25000	30	Automatic
Chevrolet	Camaro	2020	32000	22	Automatic

supplies a set of operations to analyze and extract knowledge from an information system.

Table 13.1 is an example of information system. Suppose we have a dataset of cars, where each car is described by its make, model, year, price, fuel efficiency, and transmission type.

In this information system, the universe of objects U is the set of cars, the set of attributes A is {Make, Model, Year, Price, Fuel Efficiency, Transmission}, and the valuation function V maps each car to its attribute values. Using rough set theory, we can perform various operations on this information system, such as attribute reduction, attribute selection, and rule induction. As an illustration, we may utilize attribute reduction to get rid of useless qualities and determine which ones are most crucial for categorization. Additionally, we may create classification rules for automobiles based on their characteristics using rule induction.

13.2.1.2 Decision system

A decision system is represented as a pair (U, D), where U is the universe of objects and D is the decision attribute or decision set. The attribute which creates the decision about an information system is named as decision attribute. In other words, it is the attribute that is used to classify the object into a particular category. Rule induction involves generating decision rules from the data in the decision system. A decision rule is a statement that describes the relationship between the attributes and the decision attribute. These rules can be used to make decisions about new objects based on their attributes. For the above information system (Table 13.2), if we add the decision attribute "Buy or not," then it is converted as decision system.

13.2.1.3 Indiscernibility relation

The indiscernibility relation is a fundamental concept used to determine the similarity between objects or elements in a given set. It is an equivalence relation on the set of objects in which two objects are said to be indiscernible if they have the same value for all attributes or properties under consideration. In other words, if two objects cannot be distinguished based on their attributes, they are said to be indiscernible.

Table 13.2 Decision system

Objects	**Conditional attributes**					**Decision attribute**
Make	**Model**	**Year**	**Price**	**Fuel efficiency**	**Transmission**	**Buy or not**
Honda	Civic	2015	12000	35	Manual	Yes
Toyota	Corolla	2018	15000	32	Automatic	Yes
Ford	Mustang	2014	22000	25	Manual	No
Honda	Accord	2019	25000	30	Automatic	No
Chevrolet	Camaro	2020	32000	22	Automatic	Yes

Formally, let U be a universe of discourse and let X be a subset of U. The indiscernibility relation on X is denoted by $\sim$ and defined as follows:

$\forall x, y \in X, x \sim y \Leftrightarrow \forall a \in U, a(x) = a(y)$; here $a(x)$ is the value of the attribute a for object x. These equivalence classes partition the set X into subsets of objects that are indiscernible from each other.

13.2.1.4 Set approximation

Set approximations of rough set are constructed from equivalence classes of indiscernibility relation. The lower approximation of a subset $\mathcal{A}$ of $\mathbb{X}$ is the set of all objects in $\mathbb{X}$ that are certainly in $\mathcal{A}$, while the upper approximation of A is the set of all objects in X that are possibly in $\mathcal{A}$.

Mathematically it is exhibited as:

$$\underline{\mathcal{A}}(\mathbb{X}) = \{x : [x]_A \subseteq \mathcal{A}\}$$

$$\overline{\mathcal{A}}(\mathbb{X}) = \{x : [x]_A \cap \mathcal{A} \neq \emptyset\}$$

$$BND\ (\mathcal{A}(\mathbb{X})) = \overline{\mathcal{A}}(\mathbb{X}) - \underline{\mathcal{A}}(\mathbb{X})$$

If $BND(\mathcal{A}(\mathbb{X})) \neq \emptyset$, then the target set $\mathbb{X}$ is rough, otherwise it is classical set. The process of analyzing the objects boundary region is NP hard. Rough set is one of the mathematical tools to address this issue. It does not expect additional parameters. Based on the available information, it finds the hidden patterns from the data. From these patterns, decision-makers can able to take optimal decision based on classification rules. Formation of these rules can be done through reduct and core of rough sets.

13.2.1.5 Rough membership function

Rough membership function assigns degree/dependency for each object with respect to the decision set. In the rough set theory, a set is characterized by its approximation pair, which are subsets of the universe of discourse that contain all the elements of the set with a certain degree of certainty. The rough set membership function takes as input an element of the universe of discourse and a set defined by its lower and upper approximations, and returns a value between 0 and 1. The rough membership function for an information system $\mathbb{M} = (U, F)$ and the non-empty set $G \subseteq U$ is

$$\omega_G^F(f) = \frac{|[f]_G \cap G|}{|[f]_F|} \textit{for some} f \in U$$

13.2.1.6 Reduct and core

Reduct and core are lean forms of an information system. In other words, reduct contains only the essential attributes that are necessary for distinguishing between objects in the set. The core of a set of attributes is the intersection of all reducts of that set. It represents the smallest subset of attributes that are both necessary and sufficient for distinguishing between objects in the set. In other words, the core is the set of attributes that cannot be eliminated without losing the ability to distinguish between objects in the set. Both the reduct and the core are important concepts in rough set theory, as they provide a way to reduce the dimensionality of data while preserving its essential structure. Let $A \neq \emptyset$ be attribute set and P be the binary relation on a universe U of objects. The rough set induced by A and P is defined as follows: for any X, Y subset of U, we say that X and Y are $P - \textit{indiscernible}$ if for every attribute a in A, we have (x, y) in P for all x in X and y in Y.

Given the rough set (U, P), a subset R subset of A is said to be a reduct of A if and only if R preserves the indiscernibility relation induced by P, that is, for any x, y in U, x and y are $P - \textit{indiscernible}$ with respect to A if and only if they are $P - \textit{indiscernible}$ with respect to R. The core of A with respect to (U, P) is the intersection of all reducts of A, denoted by *Core_P(A)*. In other words, the core of A is the smallest subset of A that is a reduct of A and preserves the indiscernibility relation induced by P. It is the set of attributes that are essential for distinguishing objects in U with respect to P, and cannot be eliminated without losing this ability. Applications of rough set in feature selection has been clearly examined and implemented for different datasets in [2,3].

13.3 Rough graph

Rough graph is a graphical representation in approximation space. An approximation space is a mathematical model that consists of a set of objects and a set of attributes that describe those objects. The construction of a rough graph was initially proposed by Tong He in 2006. Tong He proposed a representation of rough set graphs using a matrix-based approach, which is also known as the decision theoretic rough set model. In this approach, the rough set graph is represented by a decision table, which is a matrix that contains the attributes and values of the objects [4]. Giampiero *et al.* [5] proposed rough set-based digraph without having loops and multiple edges. They used adjacency matrix for drafting the graph that depends on the nature of the information table. Imran *et al.* [6] constructed the graph using similarity relation. In this work, they proposed orbit for addressing an

Table.13.3 Information system

Objects	Conditional attributes (c_i)			Decision attribute (D)
	BMI value (c_1)	**Blood sugar level (c_2)**	**HB level (c_3)**	**Diagnostic report (D)**
s_1	High	Normal	10	Pass
s_2	Normal	Above normal level	10	Pass
s_3	Normal	Normal	13	Pass
s_4	High	Normal	7	Fail
s_5	Normal	Above normal level	7	Fail
s_6	High	Normal	13	Fail

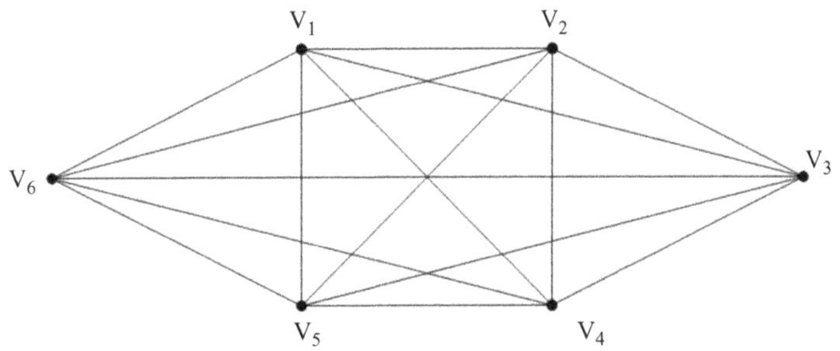

Figure 13.2 Structure of rough graph

information system. Anitha *et al.* proposed rough graph using membership values of vertices and they discussed the labeling of rough graphs [6–9]. Example for membership value-based rough graph is defined in Table 13.3.

For this information system, the graph shown in Figure 13.2 has been drawn with the condition that [10–14]

$$E = \begin{cases} Membership\ values(v_l, v_j) > 0 \Rightarrow edge\ present \\ Membership\ values(v_i, v_j) = 0 \Rightarrow no\ edge \end{cases}$$

13.4 Proposed work

13.4.1 Rough hypergraph

This section describes the process of connectivity between the rough graph and hypergraph. In particular, hypergraphs can be used to represent the lower and upper approximations of a rough set, where each node in the hypergraph corresponds to an element in the universe of discourse and each hyper-edge corresponds to a subset of elements that are indistinguishable with respect to the rough set. This representation can be used to perform various operations on rough sets such as computing set approximation of a target set and reduction of attributes from the information system. Hypergraphs are a generalization of graphs that allow edges to connect more than two vertices, enabling the representation of higher-order relationships between data points; we can also define the following terms

Cardinality: The cardinality of a hyper-edge is the number of vertices it contains. The hypergraph is said to be $k-uniform$ if every hyper-edge has k vertices.

Incidence matrix: The incidence matrix of a hypergraph is a binary matrix with rows representing the vertices and columns representing the hyper-edges. Each entry in the matrix is 1 if the vertex is in the hyper-edge, and 0 otherwise.

Hypergraph connectivity: Hypergraphs can be disconnected or connected. A hypergraph is termed as connected if there exists a path between any two vertices.

Hypergraph colorings: Hypergraphs can be colored, where vertices in the same hyper-edge are assigned the same color. A hypergraph is said to be k-colorable if it is colored with k colors such that no two vertices in the same hyper-edge have the same color.

Hypergraph matching: Hypergraphs can also be matched, where a matching is a subset of hyper-edges such that no two hyper-edges share a vertex. Finding a maximum matching in a hypergraph is a combinatorial optimization problem.

Related works

Authors	**Techniques developed**
Nabwey and Omar [15]	Implemented Hypergraph to identify minimal transversal reducts for wart data
Donglin *et al.* [16]	Proposed uncertainty vertex-weighted Hypergraph learning for the purpose of analyzing COVID 19 CT scan images
Abd-El Aziz [17]	Signalized the value of approximations using rough hypergraphs
Praba and Gomathi [18]	Employees satisfactory level has been calculated using rough hypergraph
Pawlak [19]	Rough set with flow graph is used as descriptor tool for data analyzation

13.4.2 Methodology

Hypergraph-based feature selection is a methodology for selecting relevant features from high-dimensional data using hypergraphs [20–22]. Hypergraphs are a generalization of graphs that allow edges to connect more than two vertices, enabling the representation of higher-order relationships between data points.

The methodology for hypergraph and rough set-based feature selection typically involves the following steps:

1. Find all possible reduct sets for an given information system.
2. Construct a hypergraph: The first step is to construct a hypergraph from the reduct set, where each data point is represented by a vertex, and hyper-edges connect subsets of vertices that share common features. The hypergraph is constructed based on a rough membership values for the reduct set.

3. Calculate centrality measures for the hypergraph: Next, centrality scores are determined for each vertex in the hypergraph. Each vertex in the hypergraph is given a centrality measure, which may be calculated using a variety of metrics, including degree, eigenvector, and betweenness centrality.
4. Select relevant features: Finally, the vertices with the highest centrality measures are selected as the most relevant features for the analysis. The number of features to select is determined by a predefined threshold or through cross-validation.

13.4.3 Experimental results

Data description: Screening test of autistic spectrum disorder (ASD) for adults is considered. This dataset contains 20 conditional features

($Attr_1$ *to* $Attr_{10}$→A1 to A10 scores are behavioral features, $Attr_{11}$→age, $Attr_{12}$→gender, $Attr_{13}$→ethinicity, $Attr_{14}$→jaundice, $Attr_{15}$→austim, $Attr_{16}$→country, $Attr_{17}$→recently used app, $Attr_{18}$→result, $Attr_{19}$→age desc, $Attr_{20}$→relation) and 1 decision feature (Class/ASD). It is essential to identify the most informative feature for early diagnosis. Total number of instances 704. Their descriptive statistics are defined in Figure 13.3.

Classifier model: simple naïve Bayes classifier

Correctly classified instances	6S2 (96.875%)
Incorrectly classified instances	*22* (3.125%)
Kappa statistic	0.9228
Mean absolute error	0.033
Root mean squared error	0.1386
Relative absolute error	8.3977%
Root relative squared error	31.2756%
Total number of instances	704

Classification accuracy

	TP rate	FP rate	Precision	Recall	F-Measure	MCC	ROC Area	PRC Area	Class
	0.961	0.011	0.996	0.961	0.978	0.925	0.997	0.999	NO
	0.989	0.039	0.903	0.989	0.944	0.925	0.997	0.972	YES
Weighted Avg.	0.969	0.018	0.971	0.969	0.969	0.925	0.997	0.992	

Confusion matrix

a	*b*	<— **classified as**
495	20	*a* = NO
2	187	*b* = YES

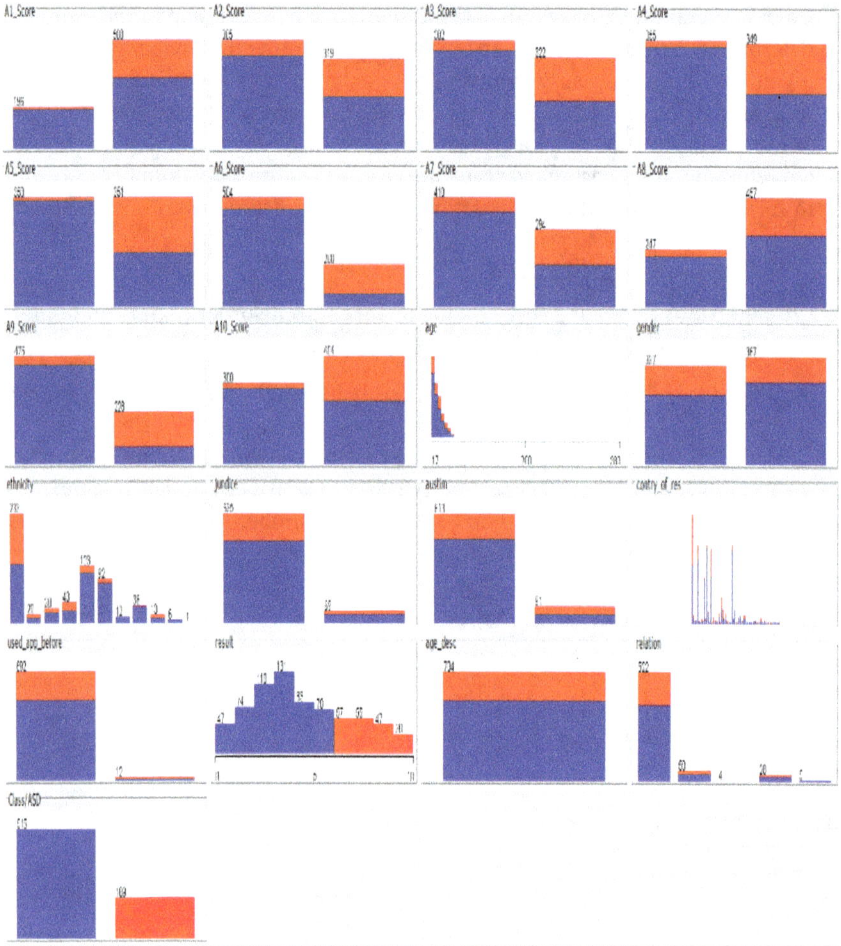

Figure 13.3 Descriptive statistics

Rough set reduct generation ($\Re_i$ – *reduct*)

$$\Re_1 = \left\{ \begin{array}{c} Attr_1, Attr_2, Attr_3, Attr_4, Attr_5, Attr_6, Attr_7, Attr_8, Attr_9, \\ Attr_{10}, Attr_{13}, Attr_{14}, Attr_{15}, Attr_{16}, Attr_{18} \end{array} \right\}$$

$$\Re_2 = \left\{ \begin{array}{c} Attr_1, Attr_2, Attr_3, Attr_4, Attr_5, Attr_6, Attr_7, Attr_8, Attr_9, \\ Attr_{10}, Attr_{12}, Attr_{14}, Attr_{15}, Attr_{16}, Attr_{18}, Attr_{19} \end{array} \right\}$$

$$\Re_3 = \left\{ \begin{array}{c} Attr_1, Attr_2, Attr_3, Attr_4, Attr_6, Attr_7, Attr_8, Attr_9, \\ Attr_{10}, Attr_{12}, Attr_{14}, Attr_{15}, Attr_{16}, Attr_{17}, Attr_{19} \end{array} \right\}$$

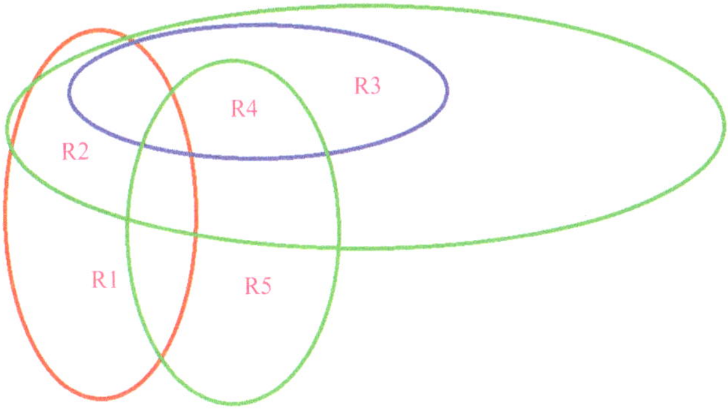

Figure 13.4 *Rough hypergraph (with respect to reduct)*

$$\mathfrak{R}_4 = \left\{ \begin{array}{c} Attr_1, Attr_2, Attr_3, Attr_4, Attr_6, Attr_7, Attr_8, Attr_9, \\ Attr_{13}, Attr_{14}, Attr_{15}, Attr_{16}, Attr_{17}, Attr_{20} \end{array} \right\}$$

$$\mathfrak{R}_5 = \left\{ \begin{array}{c} Attr_1, Attr_2, Attr_3, Attr_4, Attr_6, Attr_7, Attr_8, Attr_9, \\ Attr_{10}, Attr_{13}, Attr_{14}, Attr_{15}, Attr_{16}, Attr_{18}, Attr_{20} \end{array} \right\}$$

Rough hypergraph (Figure 13.4)
Optimal selected attributes are (12 attributes)

$$= \{Attr_1, Attr_2, Attr_3, Attr_4, Attr_5, Attr_6, Attr_7, Attr_8, Attr_9, Attr_{10}, Attr_{16}, Attr_{18}\}$$

13.5 Results and discussion

Hypergraph and rough set-based feature selection has been shown to be effective in identifying relevant features from high-dimensional data, particularly in cases where traditional feature selection methods struggle due to the presence of complex and nonlinear relationships between data points. From this experimental dataset, out of 20 attributes only 12 attributes are selected and these are considered as consistent and eminent features remaining eight attributes are inconsistent and removal of these attributes will not be useful for further diagnosis. Hypergraph theory is a rapidly growing field with ongoing research, and new results and applications are constantly being discovered.

References

[1] Zdzislaw P. Rough sets- *Theoretical Aspects of Reasoning about Data*. Springer: Dordrecht; 1991. https://doi.org/10.1007/978-94-011-3534-4.

[2] Anitha K. and Datta D., Analysing high dimensional data using Rough tolerance relation. *2022 International Conference on Electrical, Computer,*

Communications and Mechatronics Engineering (ICECCME); Maldives, Maldives; 2022, IEEE, pp. 1–5, doi:10.1109/ICECCME55909.2022.9988257.

[3] Anitha K. and Thangeswari M. Rough set based optimization techniques. *2020 6th International Conference on Advanced Computing and Communication Systems (ICACCS)*; Coimbatore, India; 2020, IEEE, pp. 1021–1024, doi:10.1109/ICACCS48705.2020.9074212.

[4] Zhou L., Nandal A., Dhaka A., *et al.* Artificial neural network dual hesitant fermatean fuzzy implementation in transportation of COVID-19 vaccine. *Journal of Organizational and End User Computing (JOEUC), IGI Global.* 2023;35(2):1–23.

[5] Chiaselotti G., Ciucci D., Gentile T. and Infusino F. Rough set theory and digraphs. *Fundamenta Informaticae*. 2017;153:291–325.

[6] Anitha K. and Komathi J. Study of fuzzy magic graphs on intuitionistic space. *International Journal of Recent Technology and Engineering*. 2019;7:273–278.

[7] Komathi J., Anitha K. and Meiyappan D. Bimagic labeling on families of graph in fuzzy intuitionistic space. *Test Engineering & Management* (ISSN: 0193-4120). 2020, 16313–16317.

[8] Nithya R. and Anitha, K. Even vertex ζ-graceful labeling on rough graph. *Mathematics and Statistics*. 2023;11(1);71–77. DOI:10.13189/ms.2023.110108.

[9] Aruna Devi, R. and Anitha, K. Construction of rough graph through rough membership function. *Mathematics and Statistics*. 2022;10(6):1344–1353. DOI:10.13189/ms.2022.100622.

[10] Molnár, B. Applications of hypergraphs in informatics: a survey and opportunities for research. *Annales Universitatis Scientiarum Budapestinensis de Rolando Eotvos Nominatae Sectio Computatorica*. 2014;42:261–282.

[11] Nasreen Kausar, M. Sajida Kousar, M., Gulistan, M., and Anitha, K. Characterization of non-associative ordered semigroups by the properties of F-ideals. *Fuzzy Information and Engineering*. 2020;12(4):490–508, DOI: 10.1080/16168658.2021.1924513.

[12] Anitha, K. Rough set theory on topological spaces. In: Miao, D., Pedrycz, W., Ślęzak, D., Peters, G., Hu, Q. and Wang, R. (eds.) *Rough Sets and Knowledge Technology*. RSKT 2014. Lecture Notes in Computer Science (2014), vol 8818. Springer, Cham; 2014. https://doi.org/10.1007/978-3-319-11740-9_7

[13] Anitha, K. and Venkatesan, P. Properties of hesitant fuzzy sets. *Global Journal of Pure and Applied Mathematics*. ISSN 0973-1768.2016;12(1):114–116.

[14] Sharma, M. K., Dhiman, N., Mishra, V. N., Mishra, L. N., Dhaka, A., and Koundal, D., "Post-symptomatic detection of COVID-2019 grade based mediative fuzzy projection". *Computers and Electrical Engineering*, Elsevier, vol. 101, 2022.

[15] Nabwey, H.A. and Omar, A. Methodology based on rough set theory and hypergraph for the prediction of wart treatment. *International Journal of Engineering Research and Technology*. 2020;13(3):552–559. ISSN 0974-3154.

[16] Dia, D., Shi, F., Yan, F., *et al.* Hypergraph learning for identification of COVID-19 with CT imaging. *Med Image Anal*. 2021;68:101910. doi: 10.1016/j.media.2020.101910.

[17] Radwan, A.E.A., Nayle, M.S. and Nasir, A.I. New rough sets properties on graph theory. *International Journal of Contemporary Mathematical Sciences*. 2012;7(25):1217–1232.

[18] Praba, B. and Gomathi, G. Hypergraphs and rough sets with their applications in decision-making problems. *New Mathematics and Natural Computation*. 2022;18(2):293–311.

[19] Pawlak, Z. Rough sets and flow graphs. in Ślęzak, D., Wang, G., Szczuka, M., Düntsch, I. and Yao, Y. (eds.). *Rough Sets, Fuzzy Sets, Data Mining, and Granular Computing*. RSFDGrC 2005. Lecture Notes in Computer Science, vol 3641, Springer, Berlin, Heidelberg; 2005.

[20] Sharma, A.K. and Ganchev, T. Breast cancer classification using CNN extracted features: a comprehensive review. In *Application of Deep Learning Methods in Healthcare and Medical Science (ADLMHMS-2020)*. Taylor and Francis: CRC Press; 2022.

[21] Nandal, A. Gamboa-Rosales, H., Dhaka, A., *et al.* Image edge detection using fractional calculus with feature and contrast enhancement. *Circuit System and Signal Processing*, Springer, 2018, vol. 37(9), pp. 3946–3972.

[22] Jakhar, S. P., Nandal, A., Dhaka, A., Jiang, B., Zhou, L., and Mishra, V. N., Fractal feature based image resolution enhancement using wavelet-fractal transformation in gradient domain. *Journal of Circuits, Systems and Computers, World Scientific*. 2023;32(8).

Chapter 14

Machine learning for neurodegenerative disease diagnosis: a focus on amyotrophic lateral sclerosis (ALS)

Ajantha Devi[1]

Motor neuron degeneration is the primary cause of severe illness amyotrophic lateral sclerosis (ALS). As with all other neurodegenerative diseases, finding drugs that can modify the course of the disease has proven challenging for a variety of reasons. However, ALS is one of the few neurodegenerative diseases for which disease-modifying drugs have been approved. Significant progress has been made in ALS preclinical models, genetics, pathology, biomarkers, imaging, and clinical readouts over the last 10–15 years. In addition, cutting-edge therapy paradigms are used in medical fields with significant unmet medical needs such as neurodegenerative diseases. These advancements have broadened our knowledge base and made it possible to identify candidate targeted treatments for ALS with a variety of modes of action. Magnetic resonance imaging (MRI) and positron emission tomography (PET) are two image-based methods that are currently used to diagnose and monitor neurodegenerative diseases. These image-based techniques provide several benefits for the detection and tracking of neurodegenerative disorders. They offer precise, quantitative readings that enable early illness identification and progression monitoring. Additionally, thcy allow for diffcrcntial diagnosis, which aids in separating several neurodegenerative diseases with comparable clinical signs. Furthermore, by offering useful indicators for illness prediction and tracking therapy response, these imaging modalities aid in the development of novel therapeutic approaches. The standardization of imaging methods, the interpretation of complicated imaging data, and accessibility to cutting-edge imaging equipment in environments with limited resources are still issues in the area of neuroimaging. According to the information in this chapter, image-based techniques like MRI and PET have become effective tools for the early detection and monitoring of neurodegenerative illnesses. These methods help with differential diagnosis and creation of innovative therapeutic approaches, and provide significant insights into the underlying disease processes. The management of neurodegenerative illnesses can be improved, which will eventually improve the quality of life for those who are

[1]Research Head AP3 Solutions, India

affected, with the help of ongoing developments in neuroimaging technology and their incorporation into clinical practice.

14.1 Introduction

A set of conditions that impact the central nervous system are referred to as neurodegenerative diseases (NDDs). They are distinguished by the gradual degeneration of neuronal structure or function of human brain as in Figure 14.1. Biomarkers specific to each illness must be identified in order to give a tool for early diagnosis before any potential therapeutics can be developed. Early stage diagnosis is crucial for the identification of candidates for clinical trials, the monitoring of therapies, and the improvement of pathogenesis [1].

For instance, although Alzheimer's disease (AD) is caused by completely distinct mechanisms, people with vascular dementia have many of the same early cognitive signs [2]. Patients with vascular dementia may be included in clinical trials for an AD medication without a biomarker specific to the illnesses, which would reduce the probability of success.

Since the brain is the main organ affected by these disorders and may be utilized to acquire useful information on structural and functional changes there, neuroimaging technologies are ideal instruments for examining this. It has already been demonstrated that the brain's metabolic processes and levels of certain compounds, such as amyloid and tau in AD [3], can be measured using positron emission tomography (PET). However, it is costly, intrusive, and the radioactive tracer restricts the kind of data that can be acquired in a single scanning session.

Magnetic resonance imaging (MRI), on the other hand, is a less costly, noninvasive tool that could provide extra or substitute biomarkers. The key to MRI's effectiveness is its ability to control the "spin" (a fundamental quantum feature of particles) of hydrogen nuclei, which mostly belong to water and to a lesser extent fat, in various ways. This implies that depending on how the spins of the nuclei have been altered, the signal that they subsequently emanate can be rendered

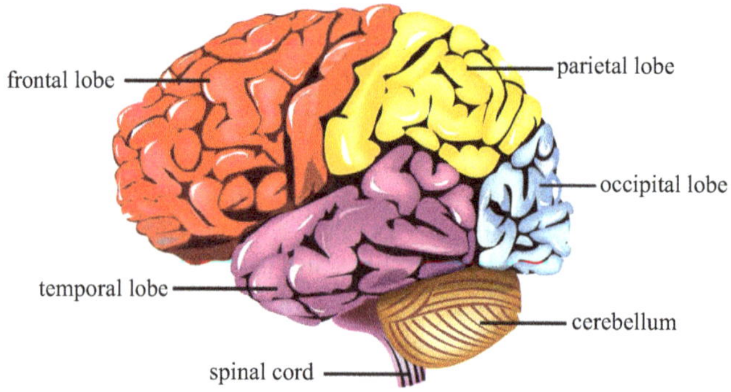

Figure 14.1 Structure of a human brain

sensitive to various tissue features. As a result, many MRI modalities [4–6] can be acquired within a single scanning session. In this chapter, we concentrate on three different imaging modalities: structural MRI, diffusion MRI, and functional magnetic resonance imaging (fMRI).

14.2 Neurodegenerative diseases

A collection of chronic, progressive illnesses known as NDDs predominantly affect the neurological system and cause the degeneration and death of neurons in certain areas of the brain or spinal cord. Numerous motor, cognitive, and behavioral symptoms are produced by these illnesses, which frequently causes serious impairment and worse quality of life for those who are affected.

There are several well-known neurodegenerative conditions, including AD, Huntington's disease (HD), Parkinson's disease (PD), and amyotrophic lateral sclerosis (ALS). The most common type of dementia, according to the Alzheimer's Association (2021) [7], is AD. It is characterized by the accumulation of abnormal proteins in the brain, which affects memory and cognitive function. HD is a genetic illness that causes a slow degradation of brain nerve cells, decreasing movement, affecting cognitive function, and generating mental symptoms, according to the Huntington's Disease Society of America, 2021 [8]. Lou Gehrig's disease, generally known as ALS, is a motor neuron disease that worsens over time and damages the nerve cells that regulate voluntary muscles, causing muscular weakness and trouble breathing, speaking, and swallowing (ALS Association, 2021) [7].

The cerebral cortex serves as the source of upper motor neurons as shown in Figure 14.2, which eventually gives rise to the corticospinal and corticobulbar tracts that terminate on cranial nerves. The anterior horns of the spinal cord's axons give rise to lower motor neurons. Lower motor neurons begin in the cranial nerve nuclei of the brainstem and the anterior horns of the spinal cord, and they terminate in muscles. The precise origins of neurodegenerative illnesses, which involve a mix of genetic, environmental, and lifestyle factors, are complicated and not entirely known. According to recent research, some of the common pathogenic processes driving neurodegenerative illnesses include protein misfolding, aggregation, and accumulation, oxidative stress, inflammation, mitochondrial dysfunction, and poor protein clearance pathways [9].

For prompt intervention and efficient therapy, neurodegenerative illnesses must be diagnosed as shown in Figure 14.3, correctly and early. Imaging techniques, such as MRI and PET, have emerged as potent tools for detecting structural, functional, and molecular changes related to NDDs, even though clinical assessment and medical history continue to play a significant role in diagnosis.

According to Sharma *et al.* [10], MRI is a non-invasive imaging method that offers precise structural information about the brain, enabling the observation of brain atrophy, anomalies in the white matter, and other structural alterations linked to neurodegenerative disorders. Contrarily, PET enables the visualization of molecular and metabolic changes in the brain, offering insightful information into the pathological mechanisms underlying NDDs like dopamine deficiency in PD and amyloid and tau protein deposition in AD [11].

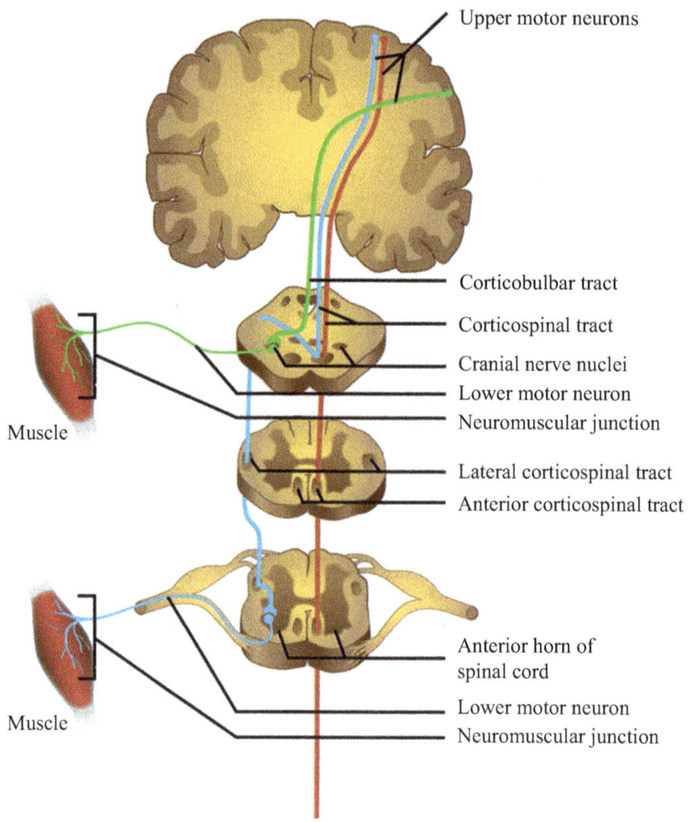

Figure 14.2 Upper motor neuron

The early identification, differential diagnosis, and monitoring of disease progression in neurodegenerative illnesses have all showed tremendous promise when using these imaging approaches. They aid in the creation of novel therapeutic methods by acting as useful biomarkers for illness prediction and tracking therapy response [12]. The standardization of imaging methods, the interpretation of complicated imaging data, and accessibility to cutting-edge imaging equipment in environments with limited resources are still issues in the area of neuroimaging. In order to ensure the acceptable and ethical use of neuroimaging methods in the diagnosis and therapy of neurodegenerative disorders, ethical issues including patient privacy, data sharing, and informed permission must also be addressed.

14.2.1 Alzheimer's disease

According to the Alzheimer's Association, 2021, AD, a neurological condition that affects millions of people worldwide, may be to blame for 60–70% of dementia cases [7]. In particular areas of the brain, the illness is defined by a gradual loss of

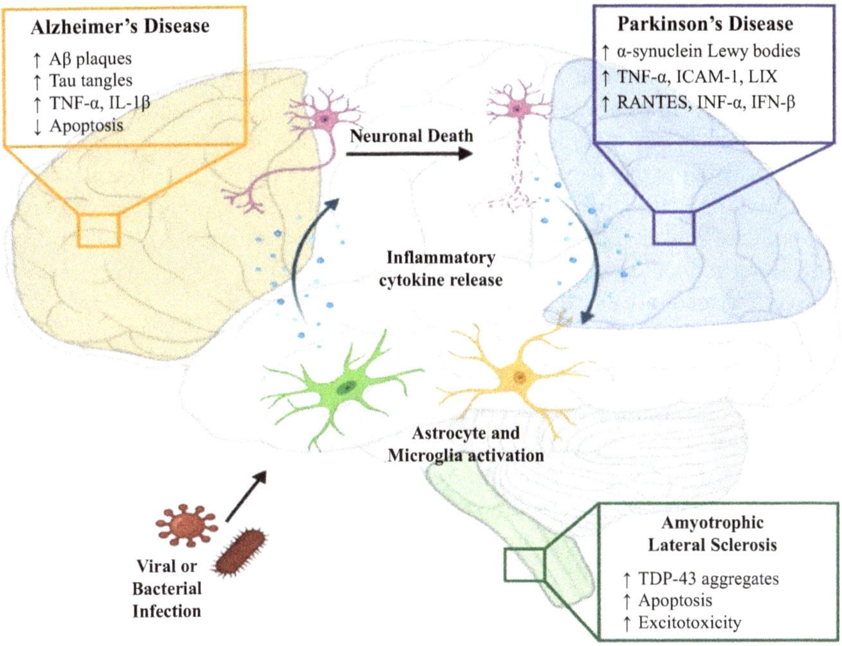

Figure 14.3 Neurodegenerative diseases

neurons, which results in cognitive deterioration, memory loss, and functional impairment.

AD is thought to be caused by a complex interaction of genetic, environmental, and lifestyle factors, while its specific etiology is yet unknown. The build-up of aberrant proteins, such as amyloid beta and tau, in the brain is one of the defining characteristics of AD [13]. These proteins create plaques and tangles, which harms and kills neurons and worsens AD cognitive and behavioral symptoms.

AD diagnosis depends heavily on clinical evaluation and medical history. However, imaging methods like PET and MRI have become effective tools for identifying structural, functional, and molecular abnormalities connected to the disease [14–16].

14.2.2 Parkinson's disease

PD is a long-term neurodegenerative condition that impairs motor function by lowering dopamine levels as a result of the death of dopaminergic neurons in the substantia nigra, a part of the brain. While the exact cause of PD is unknown, it is believed to be a combination of genetic and environmental factors, as well as the accumulation of abnormal protein aggregates in the brain [14]. Lewy bodies, which are pathological traits that are typical of PD, cause dopaminergic neuron loss.

The clinical assessment, which includes a comprehensive medical history and a neurological examination, is the main factor used to make the diagnosis of PD.

However, imaging methods like PET and MRI have grown in importance for the diagnosis and treatment of PD.

The substantia nigra and other areas of the brain affected by PD can be evaluated using MRI to determine changes in brain structure [15]. Contrarily, PET can offer functional data regarding the brain's dopamine metabolism and receptor binding, assisting in the diagnosis and management of PD [16]. PD presently has no known cure; however, there are a number of therapies that can be used to control both its motor and non-motor symptoms such as dopaminergic drugs, deep brain stimulation, and physical therapy. For those with PD, a prompt diagnosis and effective treatment are essential to maximizing outcomes.

14.2.3 Huntington's disease

An afflicted person has a 50% chance of transferring the disease-causing gene to each of their progeny, making HD an uncommon and fatal neurological condition that is inherited autosomal dominant. According to Walker [17], HD is defined by the gradual degeneration of neurons in specific brain areas, which results in a variety of motor, cognitive, and psychiatric symptoms. A genetic mutation in the huntingtin (HTT) gene is the underlying cause of HD and results in the synthesis of an aberrant HTT protein with an enlarged CAG repeat, which causes the build-up of toxic protein aggregates in the brain [18]. The neurons in the damaged brain areas eventually die as a result of these aggregates, which interfere with regular cellular activities. A thorough medical history, physical exam, genetic testing to confirm the existence of the HTT mutation, and clinical evaluation are commonly used to make the diagnosis of HD. However, HD can also be identified and monitored through neuroimaging methods like MRI.

According to Rosas *et al.* [19], an MRI can detect structural abnormalities in the brain, such as the caudate nucleus, which is frequently impacted by HD, atrophy, or shrinking of certain brain areas. MRI can shed light on the cortex and white matter, two additional areas of the brain implicated in HD disease. HD presently has no known cure; thus, symptom alleviation and supportive care are the main therapeutic goals. However, early diagnosis is necessary to enable participation in clinical trials and research projects aimed at creating new therapies, as well as to give persons with HD and their families with the proper care and support.

14.2.4 Amyotrophic lateral sclerosis

Brown and Al-Chalabi [20] assert that Lou Gehrig's illness, also known as ALS, is a neurodegenerative disorder that damages motor neurons in the brain and spinal cord, leading to the eventual loss of muscular function. The exact cause of ALS is still unknown, despite the fact that genetic and environmental factors are considered to be involved. An extensive medical history, physical examination, and electromyography (EMG) to evaluate motor neuron function are frequently used to make a clinical diagnosis of ALS. When it comes to anatomical alterations in the brain and spinal cord brought on by ALS, MRI can offer useful information. It can show atrophy or shrinking of certain areas, such the spinal cord, motor cortex, and corticospinal tracts, which are known to be impacted by ALS [21]. Additionally, an MRI might aid in ruling out illnesses like tumors or compressive lesions that may exhibit same symptoms.

14.3 The development stages of NDDs

Although the precise timing and duration of each stage might vary from person to person, neurodegenerative disorders normally progress in a predictable order of stages. Retrogenesis, cognitive impairment, and gait abnormalities are the three primary phases of NDDs.

Retrogenesis: The initial stage of NDDs includes the brain's regression in the opposite direction of how it developed. This implies that early developed brain areas are impacted first, followed by later-developed brain regions. For instance, in AD, the brainstem and other lower-level structures are impacted, later while the hippocampus and other parts of the cortex that are crucial for memory and higher-order cognitive function are initially affected.

Cognitive impairment: The second stage of NDDs is known as cognitive impairment, and it is characterized by a steady deterioration in cognitive function. A variety of symptoms, including memory loss, confusion, issues with language and communication, as well as changes in mood and behavior, characterize this stage. Cognitive impairment in AD often starts off moderate, affecting only executive function and memory, and then gets worse with time, affecting all facets of everyday life.

Gait abnormality: The third and last stage of NDDs is gait abnormality, which is characterized by a steady loss of coordination and movement. Symptoms including trouble walking, falls, and lack of balance define this stage. Gait irregularity, for instance, is a defining sign of PD that appears later in the illness course and is brought on by the degradation of dopaminergic neurons in the substantia nigra, which regulates movement.

For successful symptom management and the development of medicines that can reduce or stop disease progression, early diagnosis of NDDs [22–24] is crucial. Several diagnostic techniques, including imaging, genetic testing, and biomarker analysis, can be used to find NDDs early on, even before any clinical symptoms appear. This can assist in identifying those who are very susceptible to NDDs and enable early intervention to stop or postpone the beginning of the disease.

14.4 Neuroimages on neurodegenerative diseases

Neuroimages, pictures of the brain that directly or indirectly depict its structure and/or function, are the main source of knowledge in the area of neuroscience. Neuroimaging includes a number of methods [25] used to get these pictures, including MRI, computed tomography (CT), diffuse optical imaging, PET, as shown in Figure 14.4. As opposed to CT, which employs ionizing radiation, MRI uses magnetic fields and is said to be less intrusive. Additionally, MRI allows

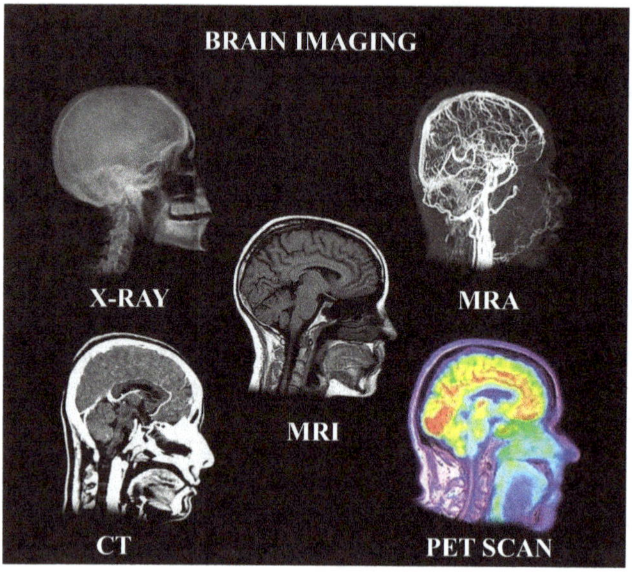

Figure 14.4 Various neuroimaging

excellent contrast viewing of many types of brain tissue and comparable spatial resolution to other methods. In order to identify aberrant volume changes that are suggestive of neurodegenerative illnesses, MRI is frequently used to evaluate specific brain areas such as the hippocampus, entorhinal cortex, or amygdala.

The two forms of MRI that are most often distinguished are structural MRI and fMRI. However, Symms *et al.* [26] contend that because the structure and function of the brain are frequently linked, the distinction between structural and fMRI is hazy. While structural MRI frequently provides static anatomical information, fMRI provides dynamic physiological information. fMRI techniques include blood oxygen level dependent (BOLD) and cerebrospinal fluid (CSF) pulsation measurements [27,28]. fMRI has been extensively utilized to study a variety of brain processes, including cognition, language, vision, and motor abilities. fMRI is an effective technology because it can map brain functions with comparatively high spatial and temporal resolution.

14.4.1 Structural magnetic resonance

The basis for how MRI [29] works is Nuclear Magnetic Resonance (NMR), which describes how proton atoms behave when exposed to fluctuating radiofrequency (RF) in a magnetic field, aligning their orientation and altering their charge. Two distinct scans generated utilizing T1 and T2 weighted sequences are shown in Figure 14.5. T2-weighted pictures distinguish between fat and water based on the spin-spin relaxation time, whereas T1-weighted images show variations in the spin-lattice relaxation time of various tissues. Since T2-weighted pictures are very sensitive to a variety of clinical disorders, they are often employed in

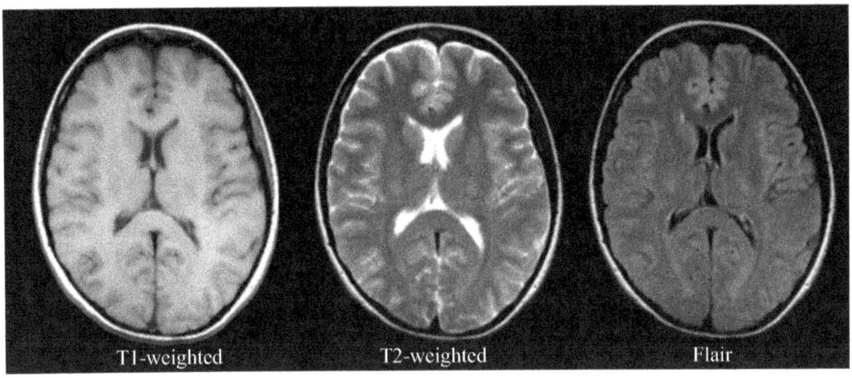

Figure 14.5 Magnetic resonance imaging

neuroimaging. The signal in T2-weighted pictures tells us how much water (or proton density) is present in each pixel. However, occasionally the addition of this information might make a diagnosis more difficult.

This restriction can be overcome, for example, by capturing two or more pictures with various echo durations or by using curve-fitting methods to a long spin echo train, which enables the generation of T2 maps independent of proton density. Similar to this, by creating a picture based on a single parameter, T1 mapping may similarly avoid changes in image intensity brought on by proton density. T1 maps can be created by combining two or more pictures with various repetition rates, flip angles, or inversion rates.

It is advised to capture additional pictures to measure and adjust for these flaws since T1 mapping is more vulnerable to RF field inhomogeneities than T2 mapping is [30]. For instance, volume measurements of the hippocampus may be performed using coronal T1 scans, which offers three-dimensional and high-resolution pictures [31]. Numerous cross-sectional and longitudinal studies that look at disorders like hippocampal sclerosis and AD as well as changes in hippocampus volume over time use these kinds of pictures as their starting point. Additionally, it has been demonstrated that both T1 and T2 imaging methods may identify variations in the iron concentration of various gray matter areas in PD patients.

14.4.2 Diffusion tensor imaging

The foundation of diffusion tensor imaging (DTI) as shown in Figure 14.6 is NMR. The effects of molecular diffusion in the NMR signal are encoded by DTI using bipolar magnetic field gradient pulses [32]. Molecular diffusion describes the random translational motion that molecules experience due to their thermal energy, also referred to as Brownian motion.

The distribution of water molecule displacement inside a diffusion MRI image voxel, which generally covers a volume of several cubic millimeters, can be statistically assessed. The molecular mobility in tissues could not be uniform in all

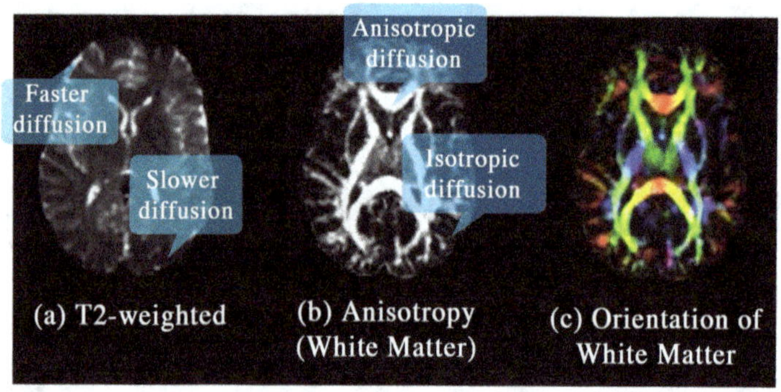

Figure 14.6 Diffusion tensor imaging (DTI)

directions because of the three-dimensional structure of diffusion. Both the physical configuration of the medium and the existence of obstructions that restrict molecule mobility in the diffusion process can contribute to this anisotropy.

Because diffusion is represented by magnetic field gradient pulses, MRI can only detect molecular displacements in the gradient's direction [33]. Differences in diffusion data can be used to detect diffusion anisotropy by changing the direction of gradient pulses. This unique capability is not achievable with conventional MRI parameters like T1 or T2. Diffusion anisotropy in the brain's white matter is caused by a specific arrangement of bundles of parallel-running, more or less myelinated axonal fibers. Diffusion proceeds more faster in the direction of the fibers than it does in the perpendicular direction, allowing for the mapping of the spatial orientation of white matter tracts. A color scale may be used to indicate the direction of the fastest diffusion, which displays the general orientation of the fibers.

Figure 14.6(a) shows a T2-weighted image where brighter areas represent faster diffusion and darker areas represent slower diffusion. Figure 14.6(b) shows a DTI image where brighter areas indicate anisotropic diffusion and darker areas indicate isotropic diffusion. Finally, Figure 14.6(c) shows a color scale reconstruction of white matter tracts, which allows for visualizing the orientation of the fibers based on the direction of the fastest diffusion.

The composition and architectural structure of brain tissue can be learned in vivo using DTI [34]. Anisotropy and diffusivity-related indices from DTI images frequently provide physiological information. Such indicators may be compared among various patient cohorts to get insightful results. For instance, fractional anisotropy, which quantifies the amount of diffusion that is directionally dependent, can be used to evaluate alterations in the white matter's microstructures that are connected to the onset of illness [35]. The same is true for mean diffusivity, which ignores the direction of diffusion in favor of the average diffusion inside a single voxel.

14.4.3 Functional magnetic resonance imaging

In clinical settings, fMRI as shown in Figure 14.7 [36] is frequently utilized to locate the expressive cortex for neurosurgery planning in situations like

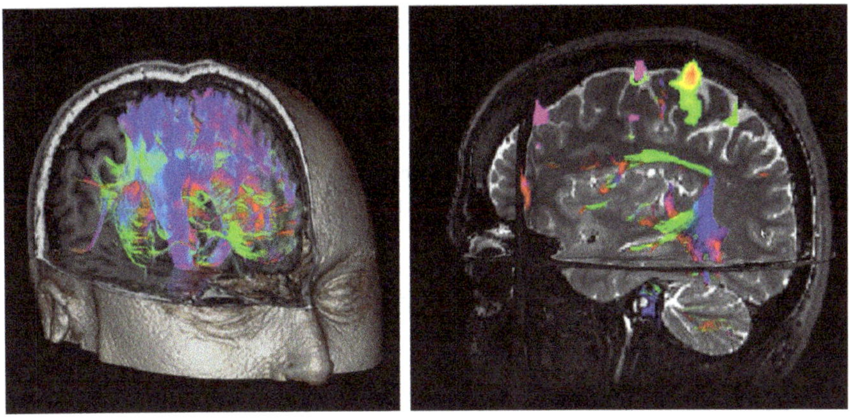

Figure 14.7 Functional magnetic resonance imaging

intractable epilepsy or tumor resections. Furthermore, substantial research has been done on the use of fMRI to spot abnormalities in brain activity linked to NDD including AD and moderate cognitive impairment (MCI) [37,38] in their early stages. Particularly, fMRI has demonstrated potential for early AD diagnosis in those with genetic risk factors and for predicting the onset of AD in people with MCI. Other investigations, including electrophysiologic ones like magnetoencephalography (MEG) and electroencephalography (EEG), have corroborated these results.

RS-fMRI has also been used to detect changes in functional connectivity in a number of neurodegenerative illnesses [39,40] that impact resting-state networks including the default mode network and others. These conditions include multiple sclerosis (MS), PD, HD, and frontotemporal dementia. Studies on functional connectivity in MS patients and verbal episodic memory tests support the idea that neurodegenerative illnesses frequently impact linked cortical networks rather than discrete brain areas.

In addition, fMRI research has demonstrated that, in addition to resting-state fMRI (RS-fMRI) studies, changes in brain activity can be seen during particular tasks in people with a variety of NDD, including ALS, AD, PD, HD, semantic dementia in the frontotemporal lobar degeneration spectrum, and HIV-positive people [41,42]. For instance, Tessitore *et al.* found that people with sporadic ALS had increased activity in the anterior putamen, a region involved in motor execution, and decreased activity in brain areas connected to the main motor and premotor cortex.

14.5 Machine learning and deep learning applications on ALS

The goal of the study by van der Burgh [43] was to divide ALS patients into three survival groups using clinical data and MRI. Data from 135 ALS patients were used in the study. The authors emphasized the effectiveness of deep neural networks for making predictions from complicated data, but they also pointed out that additional research was necessary to fully grasp how input and output factors related to ALS development.

In the research by Pires *et al.* [44], the scientists created a model to foretell when an ALS patient may require non-invasive ventilation (NIV) assistance within a specified time range (3, 6, and 12 months). The most effective model for predicting NIV support for each of the three time window values was the Random Forest model, which was chosen using a feature selection (FS) ensemble technique to choose pertinent biomarkers for each group. Age at illness beginning, disease duration, and ALSFRS (ALS Functional Rating Scale) score were other significant biomarkers that were present in 75% of the time. The scientists recognized BMI, forced vital capacity (FVC), and VC as useful biomarkers present in all groups. Instead of developing generic models that treat all patients equally, they stressed the benefit of employing customized ML models for various patient categories.

The objective of Gordon and Lerner [45] was to assess the ability of ordinal classifiers to forecast the functional deterioration in ALS patients. The PROACT dataset, which contained details from the initial and last patient visits, was used by researchers. A total of 3,772 patients in all were included in the dataset. Clinical, demographic, ALSFRS, FVC, medication, vital signs, and laboratory testing were among the different biomarkers that the researchers examined. The ten ALSFRS items, which are questions used to gauge various facets of an ALS patient's functional state, served as the study's target variables. The ALSFRS components were mapped to the patient states, allowing the researchers to connect the patient states to the development of the illness at various times throughout.

A crowdsourcing approach was employed in a different study by Kueffner *et al.* [46,47] to assess machine learning (ML) models for ALS prediction tasks. For the ALS stratification challenge, they chose 30 teams from around the globe. The teams were given access to patient information from the first three months but were only allowed to assess six of the available biomarkers. The study's intended predictions included the likelihood that the condition will develop at 12 months (as shown by a drop in the Functional Rate Scale (FRS) and the probability of survival at 12, 18, and 24 months. It was noted that the ALSFRS Bulbar subscore was a crucial predictor for differentiating between the groups. Tan *et al.* [48] used static and longitudinal biomarkers from the PROACT dataset, which comprised 2,424 patients, to address predictions of changes in the ALSFRS score and the FVC %. The goal of the project by Grollemund *et al.* [49] was to create a dimensionality reduction model that could forecast ALS patients' one-year survival probabilities. The study looked at a number of indicators such as gender, age, weight, illness duration, and ALS functional rating. The method assisted in determining whether biomarkers had a stronger or weaker link with survival prediction. Age and ALSFRS score, for instance, revealed a strong association, although gender and weight showed a weak correlation. The accepted model was seen to be a black-box approach, which hampered its interpretability.

Greco *et al.* [50] conducted another study in 2021 with the goal of identifying blood analytes that might discriminate between ALS and lower motor neuron illness patients. They also looked at how these individuals were categorized depending on their illness progression rates (high or low). The study examined 108 analytes from 41 ALS patients' blood as well as clinic, demographic, and other data. They created a support vector machine (SVM) model and chose features using the recursive feature elimination approach. Blood data can be used as prediction biomarkers, as shown by the SVM model's accuracy of 87.25% in categorizing ALS patients into high and low advancement groups using the top 16 ranked analytes.

Leo *et al.* [51] introduced a prediction model using both static and longitudinal data and dynamic Bayesian networks (DBN). The aim prediction for the study, which used data from the Portuguese ALS dataset with a sample size of 1,214, was disease progression associated with the requirement for non-invasive ventilation (NIV) assistance. A prediction model known as the stdDBN framework was created by researchers, and it makes use of stationary DBNs to forecast illness development and non-stationary DBNs to ascertain how the biomarkers vary over time in each subgroup.

In terms of accuracy, sensitivity, and area under the curve (AUC) metrics, the average results for predicting illness progression were over 80% for both groups, highlighting the potential of the suggested technique. Graphs showing the values at various time steps for each stage (before and after NIV) were created to show how the biomarkers vary throughout time.

14.6 Proposed research methodology

There are multiple phases in the proposed technique, as shown in Figure 14.8, for employing convolutional neural networks (CNNs) for ALS NDD diagnosis using brain pictures. First, it is necessary to compile a collection of brain pictures from both healthy people and people with ALS. These pictures can be acquired using a variety of imaging methods, including CT scans and MRI. The photos must be preprocessed when the dataset is acquired to make sure they are in the same format and have the same dimensions. Resizing, normalization, and augmentation may be used to achieve this.

The next step is to create and train a CNN model using the preprocessed dataset. Multiple convolutional layers, followed by pooling and dropout layers, can make up the CNN architecture. A softmax layer, which outputs the probability of a given picture belonging to either a healthy person or an ALS patient, can be the last output layer of the CNN. The loss function, which gauges the discrepancy between anticipated and real labels, will be used by the model to change its weights during training. Various optimization algorithms can be used to improve the training process.

The model may be used to categorize fresh brain pictures as belonging to healthy or ALS patients after it has been trained. Several performance metrics, including accuracy, sensitivity, specificity, and AUC, can be used to assess the model's performance. The performance on the test set must match that on the training [52,53] set in order to avoid the model being either over- or underfit to the data. Transfer learning, which uses a pre-trained model and retrains only a few of the layers with the new dataset, can also be used to fine-tune the model. Overall,

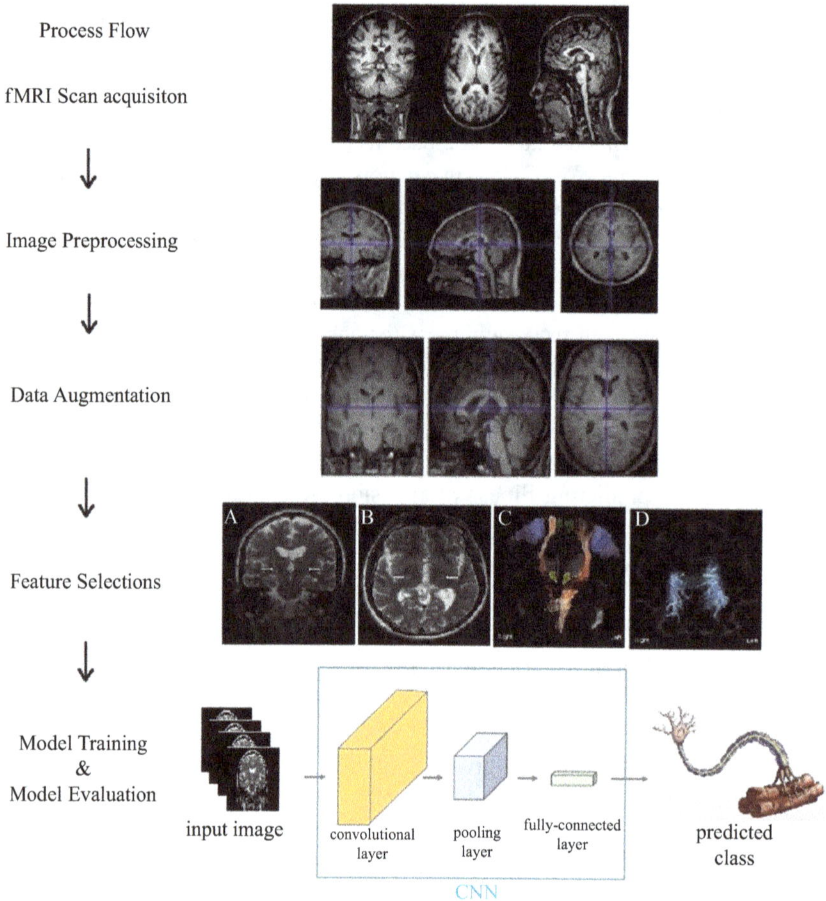

Figure 14.8 Proposed methodology to diagnose ALS using CNN

this suggested approach holds promise for a precise and quick diagnosis of ALS utilizing brain scans and CNNs.

14.6.1 Methodology flow

Data collection: The recommended process starts with data collection. In this step, brain MRI scans from both healthy people and ALS patients are gathered. The suggested CNN model uses the data from these MRI scans as its input.

Data preprocessing: To eliminate any noise or picture artifacts, the acquired MRI scans are first processed. This is done to make sure the model's input data are of a high caliber and increase the model's accuracy.

Datasets and preprocessing for ALS: With no data from other locations, the study on applying ML to predict ALS development was mostly focused on datasets from Europe and the United States. The goal of the project is to evaluate the possibility of ML treatments for ALS and the viability of combining several datasets to produce a more useful one. The PRO-ACT dataset, which includes over 10,000 samples and has been utilized in previous ALS investigations, is primarily employed in this study. However, due to its clinical trial criteria and potential underrepresentation of the actual patient population, prior investigations found flaws and biases in PRO-ACT. In order to address this, external datasets were utilized for validation in order to lower the risk of bias and enhance the evaluation of ML algorithms. The accuracy of the model may be impacted by missing values and data imbalance, and methods like missing data imputation may assist expand the sample size.

Data augmentation: To make the dataset larger, data augmentation techniques are utilized in this stage. MRI scans are subjected to random rotation, flipping, and scaling operations to enlarge the dataset. This broadens the range of pictures in the dataset and improves the generalizability of the model.

Model training: The CNN model will next be trained using the preprocessed and enhanced dataset. The CNN model is created to discover the patterns and characteristics in MRI images that set apart ALS patients from healthy people. Convolutional, pooling, and fully connected layers are only a few of the many layers that make up the CNN model.

Model evaluation: The trained CNN model is tested for accuracy and efficacy in ALS diagnosis using a different dataset of MRI images. Accuracy, precision, recall, and F1 score are among the performance indicators used for assessment.

Model optimization: In this phase, the CNN model's hyperparameters are adjusted to obtain the optimal performance. Convolutional layer hyperparameters like learning rate, number of layers, and number of filters are tuned for optimum accuracy.

Model deployment: In the suggested process, model deployment is the last stage. To diagnose ALS in patients, the trained and improved CNN model is used.

14.6.2 Approaches to predictive machine learning

Indicators of performance as well as various ML algorithms, techniques, datasets, sample sizes, and biomarkers were looked at. Even when comparing successes within the same sort of prediction, it might be challenging. The results generally showed that the capacity to forecast long-term patient functional changes using data from the initial few months, such as at 12 or 24 months, decreased significantly. Making long-term forecasts is still difficult because of the complexity and diversity of ALS.

A major danger of model overfitting is shown by the small number of samples analyzed in [46,53] experiments (135 and 41, respectively), which overshadowed the great accuracies reported by both researchers. When an algorithm overadjusts to the training data, it overfits, which results in the algorithm performing well on the training data but badly on the validation data.

Due to their "black box" character, the ML algorithms utilized in [43,49] both suffer interpretability issues. As a result, it is challenging to comprehend how input and output elements link in these investigations.

In order to evaluate the validity and reliability of the forecasts for prognostication purposes, doctors would need to understand the mechanism underlying them. Strong model performance necessitates a sizable number of biomarkers due to the intricacy of ALS. The interpretation of these models may be more difficult when black-box approaches are used. By lowering the required number of biomarkers, FS approaches may help improve model interpretability. Additionally, a number of ML frameworks, including SHAP and LIME, which are a subfield of explainable artificial intelligence, can be used to explain black-box model predictions. This is a relatively new field of study with the goal of improving the interpretability of AI.

14.6.3 Discussion on review findings

Research is crucial for a better understanding of the disease's course and to enhance the quality of life for individuals who are affected by ALS because it is a terminal and incurable condition that results in death 3–5 years after the beginning of symptoms.

Even if the research in this field is promising, a few problems still need to be resolved. The majority of the patient information included in the research originates from North America and Europe. To verify that the ML solutions can be generalized and applied to other populations, it is important to gather and analyze data from various geographic regions.

A growing number of people are turning to ML models for ALS diagnosis, which may someday displace the present gold standard of clinical evaluation utilizing the El Escorial criteria. In contrast to current ML models for ALS that have attained sensitivity and specificity values equivalent to those of the El Escorial criteria, Gouman [54] claimed that the El Escorial criteria may have poor specificity. However, a lack of training data and poor sample to feature ratios limit the models' ability to perform. Future studies should concentrate on creating models that employ multimodal data, distinguish phenotypes from disease controls within the ALS range, and identify phenotypes. The ultimate goal should be to create models that make it possible to diagnose ALS early, accurately, and independently.

In part because of programs like the Prize4Life challenge, accurate prognostic model development is currently getting a lot of attention [55]. The importance of each attribute to the course of the disease has a significant impact on how well prognostic models work. The majority of existing models are based on clinical data and laboratory testing; however, this may not be enough to properly forecast how the illness will progress. The risk of death, survival, and illness progression rates have all been reasonably estimated by more recent models despite these difficulties [55–59]. Prognostic modeling does have several substantial drawbacks, though, such as data bias, a lack of readily available data, the poor management of missing data, and low sample to feature ratios. Standardized reporting of model performance is important to enable benchmark generation, model comparisons, and repeatability.

14.7 Conclusion and future work

To improve ALS research, there is a strong need for innovative computerized risk stratification initiatives, as this area has been largely overlooked so far. Existing studies have often used simple classification procedures, limited analysis to a few groups, and focused mostly on clinical characteristics. To address this, future research should work with multimodal and longitudinal datasets and examine clustering derived from models using commonly used ALS phenotypes. By using optimized patient stratification strategies, clinically important subgroups of ALS can be identified, which can lead to enhanced clinical trial design.

There is significant potential for ML models in the field of ALS research, both academically and clinically. With the increasing availability of large datasets, multi-center projects, high-performance computing platforms, and open-source analytic tools, ML models are expected to provide more insights than traditional statistical methods. However, selecting the appropriate ML model for a specific application [60–62] requires careful consideration of the data features, model flexibility, assumptions, and limitations. Traditional ML techniques have been employed extensively in ALS research, but new models and neural network architectures have tremendous promise to advance the discipline. It's crucial to keep in mind, though that cutting-edge models like "black box" methods could face the same difficulties as tried-and-true algorithms.

Regardless of the model chosen, it is critical to thoroughly analyze data properties, assess data bias, missing data, and sample to feature ratio in order to verify the generalizability of a suggested ML model. Despite the drawbacks of outdated methodologies, data constraints and limitations are frequently disregarded in modern research. Model overfitting, which prevents the recommended model from being generalized, is the fundamental issue with present studies. Future ML studies in ALS must adhere to stringent reporting requirements, manage missing data in an efficient manner, and conduct open performance evaluations using recognized metrics. ML models will be essential for ALS research, medication discovery, and individualized patient care despite the shortcomings of present models and recent studies and the substantial improvements in the area.

References

[1] Hampel, H., O'Bryant, S. E., Durrleman, S., *et al.* (2018). A precision medicine initiative for Alzheimer's disease: the road ahead to biomarker-guided integrative disease modeling. *Climacteric*, 21(5), 406–411.https://doi.org/10.1080/13697137.2018.1509608.

[2] Sachdev, P. (2014). Vascular cognitive disorder. *Neurobiology of Aging*, 35, S244–S252. https://doi.org/10.1016/j.neurobiolaging.2014.03.033.

[3] Jagust, W. (2013). Imaging the evolution and pathophysiology of Alzheimer disease. *Nature Reviews Neuroscience*, 14(5), 319–329. https://doi.org/10.1038/nrn3461

[4] Basser, P. J., Mattiello, J., and LeBihan, D. (1994). MR diffusion tensor spectroscopy and imaging. *Biophysical Journal*, 66(1), 259–267. https://doi.org/10.1016/S0006-3495(94)80775-1.

[5] Schuff, N., Woerner, N., Boreta, L., *et al.* (2009). MRI of hippocampal volume loss in early Alzheimer's disease in relation to ApoE genotype and biomarkers. *Brain*, 132(4), 1067–1077. https://doi.org/10.1093/brain/awp007.

[6] Wu, Y. C., Alexander, A. L., and Schaefer, S. M. (2019). Magnetic resonance imaging for neurodegenerative diseases: recent developments. *Annals of Translational Medicine*, 7(20), 575. https://doi.org/10.21037/atm.2019.09.129.

[7] Alzheimer's Association. (2021). *What is Alzheimer's?* Retrieved from https://www.alz.org/alzheimers-dementia/what-is-alzheimers.

[8] Huntington's Disease Society of America. (2021). *What is Huntington's disease?* Retrieved from https://hdsa.org/what-is-hd/.

[9] Kumar, A., Singh, A., and Ekavali. (2015). A review on Alzheimer's disease pathophysiology and its management: an update. *Pharmacological Reports*, 67(2), 195–203. doi: 10.1016/j.pharep.2014.09.004.

[10] Sharma, A. K., Nandal, A., Dhaka, A., and Dixit, R. (2021). Medical image classification techniques and analysis using deep learning networks: a review. In *Health Informatics: A Computational Perspective in Healthcare*. Berlin: Springer, pp. 233–258

[11] Varrone, A., Pagani, M., and Nobili, F. (2019). Molecular imaging of neurodegenerative diseases. *Acta Neuropsychologica*, 31(6), 346–357.

[12] Sharma, A. K., Nandal, A., Ganchev, T., and Dhaka, A. (2022). Breast cancer classification using CNN extracted features: a comprehensive review. In *Application of Deep Learning Methods in Healthcare and Medical Science (ADLMHMS-2020)*. Boca Raton, FL: Taylor & Francis: CRC Press.

[13] Holtzman, D. M., Morris, J. C., and Goate, A. M. (2011). Alzheimer's disease: the challenge of the second century. *Science Translational Medicine*, 3 (77), 77sr1–77sr1.

[14] Spillantini, M. G., Schmidt, M. L., Lee, V. M. Y., Trojanowski, J. Q., Jakes, R., and Goedert, M. (1997). α-Synuclein in lewy bodies. *Nature*, 388(6645), 839–840.

[15] Sharma, A. K., Nandal, A., Dhaka, A., *et al.* (2023). HOG transformation based feature extraction framework in modified Resnet50 model for brain tumor detection. *Biomedical Signal Processing and Control*, 84, 104737, ISSN 1746-8094.

[16] Maetzler, W., Liepelt-Scarfone, I., and Berg, D. (2017). Progression of Parkinson's disease in the clinical phase: potential markers. *The Lancet Neurology*, 16(6), 504–515.

[17] Walker, F. O. (2007). Huntington's disease. *The Lancet*, 369(9557), 218–228.

[18] MacDonald, M. E., Ambrose, C. M., Duyao, M. P., *et al.* (1993). A novel gene containing a trinucleotide repeat that is expanded and unstable on Huntington's disease chromosomes. *Cell*, 72(6), 971–983.

[19] Rosas, H. D., Tuch, D. S., Hevelone, N. D., *et al.* (2008). Diffusion tensor imaging in presymptomatic and early Huntington's disease: selective white matter pathology and its relationship to clinical measures. *Movement Disorders*, 23(12), 1663–1672.

[20] Brown, R. H., and Al-Chalabi, A. (2017). Amyotrophic lateral sclerosis. *New England Journal of Medicine*, 377(2), 162–172.

[21] Turner, M. R., Kiernan, M. C., and Leigh, P. N. (2011). Talbot K. biomarkers in amyotrophic lateral sclerosis. *The Lancet Neurology*, 10(3), 249–261.

[22] Dubois, B., Hampel, H., Feldman, H. H., *et al.* (2016). Preclinical Alzheimer's disease: definition, natural history, and diagnostic criteria. *Alzheimer's & Dementia*, 12(3), 292–323.

[23] Sharma, M. K., Dhiman, N., Mishra, V. N., Mishra, L. N., Dhaka, A., and Koundal, D. (2022). Post-symptomatic detection of COVID-2019 grade based mediative fuzzy projection. *Computers and Electrical Engineering*, 101, 108028.

[24] Nandal A., Blagojevic M., Milosevic D. and Dhaka A. (2021). "Fuzzy enhancement and deep hash layer based neural network to detect Covid-19". *Journal of Intelligent & Fuzzy Systems, IOS Press*, 41(1), 1341–1351 .

[25] Kobylecki, C., Jones, M., and Wilcock, G. K. (2006). A review of imaging techniques for the diagnosis of dementia. *Dementia and Geriatric Cognitive Disorders*, 21(1), 1–13. https://doi.org/10.1159/000089638.

[26] Symms, M. R., Jäger, H. R., Schmierer, K., Yousry, T. A., and Miller, D. H. (2004). The sensitivity and specificity of MRI in detecting optic nerve lesions in multiple sclerosis. *Neurology*, 62(5), 1–4. doi: 10.1212/01.WNL.0000110310.87873.27.

[27] Ogawa, S., Lee, T. M., Nayak, A. S., and Glynn, P. (1990). Oxygenation-sensitive contrast in magnetic resonance image of rodent brain at high magnetic fields. *Magnetic Resonance in Medicine*, 14(1), 68–78. doi: 10.1002/mrm.1910140108.

[28] Woods, R. P., Cherry, S. R., and Mazziotta, J. C. (1992). Rapid automated algorithm for aligning and reslicing PET images. *Journal of Computer Assisted Tomography*, 16(4), 620 633. doi: 10.1097/00004728-199207000-00003.

[29] Dale, B. M., Brown, M. A., and Semelka, R. C. (2015). *MRI: Basic Principles and Applications*. Hoboken, NJ: John Wiley & Sons.

[30] Thompson, P. M., Hayashi, K. M., Sowell, E. R., *et al.* (2004). Mapping cortical change in Alzheimer's disease, brain development, and schizophrenia. *NeuroImage*, 23, S2–S18.

[31] Brown, R. W., Cheng, Y. C. N., Haacke, E. M., Thompson, M. R., and Venkatesan, R. (2014). *Magnetic Resonance Imaging: Physical Principles and Sequence Design*. Hoboken, NJ: John Wiley & Sons.

[32] Le Bihan, D., and Johansen-Berg, H. (2012). Diffusion MRI at 25: exploring brain tissue structure and function. *Neuroimage*, 61(2), 324–341. https://doi.org/10.1016/j.neuroimage.2011.11.006.

[33] Jakhar, S. P., Nandal, A., Dhaka, A., Jiang, B., Zhou, L., and Mishra, V. N. (2023). Fractal feature based image resolution enhancement using wavelet–

fractal transformation in gradient domain. *Journal of Circuits, Systems and Computers*, 32(2), 2350035.

[34] Kamali, A., and Kramer, L. A. Chapter 2 – diffusion tensor imaging. In Filippi M. (ed.), *MRI Techniques.* Academic Press; 2010. pp. 21–46. https://doi.org/10.1016/B978-0-12-374227-9.00002-3.

[35] Mori, S. and Zhang, J. (2006). Principles of diffusion tensor imaging and its applications to basic neuroscience research. *Neuron*, 51(5), 527–539. https://doi.org/10.1016/j.neuron.2006.08.012.

[36] Mueller, S., Wang, D., Fox, M. D., *et al.* (2013). Evaluation of treatment effect in a clinical trial of Alzheimer's disease using functional MRI and arterial spin labeling. *Journal of Alzheimer's Disease*, 36(2), 351–364.

[37] Agosta, F., Scola, E., Canu, E., *et al.* (2012). Structural brain connectome and cognitive impairment in multiple sclerosis. *Human Brain Mapping*, 33 (9), 2252–2269.

[38] Qi, Z., Wu, X., Wang, Z., *et al.* (2010). Impairment and compensation coexist in amnestic MCI default mode network. *NeuroImage*, 50(1), 48–55.

[39] Biswal, B., Yetkin, F. Z., Haughton, V. M., and Hyde, J. S. (1995). Functional connectivity in the motor cortex of resting human brain using echo-planar MRI. *Magnetic Resonance in Medicine: An Official Journal of the International Society for Magnetic Resonance in Medicine*, 34(4), 537–541.

[40] Dukart, J., Mueller, K., Horstmann, A., *et al.* (2013). Differential effects of global and cerebellar normalization on detection and differentiation of dementia in FDG-PET studies. *NeuroImage*, 70, 251–268.

[41] Gerdes, M. J., Shin, D. H. and Friston, K. J. (2013). Evidence for topological anomalies in functional brain networks from fMRI studies of Huntington's disease. *NeuroImage*, 2, 162–171.

[42] Ghafoor, U., Kim, S. H., Kim, S. E., Kim, H. K. and Chung, C. K. (2013). Brain activation during attention shifting in patients with obsessive-compulsive disorder: a functional MRI study. *International Journal of Neuroscience*, 123(1), 1–7.

[43] van der Burgh, H. K., Westeneng, H. J., Walhout, R., *et al.* (2017). Deep learning predictions of survival based on MRI in amyotrophic lateral sclerosis. *NeuroImage: Clinical*, 13, 361–369. https://doi.org/10.1016/j.nicl.2016.11.024.

[44] Pires, C., Nunes, R., Machado, S., Sarmento, A. and de Carvalho, M. (2018). Forecasting non-invasive ventilation needs in patients with amyotrophic lateral sclerosis using machine learning. *Journal of Medical Systems*, 42(12), 255. https://doi.org/10.1007/s10916-018-1102-2.

[45] Gordon, P. H. and Lerner, A. J. (2019). Using ordinal classifiers to predict functional decline in ALS: a clinical trial endpoint. *Muscle & Nerve*, 59(3), 292–296. doi: 10.1002/mus.26383.

[46] Kueffner, R., Zach, N., Barachant, A., *et al.* (2019). The DREAM ALS stratification prize4life challenge: crowdsourcing solutions for clinical trial stratification. *Nature Biotechnology*, 37(6), 634–638. https://doi.org/10.1038/s41587-019-0131-4.

[47] Kueffner, R., Zach, N., Bronfeld, M., Lasko, T. A. and Derry, A. (2019). Crowdsourced analysis of clinical trial data to predict amyotrophic lateral sclerosis progression. *Frontiers in Neurology*, 10, 508.

[48] Tang, B., Zhu, X., Zhao, J., Su, X. and Wang, X. (2019). Longitudinal disease progression modeling for amyotrophic lateral sclerosis patient subtyping. In *Proceedings of the 2019 SIAM International Conference on Data Mining*,. Philadelphia, PA: SIAM, pp. 495–503.

[49] Grollemund, V., Prudlo, J., Sanson, A. and Dupuis, L. (2020). A dimensionality reduction approach to predict amyotrophic lateral sclerosis patients' survival probabilities. *Scientific Reports*, 10(1), 1–14.

[50] Greco, G., Cascavilla, A., Monelli, M., *et al.* (2021). Blood-based biomarkers in ALS: a diagnostic approach. *Journal of Neurology, Neurosurgery & Psychiatry*, 92(5), 538–547.

[51] Leão, T., Madeira, S. C., Gromicho, M., de Carvalho, M., and Carvalho, A. M. (2021). Learning dynamic Bayesian networks from time-dependent and time-independent data: unraveling disease progression in Amyotrophic Lateral Sclerosis. *Journal of Biomedical Informatics*, 117, 103730.

[52] Vairamani, A. D. (2023). Detection and diagnosis of diseases by feature extraction and analysis on fundus images using deep learning techniques. In *Computational Methods and Deep Learning for Ophthalmology*. New York: Academic Press, pp. 211–227.

[53] Ajantha Devi, V. (2021). Cardiac multimodal image registration using machine learning techniques. In Chauhan, K. and Chauhan, R. K. (eds.). *Image Processing for Automated Diagnosis of Cardiac Diseases*. New York: Academic Press, pp. 21–33.

[54] Gouman, T. K. (2017). Machine learning applications in amyotrophic lateral sclerosis: achievements, pitfalls, and future directions. *Proceedings of the IEEE*, 105(3), 517–537. https://doi.org/10.1109/JPROC.2016.2633765.

[55] Küffner, R., Zach, N., Norel, R., *et al.* (2014). Crowdsourced analysis of clinical trial data to predict amyotrophic lateral sclerosis progression. *Nature Biotechnology*, 33(1), 51–57. https://doi.org/10.1038/nbt.3051.

[56] Ong, M. L., Tan, P., Holbrook, J. D., Koh, L. W. H. and Ong, V. (2017). Prognostic factors and survival outcome in amyotrophic lateral sclerosis: a review. *Neuropsychiatric Disease and Treatment*, 13, 2149–2155. https://doi.org/10.2147/NDT.S145616.

[57] Schuster, C., Elamin, M., Hardiman, O. and Bede, P. (2017). The segmental diffusivity profile on MRI can help differentiate ALS phenotypes. *Journal of Neurology, Neurosurgery & Psychiatry*, 88(10), 835–841. https://doi.org/10.1136/jnnp-2017-315956.

[58] van der Burgh, H. K., Westeneng, H.-J., Walhout, R., *et al.* (2017). Multicentre validation of a prognostic model for initial amyotrophic lateral sclerosis diagnosis. *Journal of Neurology, Neurosurgery & Psychiatry*, 88 (12), 1042–1045. https://doi.org/10.1136/jnnp-2017-315603.

[59] Westeneng, H.-J., Debray, T. P. A., Visser, A. E., *et al.* (2018). Prognosis for patients with amyotrophic lateral sclerosis: development and validation of a

personalized prediction model. *Lancet Neurology*, 17(5), 423–433. https://doi.org/10.1016/S1474-4422(18)30078-5.

[60] Ajantha Devi, V. and Nayyar, A. (2021). Fusion of deep learning and image processing techniques for breast cancer diagnosis. In Kose, U. and Alzubi, J. (eds.). *Deep Learning for Cancer Diagnosis*. 2021, Singapore: Springer, pp. 1–25.

[61] Sharma, A. K., Amita, N., Arvind, D., Deepika, K., Bogatinoska, D. C. and Alyami, H. (2022). Enhanced watershed segmentation algorithm-based modified ResNet50 model for brain tumor detection. *BioMed Research International*, 2022.

[62] Sharma, A. K., Amita, N., Arvind, D., and Bogatinoska, D. C. (2022). Brain tumor classification via UNET architecture of CNN technique. In *International Conference on Cyber Warfare, Security (SpacSec2021)*, Berlin: Springer, 9–11 Dec 2021. Manipal University Jaipur; pp. 18–33.

Chapter 15

Using deep/machine learning to identify patterns and detecting misinformation for pandemics in the post-COVID-19 era

Javaria Naeem[1] *and Ömer Melih Gül*[2]

The latest advancements in technology have made it simpler for individuals to communicate and share information with one another. On the other hand, this leads to the proliferation of false information due to the fact that it is difficult to determine which information is genuine. In the year 2020, when the globe was facing COVID-19, the spread of this fatal disease was referred to as an "Infodemic" by the World Health Organization. This chapter considers this issue and surveys the existing deep/machine learning-based approaches for the detection of patterns and the identification of false information pertaining to COVID-19 pandemics in the post-COVID-19 era. Using a deep learning method along with convolutional neural networks (CNN) and recurrent neural networks (RNN) models, this information identifies any false information that may be there. First, CNN is put to use in order to extract features from the text. Next, RNN is put to use in order to capture the sequence of the text. These models are trained independently from one another. The output of the CNN is used as an input in the RNN model.

15.1 Introduction

Due to technological advancement, now it's easy to access information through social media, which in turn changed the lifestyle of people. Now people prefer media platforms to get information. In 2020, more than 3.6 billion users used media. There will be a massive increase in social media users by 2025 [1]. There are so many advantages of using social media such as instant and trouble-free communication, business promotions, and online education. However, social platform has various pitfalls due to the open access. One of the well-known ones is misinformation. Misinformation is a decisive combination to the community. This is a big problem for the users of social media. Fake news is a worldwide issue

[1]Department of Big Data Analytics and Management, Bahcesehir University, Turkey
[2]Department of Computer Engineering, Bahcesehir University, Turkey

because of pandemic disasters. Social media platform allows the interaction and dispersal of unique ideas. Users of social media can spread the news and ideas by resharing the information, which leads to misinformation. Social media networking is a platform that helps users to spread fake news swiftly. Media platform is considered a strong device for the dispersion of massive unfiltered data, authorizing a deception occurrence and consequently exacerbating the chances of stratagems the public's discernment of truth by the dispersion of forgery data.

It is strenuous to control the sharing of forgery news in the current world. This indicates that ministry individuals are involved in the procreation of disinformation to a huge viewer. Thus, misinformation has touched nearly all aspects of our life. And the most daunting thing in the latest year is the dispersion of wrong data in this era of the deadly pandemic virus named Coronavirus. In 2019, the deadly virus COVID-19 was reported in China, and this virus has spread all over the world. This virus has killed various people. First, it was considered that this disease is passed on to humans from animals. This virus is transmitted from one person to another person through contact, fomites, and drops. On 14 May 2020, cases of Coronavirus rose to 4.4 million, of which 1.6 million recovered and 298,000 died. According to the latest research, various hearsay and forgery news are spreading about COVID-19. It is strenuous to discriminate false news from data whose accuracy cannot be interrogated [2]. Consequently, fake news on social platforms has increased the situation to the worst level regarding the Coronavirus pandemic.

Researchers started to find the treatment for COVID-19 but false news has escalated on social platform which experts think is a big source of threats to this pandemic. False news about health constitutes a predictable threat to people's health. Consequently, the intense growth of misinformation sharing, especially during global issues, requires large-scale research to understand the nature and aim behind the false information expansion through the social platform. Currently, limited information is known regarding soothsayers of false news increase through social media. Moreover, studies related to forgery news are finite but the data on false information are growing. It is still an unknown reason for misinformation proliferation via social platforms. The last piece of evidence has proved that false news on health problems is not new. However, the emergence of media platform which allows people to share data without any barrier has increased the procreation of fake data in the health sector [3]. There is a massive amount of forgery information and its immediate spread which results in users being mixed up, wanting accurate and true information related to the epidemic. Media, for example, Facebook, Instagram, and Twitter, have provided interaction between people across the globe. However, due to the massive information, some of the data are real and some are not real. False news leads users the wrong way. Therefore, it is necessary to find a way to recognize false information related to COVID-19.

Recognizing false news is not a simple thing; the main aim of this is to describe fake facts. Research on false news identification has used distinct models of classification. False news on media platforms has a major pessimistic social effect. As a consequence, the detection of false news on social media has fascinated researchers' engrossment. The present methods for identifying fake data use information and media content as sources for studying operations. False news is responsible for misleading users and dispersal of fake data. Currently, the detection of false information is carried out by manual checking, and suspicious cases are given to specialists for confirmation. This process is time-taking and laborious and we cannot get a rapid response. At once, a procedure for creating false is more advanced. The content of false news contains a broad scale of domains and it exchanges quickly. Various countries have developed fact-finding centers to fight the dispersion of false information, including Agence France-Presse Fact Check and PolitiFact. Corporations hired qualified individuals to analyze news data produced by massive media and try to give timely recognition of news purposeful to be false. The degree of false news on deadly diseases has become a pandemic. These reports reveal their aims and families to mistreat and take to antagonism and abomination among various people, which results in different social issues [4].

Hardly one or two studies have scrutinized the identification of Coronavirus related to false news, and present findings and awareness endeavors rely heavily on suitable experts, which can restrict the disclosure of conflicting viewpoints, avoiding separate items from having numerous annotations, thus bringing down the hassle of model training [5–7].

This research proposes a method for the identification of false news using deep learning (DL) techniques such as convolutional neural networks (CNN) and recurrent neural networks (RNN). Feature learning can be carried through using a massive amount of false and real news content to make sure the model can identify the characteristic and writing type of false information, hence attaining high detection precision.

This chapter defines a technique for recognition of fake rumors about the pandemic. We used a hybrid model by combining CNN and RNN for getting high accuracy.

15.2 Literature review

This part includes false news content and an analysis of the increased amount of studies that analyze false information related to COVID-19.

15.2.1 Difference between misinformation and disinformation

An exact definition of fake news is currently unavailable but it can be classified as "misinformation" or "disinformation." The distinction between them is that misinformation is purposefully forging and exciting rumors to provide an advantage to someone [8]. On the other hand, misinformation is the spread of invalid information and is not relevant to expert verification [9].

False information on the internet is first considered misinformation because the main aim of spreading wrong information is to defraud people. The target of giving wrong information is to cause a broad conversation between people that converts to deception that can affect belief of people while hiding the origin and

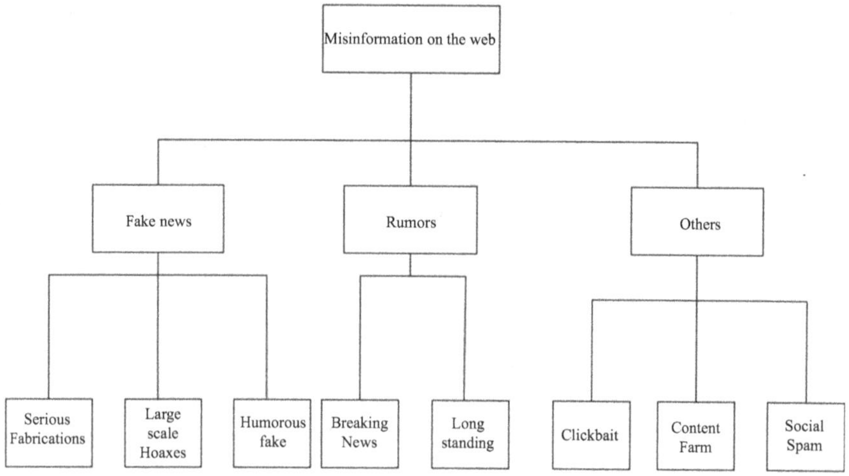

Figure 15.1 Categorization of various types of misinformation on website

genuine intention. Misinformation on the internet is categorized into rumors, false news, and others as shown in Figure 15.1.

15.2.1.1 Fake news

When information is manufactured or deliberately misrepresented with the intention of being disseminated as news, this is referred to as "fake news." Purpose of rumor is often to deceive or mislead people, to spread propaganda or promote a particular agenda, or to generate clicks and revenue through sensational headlines and stories.

Fake news can take many forms, including completely made-up stories, misleading or exaggerated headlines, manipulated images or videos, and selectively edited quotes. Additionally, it can be disseminated via a number of platforms, including social media, websites, blogs, and even conventional news sources.

Fake news is divided into three parts: numerous news, serious fabrication, and large-scale hoaxes. The most common type of fabricated news is serious fabrications that describe purposefully deceitful texts and photographs. Articles like this are typically disseminated over several social platforms. The size of the information generated is larger than that of basic news and is created to mislead people in society. Hoaxes on a large scale often originate from well-known individuals, and they tend to contain more extensive information compared to regular news [10]. Numerous web-based data such as online news, Facebook and Instagram posts and public announcements, and other social media content are considered for the detection of false information. Moreover, various websites exist for verifying content, including Polifact, Snopes, and Boomlive, which gather verified information about the accuracy of popular claims. These websites are particularly adept at detecting false information related to COVID-19 and other commonly discussed topics (Chakraborty, 2020).

Fake news can have serious consequences such as spreading misinformation and damaging people's reputations or causing social unrest. It can also reduce the credibility of legal sources and destroy public trust in social platforms.

To prevent the dissemination of false information, it's crucial to adopt a critical approach toward the sources of information we consume and verify the accuracy of claims before sharing them. Moreover, trustworthy news outlets implement fact-checking protocols to ensure the accuracy of their reporting, making them a dependable source of information.

15.2.1.2 Rumors

Rumors are pieces of information or stories that are passed around informally, often through word of mouth, and typically without any evidence to support them. They can be true, false, or somewhere in between, and can take many different forms, including gossip, urban legends, and conspiracy theories.

Rumors often spread quickly and can have serious consequences, particularly when they relate to public figures, events, or issues. They can cause harm to individuals, communities, and organizations, and can fuel social unrest and conflict. One of the challenges of dealing with rumors is that they are often difficult to verify or refute, and can persist even in the face of evidence to the contrary. This is partly because rumors can tap into people's fears, anxieties, and prejudices, and can be more emotionally compelling than facts.

There is no precise definition for rumors circulating online. Currently, the prevailing definition is "information that is publicly announced but lacks verification and displays signs of manipulation" [11]. Subsequent definitions of rumors have described them as stories that are "told with a significant degree of certainty, but are challenging to confirm and may incite fear even when they seem credible" [12]. Studies have classified rumors into two categories: long-standing and breaking news. Long-standing rumors persist in the media for a long time but lack verification and cannot be fully authenticated. Conversely, breaking news rumors emerge from unexpected events and spread rapidly. Certain individuals intentionally disseminate false information to cause confusion or misunderstanding. As these rumors are often convincing and misleading, they can spread easily and pose a risk to public safety. Such rumors should be identified and stopped from further spread [9]. To address rumors, it's essential to approach information without bias and question claims that lack supporting evidence. It's also vital to consult reliable sources of information to verify the accuracy of claims.

15.2.1.3 Others

The use of captivating headlines or outlines that are distributed on social networks and are intended to entice people to click through to the material that is not precisely represented by the outline is an example of the practice known as "clickbait" which plays a major way in the dissemination of false information [13].

"Spam" is an email sent to a user with attractive headings which convince a user to download or open files. Spam is to get personal information [14].

A "content farm" is a website that publishes articles of poor quality using a variety of ways, with the goal of rapidly generating a large amount of traffic in order to further the website's financial interests [15]. These articles divert people and mostly include false information. These articles get heavy traffic and come on the highest rank on search engine optimization because of which people easily get distracted.

Out-of-context quotes involve using a quote from someone that has been taken out-of-context in order to support a particular argument or viewpoint. By only using part of a quote, the original meaning may be distorted or misinterpreted.

Conspiracy theories involve promoting unfounded or baseless claims that are often designed to appeal to people's fears or prejudices. Conspiracy theories may involve complex and convoluted explanations that are difficult to disapprove of and can be used to support a range of different viewpoints.

False flags involve creating a fake incident or even to blame it on someone else, often with the intention of justifying a particular action or policy. False flag attacks may be used to create a sense of crisis or to justify increased security measures.

Astroturfing involves creating the appearance of a grassroots movement or campaign in order to promote a particular agenda or viewpoint. Astroturfing may involve using fake social media accounts or paying people to post or comment on articles in support of a particular viewpoint.

Echo chambers involve surrounding oneself with like-minded individuals or sources of information which can reinforce existing beliefs and make it harder to consider alternative viewpoints.

15.2.1.4 COVID-19 disinformation

During the COVID-19 pandemic, false information proliferated on an unprecedented scale. Consequently, researchers have examined the connection between social media platforms and the spread of misinformation during the epidemic [2,16]. The volume of misinformation related to COVID-19 has soared on the internet. Fact-check finds out that the messages are not fully fictional, information is mixed from different sources in a way to show wrong information for example, adding a forged caption to an original picture, thus remarkably surging the trustworthiness of the information.

On a regular basis, misleading material concerning COVID-19 is disseminated across various social platforms like YouTube, Facebook, and Instagram. These social media businesses have begun to implement the strategy in order to identify and remove false information; however, some fake data still appear on these sites. Fake information about COVID-19 is divided into different categories in Table 15.1.

15.2.2 Detection of fake news

Over the past years, false information has outspread speedily on media platforms and the prediction of fake news has become hard. The latest research investigates the difference between wrong and true information, using ML and DL.

Table 15.1 Types of COVID-19 forgery information [17]

Category	Description
Transmission avoidance	Use of particular method to stop or detect COVID-19
Virus origins	Is COVID-19 natural or made by men?
Home remedies	Taking particular food to cure the virus
Virus characteristics	Characteristics of COVID-19
Uncontrolled transmission	Which countries and cities are facing uncontrolled COVID-19
Conspiracy theories	Conspiracies about scientific data
Medicines and vaccines	Making of COVID-19 vaccines
Discriminatory views	Discrimination between groups due to COVID-19

Natural language processing (NLP) technique, for instance, word embeddings and sentimental analysis, can be used for analyzing text of news articles and determining whether they contain misleading or false information. For example, a neural network could be trained to identify patterns of language that are commonly associated with fake news articles, like using emotional language or the repetition of certain phrases. Another method called metadata analysis can also be used to analyze metadata associated with news articles such as the source of the article, the date it was published, and the social media engagement it has received. By analyzing these factors, a neural network can determine whether the news article is likely to be trustworthy or not.

Certain scholars have observed that satirical language can be detected as a means of identifying misinformation, utilizing a single support vector machine (SVM). SVM also gives a good result with an accuracy rate of 97% for finding misinformation in other than the English language [18]. In another research, SVM is used for the detection of fake information on Twitter [19]. Some of the research used random forest tree classifiers to find fake news and some used random forest and term frequency-inverse document frequency to find wrong political data on social platforms with an accuracy of 85% [20]. A method is proposed by Hua *et al.* in which an amalgamation of sentence features, category tags, parts-of-speech tags, and databases of vocabulary attributes is gathered for finding fake information from posts on social platforms [13].

Furthermore, the COVID-19 pandemic has led to heightened ethnic tensions as a result of the "infodemic" of false information circulating online that targets specific groups for discrimination, causing significant concern among academic and government institutions at the national level. Several studies have proposed an approach for analyzing data related to COVID-19 [21,22], which involves utilizing three transformer models—BERT (bidirectional encoder representations transformers), ALBER, and XLNET (auto agression transformer) for identifying false news. The F1-score for this method was found to be 98.55%. Ref. [23] proposed a method for the detection of fake news by using different DL techniques. A model was developed to analyze text in both English and Chinese, and results were compared based on distinct text features. This method involves utilizing the short-term memory model, bidirectional long and short memory model, and gated recurrent unit. The BiLSTM model yielded the most accurate detection results, with 94% accuracy for short English sentences and 99% for long English sentences, while accuracy for Chinese text was found to be 82% [11]. Patwa *et al.* proposed a method by using binary classification for finding fake and real news. This method also includes other four algorithms—support vector machine, logistic regression, decision tree, and gradient boost [1].

In another research, information fusion is done to get real information from Government, health, and broadcasting websites. However, false information is gathered from social platforms. Thirty-nine features were generated from multimedia content and used to find false information about Coronavirus by implementing DL techniques. This method increased the results from 59% to 86% [24]. Researchers in 2021 worked on CTF, COVID-19 dataset of Twitter false news. In addition, they proposed the cross-SEAN model, which was a cross-stitch-based semi-supervised end-to-end neural attention model. This approach acknowledged the massive volume of unlabeled data. The cross-SEAN method recognizes fake news by getting knowledge from external data [25]. In 2021, the BERT model was suggested for detecting misinformation related to social bot activities during the COVID-19 era. Researchers proposed the BERT technique to find fake information. They used the transfer learning method to find accounts of bot in a dataset of COVID-19 [8].

Research work by Mahbub *et al.* demonstrates the misinformation by including a psycho-linguistic feature which helped to find out the difference between the types of misinformation and it shows how this feature increases the detection rate of false information of COVID-19 on Twitter [26]. In 2021, [27] proposed a method for fake news detection using two approaches: The first one is fine-tuning language model based on a transformer with powerful loss functions; the second one is the removal of deleterious training instances by controlled calculation. With the first method, accuracy was 98% and after cleaning the data the accuracy improved to 99%, which proves the importance of generalization models. In the same year, the researcher evaluated the pre-trained transformer-based model for the detection of COVID-19 misinformation. The study that is being offered has dealt with a variety of transformer and recurrent models, along with certain word-implanted models that are related. In addition to this, the effectiveness of the model is evaluated by substituting a different loss function for the traditional loss function in the analysis. It showed that the use of related language models and personalized loss functions improves the result Hande *et al.* and Tashtoush *et al.* [28,29] proposed a methodology for the recognition of COVID-19 misinformation on social media networks using DL techniques such as CNN, LSTM, bi-directional LSTM, and the fusion of CNN and LSTM. This framework can automatically detect and classify the fake data of COVID-19 on social media networks. Models were trained and tested for finding false information using a dataset of COVID-19 which includes 21,379 real and false news data. The real news was assembled from international and authentic websites, for example World Health Organization, the United Nations Children's Fund, and the International Committee of the Red Cross.

On the other hand, fake information is gathered from various fact-checking sites. The result showed that CNN is the best model for the detection of fake news. Bangyal *et al.* [30] proposed various DL methods for the detection of misinformation on COVID-19. The first step was the data processing which includes the removal of noise, tokenization, replacement of null values and stemming. For the representation of data, a model is applied with inverse document frequency and term frequency. In the last step, measuring and estimation of eight machine learning (ML) methods such as random forest, K-nearest neighbors, logistic regression, support vector machine, decision tree, Adaboost, naive Bayesian, and neural networks, and four algorithms of deep learning CNN, RNN, LSTM, and GRU were used.

15.3 Proposed approach

The research about misinformation needs lots of experimentation by utilizing ML methods on a large scale of datasets. New methods get a deep-down understanding of the type of false information and the mechanism of fake news spreading all over the world. This chapter includes a method of identifying fake using a hybrid model which includes deep learning methods CNN, RNN, and RF. Hybridization of these algorithms improves the performance of the rumor recognition model.

Nowadays, bio-inspired algorithms are commonly used in various fields, and among them, neural networks are a widely applied example. Recently, many new techniques in bio-inspired computing are introduced for the detection of misinformation, for example detection and pinpointing fraudulent users [31] and opinion mining [32].

For scheduling, optimization, and searching problems, bio-inspired computing methods are used; however, neural networks are mainly used for classification and prediction. In this chapter, neural network algorithms are used as a classification for fake information identification. This, however, does not limit the potential of other bio-inspired techniques to address a wide range of problems that involve the propagation of information through different channels. The convolutional neural network has been proven to be tremendously effective for the identification of fake pictures and fact-checking by the implication of face recognition, object detection, classification, and characterization [33].

15.3.1 Neural networks

ML algorithms called neural networks are inspired by the structure and function of the human brain. These algorithms are composed of layers of interconnected processing nodes, also known as neurons. The neurons receive inputs, perform calculations, and generate outputs that are passed to other neurons in the network. Neural networks are a ML approach that mimics the structure and function of human brain, utilizing interconnected processing units or "neurons" arranged into layers. During training, the connections between the neurons can be modified to improve the network's performance on a specific task. These networks can model intricate and nonlinear relationships between input and output, making them ideal for complex tasks such as image and speech recognition. Additionally, neural networks can handle computationally intensive jobs as they can perform multiple computations simultaneously. These networks are robust to noise and can handle missing or incomplete data. This makes them useful for tasks such as data cleaning and data imputation. Neural networks can adapt to changing environments or data distributions. This makes them useful for tasks such as online learning and dynamic environments. These networks can generalize well to new, unseen data. This makes them useful for tasks such as prediction and classification, where the goal is to make accurate predictions on new data.

When processing data, neural networks typically receive input in the form of matrices or numerical vectors, which are subjected to various calculations such as additions and multiplications, to adjust the network's parameters. As there are multiple parameters involved, changing them can have a significant impact on the network's output, for example, loss function, weights, and biases which are used in the initial step and activation function at every layer.

The aim of activation functions on the layer is for getting the output for the next layer from the input data and weights and biases. The last layer is the output which provides results. There is a function which is called the loss function which helps to find the error between the calculated and required result. The loss function acts as an optimizer, and weights are arbitrarily initialized and then trained using an optimization method.

This fake information identification method lies in the category of binary classification. The output shows that news or information is either true or fake.

In a neural network, input data can be fed in a vector or matrix format to the network. Each word in the text is defined as a unique vector. These word vectors are called word embeddings. Word embeddings that are trained generally belong to a specific domain for finding the statistical relationship between all the words present in the collection. Rather than training word embeddings, it is better to use pre-trained word embedding which is available publicly. The extremely liked pre-trained word embeddings are Word2vector supplied by Google and GloVe. Words if they match the word from the text are kept; otherwise, they are eliminated [34].

15.3.2 Convolutional neural network

CNN is the most used algorithm. If we compare CNN with the previous algorithms, the fact that CNN can identify similar characteristics on its own, without the assistance of a person, is the platform's primary strength. CNN has been used in various fields including face recognition, computer vision, and speech recognition. The design of CNN was influenced by human and animal brain neurons [35].

CNN includes the multiplication of matrices that gives output to the next layer for proceeding training method. This process is called convolutional. The CNN is so named due to its reliance on convolutional operations. In NLP, individual words in a text are referred to as word vectors. These word vectors are utilized to train CNN with training process typically involving the specification of a kernel size and the number of filters. One-dimensional structures are commonly used in the field of text classification and NLP. In the CNN model, a filter of the determined measurement of windows iterates across the training data, which doubles the input and produces an output that saves in the output array. The array which shows the output is a feature map. By this method, the feature is identified from input data. The measurement of the filter is identified as kernel size and the amount of filter defines the amount of feature map for use. With the help of this method, CNN can derive the local features. There are the following steps involved in CNN:

1. Convolutional layer
   The first layer in a CNN is typically a convolutional layer, which consists of a set of learnable filters that slide over the input image. Each filter is applied to a small patch of the output image, and the output of the filter is a feature map that highlights the presence of certain local features such as edges, corners, or blobs.
2. Pooling layer
   Following each convolutional layer, a pooling layer is usually applied to reduce the spatial resolution of the feature map and enhance the network's robustness against minor input image shifts. It is customary to include pooling layers after every convolutional layer, with max pooling being the most common type. Max pooling selects the features map region with the highest value and sends it to the next layer with multiple values.
3. Fully connected layer
   Once the input image has been processed by several convolutional and pooling layers, the output is flattened into a vector and passed through one or more fully connected layers. These layers are similar to those used in other neural networks, and they are designed to learn a nonlinear mapping between the input and output.
4. Output layer
   A SoftMax layer is frequently employed as the final layer in a CNN to estimate the probability of each class based on the input image. The predicted class is determined by identifying the class with the highest probability.

15.3.3 Recurrent neural network

RNN includes the ordered processing of input for learning. During the preparation of data, a sequence is followed: the output of every single step is used as an input for the next step which is why it is called recurrent. This is carried out by remembering the previous output. This creates a dependency on training data.

In the field of NLP, it is common to recommend multiple informative articles for learning purposes that are related to each other instead of learning them individually. Recurrent neural network models typically include layers that incorporate memory cells, which can come in various types such as LSTM. The LSTM cell state is responsible for carrying information in sequence along with the current word vector. This ensures that information is retained throughout the ordered operation. LSTM models utilize weights and three types of gates during the learning process, namely as input gate, an output gate for value prediction, and a forget gate for eliminating redundant data. The following steps are usually involved in an RNN:

1. Recurrent layer
   The first layer in an RNN is typically a recurrent layer, which consists of a set of learnable weights that are applied to the input data at each time step. The output of the recurrent layer is a hidden state that encodes details about the previous inputs and the current input.
2. Activation function
   Once the input has been processed by the recurrent layer, an activation function is usually applied to the output to introduce nonlinearity into the network. The most common activation function used in RNNs is the hyperbolic tangent (tanh) function, which squashes the output between -1 and 1.
3. Fully connected layer
   After the activation function, the output is passed through one or more fully connected layers that are similar to those used in other neural networks. These layers are designed to learn a nonlinear mapping between input and output.
4. Output layer
   The final layer in an RNN is usually a SoftMax layer, which computes the probability of each class given the input sequence. The class with the most probability is taken as the predicted class.

15.3.4 Random forest

Random forest is a widely used machine learning approach that can be applied to both classification and regression tasks. This technique is based on ensemble learning, which involves the combination of multiple decision trees to make predictions. The fundamental idea behind a random forest is to construct many decision trees using various subsets of the training data and different subsets of the features, and then aggregate their predictions. The following are the primary steps involved in building a random forest.

1. *Randomly sample the training data:* The first step in building a random forest is to randomly sample the training data with replacement. This is called bootstrapping. Bootstrapping creates multiple datasets of the same size as the original dataset, but each dataset is slightly different because it contains some duplicate observations and some missing observations.
2. *Build a decision tree:* Next, for each bootstrapped dataset, we build a decision tree using a random subset of the features. This is done to prevent any one feature from dominating the model. Typically, we randomly select the square root of the total number of features, but we can also experiment with different values.
3. *Repeat step 2 many times:* Repeat step 2 again and again, typically hundreds or even thousands of times, to create a large number of decision trees.
4. *Combine the decision trees:* Finally, to make a prediction, we combine the predictions of all the decision trees by taking their average (for regression) or by taking a majority vote (for classification).

Random forest has various advantages:

- Random forest is one of the most accurate machine learning algorithms available. It typically performs well even with complex datasets that have many features.
- Random forest is relatively easy to use compared to other algorithms. There are only a few hyperparameters to tune, and the default values often work well.
- Random forest is robust to outliers because it averages the predictions of many decision trees, which helps to reduce the impact of outliers.
- Random forest can handle the missing data because it uses bootstrapping to create multiple datasets with slightly different observations.
- Random forest can handle unbalanced datasets because it uses a majority vote to make a prediction, which helps to reduce the impact of minority classes.

15.3.5 Hybrid CNN-RNN-RF model

This proposed model includes the combination of CNN, RNN, and RF. Both RNN and CNN extract features from the preprocessed text. The method used for the feature extraction is word embeddings from RNN and a combination of word embeddings and n-grams from CNN. Then select the most important features from RNN and CNN using methods such as feature importance scores. After that, predictions of the RNN, CNN, and RF models are combined using an ensemble model. The selected feature will go as an input to RF.

The selected feature is utilized as an input for the RF model. A hybrid approach combining CNN and RNN is shown to be effective in numerous classification and regression tasks due to their ability to capture local and sequential data patterns.

Along with advantages, there are disadvantages too. For example, finding optimal parameters and datasets is such a hectic thing; the need for huge training datasets and the absence of interpretability have a major impact on performance.

15.4 Methodology

The implementation of this hybrid method is in python and the performance is evaluated on two fake information datasets. The following sections define the detail of the datasets and describe the implementation.

15.4.1 Datasets

In this chapter, four databases of COVID-19 fake information are used which are public. (1) COVID19- FNIR, (2) COVID19FN, (3) Fake news database, (4) COVID19 news dataset both fake and real.

A. COVID19-FNIR
   This dataset contains 7,588 news about the Coronavirus (COVID-19). The data are gathered from USA, India, and Europe in February 2020. Moreover, real information was collected from Twitter, and wrong information from Poynter.

B. COVID19FN
   This dataset consists of four categories which include title, country, text and label, publish date, and URL of the source. Articles were gathered from Poynter site and other sites.
C. Misinformation database
   This dataset consists of six attributes which include ID, date, text, subject, title, and label. Approximately 31,400 articles were gathered from social sites.
D. Fake and real information dataset
   This dataset consists of 16,988 true and false news. It was gathered from two sources, i.e., CBC and web portals.

15.4.2 Data-cleaning

Data cleaning includes five steps which are as follows:

Text parsing

In the text parsing step, a tokenization method is used for splitting for analysis.

Data cleaning

In the cleaning of data, regular expression is used for extracting English letters, numbers and a mix of both. This method helps reduce the noisy data.

Part of speech tagging

In the part of speech method, each word is assigned according to the part of speech as noun, verb, or adjective.

Removal of stop words

All the words which are the same are removed such as "the," "and," and "a."

Stemming

In this method, the replacement process is applied in order to replace each single word with it root; this way redundancy is removed from the word. Moreover, English stemmer is utilized in order to reduce the text in datasets.

15.4.3 Feature extraction method

During the process of feature extraction, two subphases are applied. During these subphases, DL and ML techniques of word embedding and term frequency-inverse document frequency are utilized.

1. In ML, for assigning the weights to the text in preprocessing phase TF-IDF is used. It helps in determining the frequencies of words in the text.
2. In DL, word embedding is used to convert words into vector form. The same words have the same vectors. The glove technique is a famous unsupervised learning method to produce vector presentation.

15.5 Proposed method

In this part, the framework of our proposed method is explained. Figure 15.2 shows the group of layers of method which works in a sequence for finding the

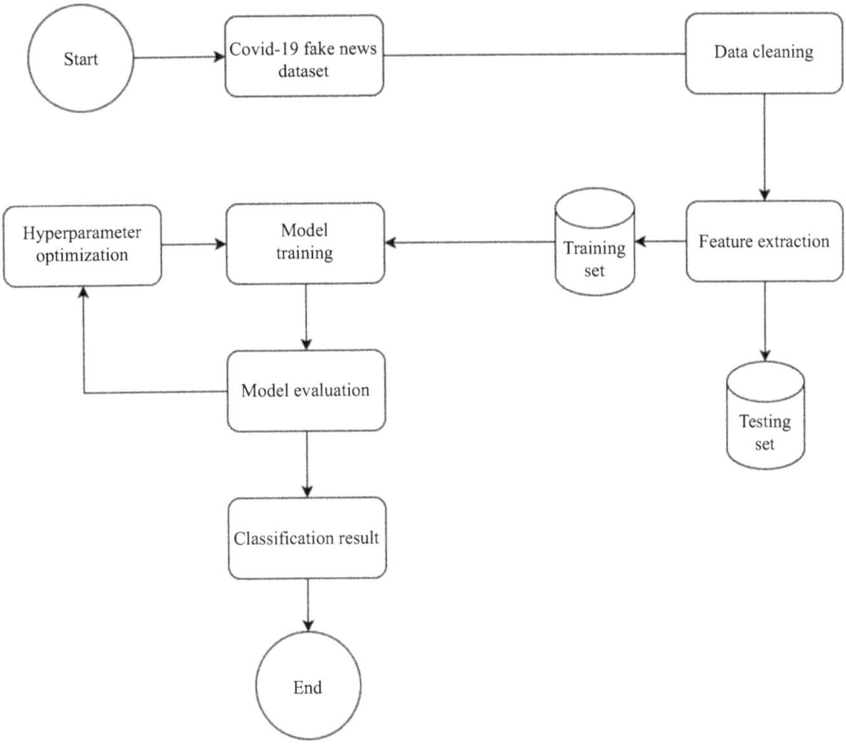

Figure 15.2 Data fusion

identification of misinformation. Moreover, we used hyper-optimization for improving the performance of the model. The description of layers is explained as follows:

1. Embedding layer:
   In this layer, unique words are converted into the vector. Kera library is used that contains three parameters. The input-dim consists of the size of the vocabulary and the output-dim consists of the input sequence length.
2. Drop-out layer:
   A drop-out layer is used to avoid overfitting and it helps to lower complexity. This layer takes the output from the embedding layer. For configuration, the value is assigned in the range of 0.1–0.9.
3. Convolutional layer
   This layer uses the output of the drop-out layer as input. The convolutional layer has two sections: filter and feature map. In the first section, filtering is applied to input words that provide a map with features. Features describe the design of input. An activation function named ReLU was implemented for the recognition of features in the news.

4. Pooling
   This layer involves the max operation for lowering the features inside the mapping procedure of the feature, especially setting a high-value help in fetching the important trait.
5. LSTM
   LSTM is a kind of RNN which is specifically used for forecasting. This layer is used for joining the two models, namely CNN and RNN.
6. Flatten layer
   In this layer, data are transformed into one-dimension, and the result of the layer is used in the next layer.
7. Dense layer
   This is an intensely linked RNN layer. It includes a few parameters such as activation, kernel, input, and bias. The input variable describes the data of the input. The kernel describes the weight of data and activation describes the activation function.
8. Output layer
   This layer takes the output of flattened layer for getting the concluding output.

15.6 Comparison of models

Various models have been correlated with the proposed method:

1. There are seven ML models that are compared with the presented model, which includes NB, SVM, LR, KNN, DT, PCA, and K-means.
2. LSTM which is a kind of RNN has been compared. This includes five layers: embedding, hidden, dropout, flatten, and output layer. Embedding layer is the same as used in the proposed model. The hidden layer is present in LSTM in which a weight adjusting method is used with a regular rate for 12. The next layer is dropout; this layer eradicates the overfitting problem and clarifies the model. In this layer, value is set in the range of 0.1–0.9. The next layer is the flattened layer; it transforms the whole text into a vector form. The last layer is the output layer; it takes in the output generated by the flattened layer to give final result, which shows the real and fake news classified results.
3. CNN contains embedding, convolutional, pooling, flatten, dense, and output layers.
4. RF includes random sampling, random feature sampling, decision tree creation, ensemble learning, and prediction.

15.6.1 Hyperparameter optimization method

Hyperparameter optimization is used for optimizing automatically the hyperparameters for ML and DL models. For the selection of the best-fitted values, hyperparameter tuning is used. It increases the performance of the models.

15.6.1.1 Optimization of the ML model

For the optimization of the ML model, a grid search approach is used. Initial values have been defined for individual hyperparameters. The model goes through the values and chooses the best one, which helps the model get a high level of accuracy. Moreover, the K-fold cross-validation method is utilized in which datasets are equally split up into K-fold. K-1 is for preparation and the remaining K-fold is for testing. This method is repeated up until the procedure outstretch that every single fold has been utilized for testing. In the end, classifier is tested on the basis of precision within ten-fold.

15.6.1.2 Optimization of the DL model

A built-in python library is used named Hyperopt, in which hyperparameters are dispersed asynchronous. Hyperopt is a free library for AutoML.

15.6.2 Evaluation benchmarks

Four common ML metrics are considered, namely accuracy, recall, F1-score, and precision. Equations (15.1)–(15.4) show the formula for calculation, in which TP is true positive, TN is true negative, FP is false positive, and FN is false negative.

1. Accuracy is a widely used benchmark for ML and DL frameworks. It calculates the percentage of accurately forecasted observations. The formula for calculating accuracy is

$$\text{accuracy} = \frac{\text{TP} + \text{TN}}{\text{TP} + \text{FP} + \text{TN} + \text{FN}} \tag{15.1}$$

2. The other benchmark is precision which shows the ratio of TP. The formula for calculating precision is

$$\text{precision} = \frac{\text{TP}}{\text{TP} + \text{FP}} \tag{15.2}$$

3. The third benchmark is recall, which is used to find out the sum of positive results from true class. The formula for calculating recall is

$$\text{recall} = \frac{\text{TP}}{\text{TP} + \text{FN}} \tag{15.3}$$

4. The last is the F1 score. It describes the trade-off between recall and precision. It gives the weighted average. The formula for calculating F1 score is as follows:

$$\text{F1 score} = \frac{2.\text{precision}.\text{recall}}{\text{precision} + \text{recall}} \tag{15.4}$$

15.7 Future work

In this chapter, three CNN, RNN, and RF models were used for finding the fake news about COVID-19. Our datasets are limited to text only. In the future, we will include emoji analysis to analyze whether the text is positive or negative; moreover, we will include images and perform analyses of fake images.

In the future, we will address the concerns regarding the safety of healthcare systems. Combating hostile cyberattacks is extremely important for essential infrastructures in medical systems such as those that store personal health records in a secure environment. Lately, a number of academics have proposed a solution using DL called RF fingerprinting as a means of overcoming malicious attacks, particularly spoofing attacks [34–36]. The research presented in [36–38] suggests a fine-grained data augmentation strategy as a means of enhancing the performance of a DL-based RF fingerprinting method. In addition, the blockchain may be utilized to maintain the confidentiality of medical records, making it a potentially useful tool for fascinating applications like improving driver safety and transparency [39]. Hence, we can propose data augmentation in deep learning algorithms for the detection of patterns and the identification of false information pertaining to COVID-19 pandemic in the post-COVID-19 era.

15.8 Conclusion

In this chapter, the combination of three models, CNN, LSTM, and RF, has been proposed, which helps in detecting COVID-19 misinformation. Both CNN and LSTM are used for extracting the features and after extraction features were used by the RF random forest model for improving the accuracy of the method.

References

[1] Patwa, P., Sharma, S., Pykl, S., *et al.* (2021). Fighting an infodemic: Covid-19 fake news dataset. In *Combating Online Hostile Posts in Regional Languages during Emergency Situation: First International Workshop, CONSTRAINT 2021, Collocated with AAAI 2021, Virtual Event, February 8, 2021, Revised Selected Papers 1,* Springer International Publishing, pp. 21–29.

[2] Huynh, T. L. (2020). The COVID-19 risk perception: a survey on socioeconomics and media attention. *Economics Bulletin*, *40*(1), 758–764.

[3] Waszak, P. M., Kasprzycka-Waszak, W. and Kubanek, A. (2018). The spread of medical fake news in social media–the pilot quantitative study. *Health Policy and Technology*, *7*(2), 115–118.

[4] Velasquez, N., Leahy, R., Restrepo, N. J., *et al.* (2021). Online hate network spreads malicious COVID-19 content outside the control of individual social media platforms. *Scientific Reports*, *11*(1), 11549.

[5] Vraga, E. K. and Bode, L. (2020). Defining misinformation and understanding its bounded nature: using expertise and evidence for describing misinformation. *Political Communication*, *37*(1), 136–144.

[6] van Der Linden, S., Roozenbeek, J. and Compton, J. (2020). Inoculating against fake news about COVID-19. *Frontiers in Psychology*, *11*, 566790.

[7] Elhadad, M. K., Li, K. F. and Gebali, F. (2020). Detecting misleading information on COVID-19. *IEEE Access*, *8*, 165201–165215.

[8] Heidari, M., Zad, S., Hajibabaee, P., *et al.* Bert model for fake news detection based on social bot activities in the covid-19 pandemic. In *2021 IEEE 12th Annual Ubiquitous Computing, Electronics & Mobile Communication Conference (UEMCON)*. IEEE, (2021), pp. 0103–0109.

[9] Bondielli, A. and Marcelloni, F. (2019). A survey on fake news and rumour detection techniques. *Information Sciences*, *497*, 38–55.

[10] Rubin, V. L., Chen, Y. and Conroy, N. K. (2015). Deception detection for news: three types of fakes. *Proceedings of the Association for Information Science and Technology*, *52*(1), 1–4.

[11] Chen, M. Y., Lai, Y. W. and Lian, J. W. (2022). Using deep learning models to detect fake news about COVID-19. *ACM Transactions on Internet Technology*, *23*(2), 1–23.

[12] DiFonzo, N. and Bordia, P. (2007). Rumor, gossip and urban legends. *Diogenes*, *54*(1), 19–35.

[13] Hua, W., Wang, Z., Wang, H., Zheng, K. and Zhou, X. (2016). Understand short texts by harvesting and analyzing semantic knowledge. *IEEE Transactions on Knowledge and Data Engineering*, *29*(3), 499–512.

[14] Silverman, C. (2015). Lies, damn lies and viral content.

[15] Shu, K., Mahudeswaran, D., Wang, S., Lee, D. and Liu, H. (2020). Fakenewsnet: a data repository with news content, social context, and spatiotemporal information for studying fake news on social media. *Big Data*, *8* (3), 171–188.

[16] Hou, Z., Du, F., Jiang, H., Zhou, X. and Lin, L. (2020). Assessment of public attention, risk perception, emotional and behavioural responses to the COVID-19 outbreak: social media surveillance in China. *MedRxiv*, 2020, pp. 1–2.

[17] Chakraborty, T., Shu, K., Bernard, H. R., Liu, H. and Akhtar, M. S. (eds.) (eds.). (2021). *Combating Online Hostile Posts in Regional Languages during Emergency Situation: First International Workshop, CONSTRAINT 2021, Collocated with AAAI 2021, Virtual Event, February 8, 2021, Revised Selected Papers*(Vol. 1402). Springer Nature, pp. 16–17.

[18] Ahmed, H., Traore, I. and Saad, S. (2017). Detection of online fake news using n-gram analysis and machine learning techniques. In *Intelligent, Secure, and Dependable Systems in Distributed and Cloud Environments: First International Conference, ISDDC 2017, Vancouver, BC, Canada, October 26–28, 2017, Proceedings 1*. Springer International Publishing, New York, vol. 10618, pp. 127–138.

[19] Nikam, S. S. and Dalvi, R. (2020). Machine learning algorithm based model for classification of fake news on twitter. In *2020 Fourth International*

Conference on I-SMAC (IoT in Social, Mobile, Analytics and Cloud)(I-SMAC). IEEE, India, pp. 1–4.

[20] Jehad, R. and Yousif, S. A. (2020). Fake news classification using random forest and decision tree (j48). *Al-Nahrain Journal of Science, 23*(4), 49–55.

[21] Sharma, M. K., Dhiman, N., Mishra, V. N. "Post-symptomatic detection of COVID-2019 grade based mediative fuzzy projection". *Computers and Electrical Engineering*, Elsevier, vol. 101, 2022.

[22] Zhou, L., Chaudhary, S., Sharma, M. K., Dhaka, A., and Nandal, A. (2023). Artificial neural network dual hesitant fermatean fuzzy implementation in transportation of COVID-19 vaccine. *Journal of Organizational and End User Computing (JOEUC), IGI Global, 35*(2), 1–23.

[23] Gundapu, S. and Mamidi, R. (2021). Transformer based automatic COVID-19 fake news detection system. V3, 1–12. *ArXiv Preprint ArXiv:2101.00180.*

[24] Iwendi, C., Mohan, S., Ibeke, E., Ahmadian, A. and Ciano, T. (2022). Covid-19 fake news sentiment analysis. *Computers and Electrical Engineering, 101*, 107967.

[25] Paka, W. S., Bansal, R., Kaushik, A., Sengupta, S. and Chakraborty, T. (2021). Cross-SEAN: a cross-stitch semi-supervised neural attention model for COVID-19 fake news detection. *Applied Soft Computing, 107*, 107393.

[26] Mahbub, S., Pardede, E. and Kayes, A. S. M. (2022). COVID-19 rumor detection using psycho-linguistic features. *IEEE Access, 10*, 117530–117543.

[27] Bang, Y., Ishii, E., Cahyawijaya, S., Ji, Z. and Fung, P. Model generalization on COVID-19 fake news detection. In *Combating Online Hostile Posts in Regional Languages during Emergency Situation: First International Workshop, CONSTRAINT 2021, Collocated with AAAI 2021, Virtual Event, February 8, 2021, Revised Selected Papers 1*. Springer International Publishing; 2021. pp. 128–140.

[28] Hande, A., Puranik, K., Priyadharshini, R., Thavareesan, S. and Chakravarthi, B. R. Evaluating pretrained transformer-based models for COVID-19 fake news detection. In *2021 5th International Conference on Computing Methodologies and Communication (ICCMC)*. IEEE, 2021, pp. 766–772.

[29] Tashtoush, Y., Alrababah, B., Darwish, O., Maabreh, M. and Alsaedi, N. (2022). A deep learning framework for detection of COVID-19 fake news on social media platforms. *Data, 7*(5), 65.

[30] Bangyal, W. H., Qasim, R., Rehman, N. U., *et al.* (2021). Detection of fake news text classification on COVID-19 using deep learning approaches. *Computational and Mathematical Methods in Medicine, 2021*, 1–14.

[31] Aswani, R., Kar, A. K. and Vigneswara Ilavarasan, P. (2018). Detection of spammers in twitter marketing: a hybrid approach using social media analytics and bio inspired computing. *Information Systems Frontiers, 20*, 515–530.

[32] Yadav, A. and Vishwakarma, D. K. (2020). A comparative study on bio-inspired algorithms for sentiment analysis. *Cluster Computing*, 23, 2969–2989.

[33] Dhillon, A. and Verma, G. K. (2020). Convolutional neural network: a review of models, methodologies and applications to object detection. *Progress in Artificial Intelligence*, *9*(2), 85–112.

[34] Nasir, J. A., Khan, O. S. and Varlamis, I. (2021). Fake news detection: a hybrid CNN-RNN based deep learning approach. *International Journal of Information Management Data Insights*, *1*(1), 100007.

[35] Alzubaidi, L., Zhang, J., Humaidi, A. J., *et al.* (2021). Review of deep learning: concepts, CNN architectures, challenges, applications, future directions. *Journal of Big Data*, *8*, 1–74.

[36] Comert, C., Gul, O. M., Kulhandjian, M., *et al.* Secure design of cyber-physical systems at the radio frequency level: machine and deep learning-driven approaches, challenges and opportunities. In: Traore, I., Woungang, I. and Saad, S. (eds.) In *Artificial Intelligence for Cyber-Physical Systems Hardening. Engineering Cyber-Physical Systems and Critical Infrastructures*, vol 2. Springer: Cham; 2023.

[37] Comert, C., Kulhandjian, M., Gul, O. M., *et al.* Analysis of augmentation methods for RF fingerprinting under impaired channels. In *Proceedings of the 2022 ACM Workshop on Wireless Security and Machine Learning,* 2022, May, pp. 3–8.

[38] Gul, O. M., Kulhandijan, M., Kantarci, B., Touazi, A., Ellement, C. and D'Amours, C. On the impact of CDL and TDL augmentation for RF fingerprinting under impaired channels. *48th Wireless World Research Forum (WWRF 2022)*, 7–9 November 2022, UAE, pp. 1–6.

[39] Gul, O. M., Kulhandjian, M., Kantarci, B., Touazi, A., Ellement, C. and D'Amours, C. Fine-grained augmentation for RF fingerprinting under impaired channels. In *2022 IEEE 27th International Workshop on Computer Aided Modeling and Design of Communication Links and Networks (CAMAD)*. IEEE, 2022, November, pp. 115–120.

Chapter 16

Integrating medical imaging using analytic modules and applications

Amit Kumar[1], *Kanchan Rani*[1], *Satendra Kumar*[1], *Ajeet Singh*[1] *and Arpit Kumar Sharma*[2]

The proposed framework for integrating various medical imaging modalities using analytical modules is described in the abstract for integrating medical imaging using analytical modules. The framework's objectives include enhancing clinical workflow, making treatment planning easier, and improving diagnostic accuracy. The suggested system extracts useful elements from medical photos and incorporates them into a coherent diagnostic report using a combination of machine learning, image processing, and data analytics techniques. The framework is meant to be flexible, scalable, and compatible with current clinical systems. Better patient outcomes could result from the proposed framework's potential to considerably enhance the accuracy and effectiveness of medical imaging and diagnosis.

16.1 Introduction

Integrating analytical modules with medical imaging is an effective method for deciphering and analyzing complex medical images. In order to obtain useful information from the images, this technique combines cutting-edge imaging technologies including magnetic resonance imaging (MRI), computed tomography (CT), and positron emission tomography (PET) scans with powerful analysis tools (Figure 16.1). Analytical modules enable the automated processing of medical pictures, making it easier to identify and characterize anomalies and quantify disease development. Healthcare professionals can increase diagnostic precision [1], pinpoint viable treatments, and keep track of patient outcomes by integrating medical imaging with analytic modules. This strategy could completely change the area of radiology, representing a huge advancement in medical imaging technology. Artificial intelligence (AI) and machine learning are becoming helpful

[1]Department of CSE, Moradabad Institute of Technology, India
[2]Department of Computer and Communication Engineering, Manipal University, India

	CT	MRI	PET	SPECT	Ultrasound
Example	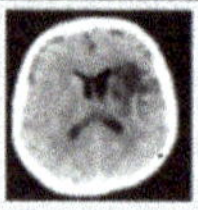	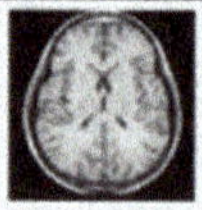	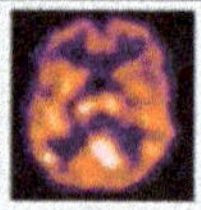	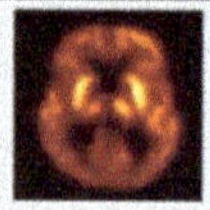	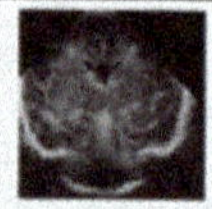
Main characteristic	Scan body organs using X-rays and produce a series of cross-sectional based images via the computer	Produce "slices" that represents the human body through applying magnetic signals	Nuclear imaging technique example where the tracers are used in diseases diagnosis	A non-invasive based technique where cross-sectional images of radiotracer within the human body are structured	Sound waves based technique that possesses a high temporal frequency and which is capable of producing quantitative and qualitative diagnostic information through a set of comprised methodologies

Figure 16.1 Various techniques of medical imaging [7]

technologies in the field of medical imaging. Radiologists can benefit from AI systems' ability to examine massive amounts of medical picture data, spot trends, and help them make quicker and more precise diagnoses. AI algorithms, for instance, can support surgical planning, detect and categorize tumors, evaluate the course of diseases, forecast treatment outcomes, and much more.

16.2 Applications of medical imaging

16.2.1 Radiology and diagnostic imaging

To identify and track a number of illnesses, a wide variety of medical imaging techniques are used in radiology and diagnostic imaging. Deep learning and AI have changed the way to view and analyze medical images. Some important radiology and diagnostic imaging applications are presented here [2].

16.2.1.1 Image interpretation

To achieve higher accuracy and efficiency, radiologists can use deep learning and AI algorithms. For example, machine learning models can automatically detect and concentrate abnormalities in X-rays, MRI images, and CT scans [3]. This can help accelerate the diagnosis process and improve the early detection of diseases.

16.2.1.2 Computer-aided detection/diagnosis

By using AI and deep learning models, CAD systems analyze medical images and highlight potential abnormalities for further evaluation by radiologists. These systems can indicate suspicious regions, such as masses or tumors, aiding radiologists in making more accurate diagnoses [4].

16.2.1.3 Quantitative imaging

Automated quantitative measurement extraction from medical images is made possible by AI techniques. By examining periodical scans over time, deep learning systems, for instance, can estimate tumor volume, gauge the course of the disease, or monitor patient outcomes. This data can help to keep track of patients' results and treatment of patients [5].

16.2.1.4 Image reconstruction and enhancement

Deep learning techniques have been used to increase low-dose imaging and improve image quality. AI systems use large datasets to rebuild high-resolution images from lower-resolution scans. However, acquired images are noisy and incomplete. To overcome these issues, a variety of image reconstruction and enhancement methods are applied. Some common methods are Fourier transform reconstruction, parallel imaging, compressed sensing, image enhancement filter, and machine learning-based techniques such as convolutional neural networks (CNN) [5].

16.2.1.5 Workflow optimization technique

AI systems can help radiology workflows run more smoothly by automating repetitive operations. AI algorithms, for instance, based on priority, reduce the waiting time for urgent situations. On the other hand, these techniques can help in organizing the report, extracting information, and creating structured reports for better communication.

16.2.1.6 Radiomics

Radiomics is a method that brings out a huge [6] number of attributes from medical images by using advanced feature analysis. Data derived from radiomic processes, such as intensity, texture, shape, and wavelength, can be retrieved from medical images. Radiomics exercise has several steps involved such as image acquisition, segmentation, feature extraction, feature selection, and data analysis (Figure 16.1).

16.2.2 Pathology

Pathology, a branch of medical imaging, is the study and analysis of images created from tissue samples or specimens. Pathologists use microscopes to examine tissue sections. Pathologists play an important role in the diagnosis of illness, identifying anomalies, and direction of treatment. Physical inspection of samples is still required but AI system and deep learning have created a new chance. There is a strong connection between pathology and medical imaging.

16.2.2.1 Digital pathology

Digital pathology is a field of pathology that focuses on digital imaging technology. Digital pathology plays an important role in clinical pathology. Pathologists can use this technique to improve the efficiency and accuracy of the diagnosis process. Deep learning is a tool to assist pathologists in clinical diagnosis.

16.2.2.2 Whole slide imaging

As pathology is going under transformation, digital pathology is becoming more and more important. Whole slide imaging (WSI) is a special utility in pathology that converts glass slides into digital slides.

16.2.2.3 AI-assisted pathology

A computer system is said to have AI if it is able to learn and solve problems in the same way that a human can. Human intelligence is needed for many jobs, and AI should be able to handle them all, including visual perception, decision-making, and communication. The relatively new topic of AI-based computational pathology has shown considerable promise in improving the precision and accessibility of high-quality health treatment for patients across a wide range of specialties. Using AI technologies can improve the pathologic diagnosis, classification, prediction, and prognosis of diseases because of the massive amount of data generated throughout the patient care lifecycle [8].

16.2.2.4 Image analysis and quantification

Medical images have several attributes. These attributes should be quantified and analyzed for better image quality. CNN has become an important tool in medical imaging [9].

16.2.2.5 Computational pathology

Utilizing AI and machine learning techniques, computational pathology combines pathology data with other clinical data, including genomes, radiomics, and patient outcomes. Computational pathology tries to create prediction models for illness diagnosis, prognosis, and individualized therapy plans by merging various data sources [10].

16.2.3 Cardiology

16.2.3.1 Telehealth

Cardiologists may now monitor patients from a distance, offer consultations, and manage chronic illnesses thanks to telemedicine, which has grown significantly in popularity in recent years. It makes it possible for patients, especially those in rural locations or with limited mobility, to get healthcare promptly [11].

16.2.3.2 Wearable technology

The heart rate, activity level, sleep patterns, and even aberrant heart rhythms can all be monitored by these devices. They provide helpful data that can aid in monitoring heart health and identifying early warning signs. Smart watches and fitness trackers are two examples of wearable technology that have gained popularity [12].

16.2.3.3 Artificial intelligence

Cardiology has greatly benefited from AI. Large volumes of patient data, including medical records, imaging, and genetic data, can be analyzed using machine learning algorithms to help with risk assessment, early detection, and individualized treatment strategies. In analyzing ECG data, forecasting cardiovascular events, and assisting in picture analysis for better diagnosis, AI systems have shown its potential.

16.2.3.4 Cardiology intervention

The way that cardiovascular illnesses are treated has been revolutionized by interventional cardiology techniques. Trans-catheter aortic valve replacement, stenting, and other minimally invasive procedures reduce the need for open heart surgery and enhance patient outcomes [13].

16.2.3.5 Cardiac imaging

The ability to see the anatomy and operation of the heart has improved thanks to advancements in cardiac imaging technologies including echocardiography, cardiac MRI, and CT. These methods support heart disease diagnosis, inform therapy choices, and track the success of interventions [14].

16.2.3.6 Genetic analysis

Genetic testing has gained importance in cardiology, especially for inherited heart disorders. Early detection and focused management for those who are at risk and their families help discover genetic abnormalities that may predispose people to particular heart disorders [15].

16.2.4 Neuroimaging

16.2.4.1 Functional magnetic resonance imaging

The commonly used neuroimaging method known as functional magnetic resonance imaging (fMRI) is capable of evaluating the activity of the brain by tracking variations in blood oxygenation levels. Our comprehension of the connection and operation of the brain has transformed. The application of fMRI in cognitive neuroscience research permits the finding of areas of the brain responsible for an array of cognitive operations, especially language processing, memory, attention, and emotion [16].

16.2.4.2 Diffusion tensor imaging

Using the aid of diffusion tensor imaging, investigators are currently able to gain insight into how the brain develops, how atypical connections in neurological conditions affect these diseases, and how these disorders develop later on. The assessment of water particle diffusion in brain cells is an area of expertise of the diffusion tensor imaging MRI techniques. White matter pathways can be identified, which reveals the links among the building blocks of the brain [17].

16.2.4.3 Positron emission tomography

The metabolic and biochemical activity of the brain is tracked using radioactive tracers in this form of measurement [18].

16.2.4.4 Electro-encephalography

Electro-encephalography (EEG) analyses the electrical signals of the brain using electrodes on the scalp. Research on the brain waves and patterns related to various mental states and neurological disorders is conducted using this method [19].

16.2.4.5 Magneto encephalography

Some important applications of magneto encephalography (MEG) are HTR, Functional brain mapping, and source localization. Basically, MEG focuses on the connectivity of brain components with each other and how they communicate with each other [20].

16.2.5 Ophthalmology imaging

Clinical evaluation, diagnosis, treatment, and ongoing monitoring of a wide range of anterior and posterior segment diseases rely heavily on ophthalmic imaging technologies. The morphological and functional evaluation of ocular structures has been greatly enhanced by the introduction and development of imaging technologies such as anterior and posterior segment optical coherence tomography (OCT), OCT angiography, retinal photography, specular and confocal microscopy, corneal topography, and ocular ultrasound. Biomarkers and predictors for high-incidence ocular and neurological illnesses have also been discovered thanks to multimodal imaging.

Researchers are increasingly interested in the use of AI and deep learning in ophthalmology because it has shown significant improvement in the accuracy and sensitivity of detecting common eye diseases like age-related macular degeneration, diabetic retinopathy, and glaucoma, and thus has the potential to become an indispensable tool in the prevention, early detection, treatment, and monitoring of these conditions.

16.2.5.1 Optical coherence tomography

Non-invasive OCT creates high-resolution cross-sectional pictures of tissue. Because of this, it is particularly helpful in areas where a biopsy cannot be performed, such as the human eye. Due to their shared operational principles, OCT is frequently compared to medical ultrasound. Both MRI and ultrasound use the wave's echo off the tissue structure to create an image of the organ or tissue. Analysis of the delayed back-reflected waves can provide the depth at which the reflection occurred. Near-infrared light, like that used in OCT, travels at a significantly higher speed than ultrasound. It is not possible to directly measure the delays of the back-reflected waves; therefore, a reference measurement is utilized. The interferometer sends some of the light to the sample and some to a reference arm whose length is already known [21].

16.2.5.2 Fundus photography

Ocular fundus examinations are typically documented using fundus photography. Using a fundus camera, fundus photographs are often taken in a medical facility. However, fundus cameras are not always readily available, even in settings where they would be most useful such as hospital emergency rooms or patient rooms. Alternative methods of documenting retinal pictures include indirect ophthalmoscope-mounted contact and noncontact fundus cameras and handheld contact and noncontact fundus cameras; however, their availability and high cost prevent their routine usage in most ophthalmology practices. Smartphones with high-resolution cameras are finding growing applications in ophthalmology, particularly for studying the anterior region and retina [22].

16.2.5.3 Corneal topography

Imaging of the cornea is crucial in the detection and treatment of many eye conditions. Corneal topography is used to define the cornea's anterior surface and its overall shape. The Placido disc is the foundation of most corneal topographical systems, and it is used to evaluate rings reflected off the cornea. Placido disc technology cannot be used to characterize the posterior corneal surface. Corneal ectasia can be diagnosed with the use of imaging of the back of the cornea. Tomographers, as opposed to corneal topographers, reconstruct the anterior region in three dimensions and reveal data concerning corneal thickness. Scheimpflug imaging is a popular method of corneal tomography [23]. The anterior and posterior corneal surfaces are located using cross-sectional pictures captured by a revolving Scheimpflug camera. Among corneal topography's many medical applications are the detection of corneal ectasia, the evaluation of corneal astigmatism, and the preparation for refractive surgery.

16.2.5.4 Virtual and augmented reality

The use of virtual reality (VR) and augmented reality (AR) in healthcare has steadily increased over the past few years. In addition, radiographic images are crucial in making diagnoses and arranging surgical procedures. Since most AR systems rely on user-generated visual input, ophthalmology and AR are inextricably intertwined fields. This means that ophthalmology is a particularly fruitful field for AR to explore. For individuals who cannot undergo conventional medical or surgical treatment, AR has great promise as an alternate or adjuvant option that is both non-invasive and convenient. AR has been used in ophthalmology for the purpose of education and clinical assistance other than treatment [24].

16.3 Key aspects of integrating medical imaging

In order to increase decision-making, facilitate communication among healthcare practitioners, and improve patient care, integrating medical imaging [25] entails integrating imaging data and technologies into the larger healthcare system. Incorporating medical imaging involves the following important factors [26,27].

16.3.1 Interoperability

Interoperability is the capacity of different systems to exchange and share information with one another. Care is increasingly being provided by a network of different institutions. Also, views, specialized care delivery, continuity of treatment, and unforeseen changes in care are just few of the reasons patients travel between health systems. There are many ways in which healthcare interoperability helps with the aforementioned problems such as providing a means for the exchange of patient test results and treatment plans as part of a transaction reducing the burden of extra administrative work on doctors that can help them stay focused on patient care. Cost savings in manual processes and fax transmissions make data-aggregation resources available to medical professionals and hospitals.

16.3.2 Picture archiving and communication system

To safely save and digitally transfer electronic images and therapeutically important reports, medical facilities often employ a technology called picture archiving and communication system (PACS). When you use PACS, you would not have to worry about securely filing, storing, retrieving, or sending any of your films, reports, or other sensitive data by hand. Instead, PACS software, workstations, and mobile devices can securely access medical records and images stored in off-site servers from virtually anywhere in the world. As the use of data analytics on medical image data becomes more commonplace, medical image storage solutions like PACS will become increasingly crucial.

16.3.3 Electronic health records

A patient's paper chart has been replaced by an electronic health record (EHR). EHRs are contemporaneous, patient-centric records that enable authorized individuals immediate, encrypted access to patient data. While it is true that an EHR contains patients' medical and treatment histories, such a system is designed to incorporate more than just the conventional clinical data obtained in a provider's office. EHRs play an important role in health information technology because they can store a patient's medical history, diagnoses, medications, treatment plans, immunization dates, allergies, radiology images, and test results.

16.3.4 Decision support systems

By combining decision support systems with medical imaging, healthcare professionals can receive help from these systems in interpreting images, providing clinical recommendations, and making diagnostic recommendations based on algorithms and AI. Accuracy and effectiveness in diagnosis can both be enhanced by these technologies.

16.3.5 Telemedicine and remote access

Integrating medical imaging with telemedicine platforms enables healthcare professionals to remotely see and interpret images, enabling consultations and collaborations across many locations. Particularly in underdeveloped areas, it improves access to knowledge and facilitates prompt decision-making.

Integrating medical imaging with data analytics and AI technologies can allow automated analysis, picture identification, and quantification. In the areas of abnormality identification, risk assessment, and treatment planning, AI algorithms can be helpful.

16.3.6 Clinical workflow optimization

By incorporating medical imaging [8] into clinical workflows, procedures are made more efficient and manual effort is decreased. To automate operations, keep track of patient imaging orders, and promote prompt reporting and communication, this may entail connecting imaging systems with radiology information systems and other hospital administration systems.

Working together with physicians, IT experts, healthcare executives, and radiologists is necessary for the integration of medical imaging. Better outcomes, more efficient processes, better patient care, and higher diagnostic accuracy can all result from it. However, to guarantee its effective installation and utilization, it necessitates careful planning, adherence to data privacy and security requirements, continuing maintenance, and training.

16.4 Analytic modules in integrating medical imaging

While many specialized image analysis techniques remain primarily in the hands of their authors and collaborators, a number of imaging frameworks have arisen to meet the needs of research groups in terms of functionality and visualization capabilities. The effort required to integrate these methods and close the gap between them and the technology suitable for clinical research and patient trials in the field of image analysis is enormous.

16.4.1 Image segmentation

Segmentation is a crucial process in the field of Computer Vision. Segmentation of a picture is the process of breaking it up into different sections that have a common characteristic. This clustering activity requires the use of narrow criteria, such as color or texture. Pixel-level categorization is another name for this procedure. In other words, it entails breaking up pictures (or video frames) into smaller pieces that may be processed separately.

16.4.2 Registration and fusion

The goal of picture registration is to make images spatially coherent with one another. Two photos are needed as input to this procedure; the original is called the reference image, and the image to be aligned is called the sensed image. The information gained from this stage can be used in further analysis procedures such as image fusion. Instead of using individual photos for each step, image fusion uses many input images to create a single, more detailed, and accurate final image.

16.4.3 Quantitative analysis

The use of numeric values allows for accurate monitoring of treatment outcomes and the establishment of baselines for image-based claims.

16.4.4 Texture analysis

In modern medical practice, digital images are used for diagnosis. In general, texture analysis is a method for assessing the location and strength of signal elements in digital images, such as pixels, and the intensity of their grayscale values. In reality, the pixel distribution in an image is analyzed mathematically to determine texture features, which in turn characterizes the texture type and, by extension, the underlying structure of the objects seen in the image. Texture analyses can be broken down into four broad categories based on the techniques used to assess the connections between pixels: structural, model-based, statistical, and transform.

16.4.5 Deep learning and AI algorithms

AI has the ability to significantly alter the medical imaging industry. To weave design solutions that account for ethical and regulatory requirements and develop AI-based algorithms that improve outcomes, quality, and efficiency, a firm grounding in the principles and application of radionics, artificial neural networks, machine learning, and deep learning is essential. Implementing AI technologies in a way that is both ethical and sustainable requires taking a more comprehensive view of their potential uses, benefits, and risks.

16.4.6 Visualization and 3D reconstruction

One way to think of 3D reconstruction is as the “rebuilding” of something that has been destroyed. However, in the field of medical imaging, data acquisition methods that essentially “tear” data apart (or capture the components of data one by one) are typically required in order to observe the underlying features. Once all the data have been pieced back together, it is crucial to be able to display it or visualize it.

16.5 Algorithm for integrating medical imaging using analytic modules

Using analytical modules to integrate medical imaging entails the processing and analysis of medical pictures within the framework of a larger healthcare system. The following are some essential methods for fusing analytical modules and medical imaging.

16.5.1 Data preprocessing algorithms

These algorithms are employed to get medical imaging data ready for analysis. To maintain uniform data quality and format across several modalities or time points may require activities like image denoising, image normalization, picture registration, or image resampling.

16.5.2 Segmentation algorithms

To recognize and delineate particular features or regions of interest within medical pictures, segmentation algorithms are used. For precise and automatic segmentation, these algorithms can make use of a number of techniques, including thresholding, region expanding, graph cuts, or machine learning-based methods like CNNs.

16.5.3 Feature extraction methods

To extract pertinent information from medical images for additional analysis, feature extraction methods are used. These algorithms employ strategies like shape analysis, intensity-based features, texture analysis, or statistical characteristics. The traits that are extracted offer quantifiable data for processing and decision-making later on.

16.5.4 Machine learning techniques

Incorporating medical imaging data requires the use of machine learning techniques. On the basis of labeled datasets, these algorithms can be trained to categorize photos, find anomalies, forecast treatment outcomes, or aid in diagnosis. Support vector machines (SVM), random forests, or deep learning techniques like CNNs are a few examples of machine learning methods utilized in medical imaging.

16.5.5 Fusion algorithms

Fusion [28] algorithms combine data from many imaging modalities or data sources to present an all-encompassing picture of the patient's state. These algorithms combine and align data from several sources, making it possible to analyze the data more comprehensively. Image registration, voxel-based fusion, and feature-level fusion methods are some examples of fusion techniques.

16.5.6 Decision support algorithms

To assist healthcare professionals in making decisions, decision-support algorithms link imaging data with clinical data. These algorithms can be used to help with diagnosis, treatment planning, or prognosis prediction. They can also include machine learning models, clinical recommendations, and expert knowledge. They support the use of imaging data for making decisions that are supported by facts.

16.5.7 Visualization algorithms

Medical imaging data are presented and interpreted in a relevant and interactive fashion using visualization algorithms. These algorithms may use 3D visualization techniques, multi-planar reconstruction, volume rendering, or surface rendering. Visualization algorithms support the interpretation and dissemination of imaging results to patients and healthcare professionals.

Together, these algorithms function as analytical modules to gather important data, support the planning of diagnoses and treatments, and provide seamless integration of medical imaging data across the entire healthcare system.

16.6 Conclusion

In summary, integrating medical imaging with analytic modules is a promising strategy that has the potential to revolutionize the healthcare industry. This integration makes it possible to do an in-depth analysis, precise diagnosis, and individualized treatment planning by fusing a variety of imaging modalities and utilizing cutting-edge analytical tools.

Data preprocessing, algorithm creation, and module integration are essential components of medical imaging integration. By tackling issues including noise reduction, picture registration, segmentation, and normalization, data preprocessing methods are essential in improving the consistency and quality of medical images.

It is crucial to create specific algorithms designed for integrating medical imaging. These algorithms ought to be able to work with a variety of imaging modalities, extract pertinent features, and provide effective data fusion. Deep learning, image fusion, and multi-modal analysis techniques are frequently used to extract valuable insights from combined medical pictures.

To offer a thorough evaluation of a patient's condition, module integration combines the results of various analytic modules. To ensure accurate interpretation and decision-making based on the integrated results, effective communication and coordination between modules are essential.

To fully realize the therapeutic advantages and efficacy of these integrated techniques, issues including interoperability, standardization, and the requirement for extensive validation studies still need to be addressed.

In general, using analytical modules to integrate medical imaging has significant promise for improving healthcare outcomes. To progress this discipline and realize its full potential for bettering patient care and outcomes, the medical imaging and data analytics communities must continue to engage in research, innovation, and collaboration.

Machine learning studies must pay close attention to data pretreatment. It is done before building a learning model so that accurate input data can be gathered. Data preprocessing is a crucial stage in any machine learning research project, necessitating not only an in-depth familiarity with the data but also an in-depth understanding of the problems that may arise with the data and a knowledge-based strategy for fixing those problems.

Integrating medical imaging algorithms ought to be able to work with a variety of imaging modalities, extract pertinent features, and provide effective data fusion. Deep learning, image fusion, and multi-modal analysis techniques are frequently used to extract valuable insights from combined medical pictures.

To offer a thorough evaluation of a patient's condition, module integration combines the results of various analytic modules. To ensure accurate interpretation and decision-making based on the integrated results, effective communication and coordination between modules are essential.

The use of analytical modules to integrate medical imaging has the potential to advance medical research, increase the accuracy of diagnoses, and enable individualized treatment plans. To fully realize the therapeutic advantages and efficacy of these integrated techniques, issues including interoperability, standardization, and the requirement for extensive validation studies still need to be addressed.

In general, using analytical modules to integrate medical imaging has significant promise for improving healthcare outcomes. To progress this discipline and realize its full potential for bettering patient care and outcomes, the medical imaging and data analytics communities must continue to engage in research, innovation, and collaboration.

References

[1] Sharma, A. K., Nandal, A., Dhaka, A. and Dixit, R. (2020). "A survey on machine learning based brain retrieval algorithms in medical image analysis," *Health and Technology*. 10, 1359–1373. https://doi.org/10.1007/s12553-020-00471-0.

[2] Gong, H., Hu, Q., Walther, A., *et al.* (2020). "Deep-learning-based model observer for a lung nodule detection task in computed tomography." *Journal of Medical Imaging*, 7(4), 042807. https://doi.org/10.1117/1.JMI.7.4.042807.

[3] Sharma, A. K., Nandal, A., Dhaka, A. and Dixit, R. (2021). "Medical image classification techniques and analysis using deep learning networks: a review.": In Patgiri, R., Biswas, A. and Roy, P. (eds.). *Health Informatics: A Computational Perspective in Healthcare. Studies in Computational Intelligence*, vol. 932. Singapore: Springer. https://doi.org/10.1007/978-981-15-9735-0_13.

[4] Nishikawa, R. M. (2007). Current status and future directions of computer-aided diagnosis in mammography. *Computerized Medical Imaging and Graphics, 31*(4–5), 224–235.

[5] Sharma, A. K., Nandal, A., Dhaka, A., Koundal, D., Bogatinoska, D. C. and Alyami, H. (2022). "Enhanced watershed segmentation algorithm-based modified ResNet50 model for brain tumor detection." *BioMed Research International*, 2022, Article ID 7348344, 14 pages. https://doi.org/10.1155/2022/7348344

[6] Mayerhoefer, M. E., Materka, A., Langs, G., *et al.* (2020). Introduction to radiomics. *Journal of Nuclear Medicine*, 61(4), 488–495.

[7] Pham, D. L., Xu, C. and Prince, J. L. (2000). Current methods in medical image segmentation. *Annual Review of Biomedical Engineering*, 2(1), 315–337.

[8] Sharma, A. K., Nandal, A., Ganchev, T. and Dhaka, A. "Breast cancer classification using CNN extracted features: a comprehensive review," In *Application of Deep Learning Methods in Healthcare and Medical Science (ADLMHMS-2020)*. Taylor and Francis: CRC Press; 2022.

[9] Fredborg, M., Rosenvinge, F. S., Spillum, E., Kroghsbo, S., Wang, M. and Sondergaard, T. E. (2015). Automated image analysis for quantification of filamentous bacteria. *BMC Microbiology*, 15, 1–8.

[10] Cui, M. and Zhang, D. Y. (2021). Artificial intelligence and computational pathology. *Laboratory Investigation*, 101(4), 412–422.

[11] Sharma, M. K., Dhiman, N., Mishra, V. N., *et al.* "Post-symptomatic detection of COVID-2019 grade based mediative fuzzy projection", *Computers and Electrical Engineering*, Elsevier, vol. 101, 2022.

[12] Wu, M. and Luo, J. (2019). "Wearable technology applications in healthcare: a literature review." *Online Journal of Nursing Informatics*, 23. https://www.himss.org/resources/online-journal-nursing-informatics.

[13] Topol, E. J. and Teirstein, P. S. (2015). *Textbook of Interventional Cardiology E-Book*. London: Elsevier Health Sciences.

[14] Kellman, P. and Arai, A. E. (2012). Cardiac imaging techniques for physicians: late enhancement. *Journal of Magnetic Resonance Imaging*, 36(3), 529–542.

[15] Murdock, D. R., Venner, E., Muzny, D. M., *et al.* (2021). Genetic testing in ambulatory cardiology clinics reveals high rate of findings with clinical management implications. *Genetics in Medicine*, 23(12), 2404–2414.

[16] Sharma, A. K., Nandal, A. and Dhaka, A. (2022). "Brain tumor detection and classification using machine learning techniques." *RAMSACT 2022*, Springer, Manipal University Jaipur, India, May 2022.

[17] Lope-Piedrafita, S. (2018). "Diffusion tensor imaging (DTI)." In: García Martín, M. and López Larrubia, P. (eds.). *Preclinical MRI. Methods in Molecular Biology*, vol. 1718. New York, NY: Humana Press. https://doi.org/10.1007/978-1-4939-7531-0_7.

[18] Lee, J. S. (2020). A review of deep-learning-based approaches for attenuation correction in positron emission tomography. *IEEE Transactions on Radiation and Plasma Medical Sciences*, 5(2), 160–184.

[19] Belkhiria, C. and Peysakhovich, V. (2020). Electro-encephalography and electro-oculography in aeronautics: a review over the last decade (2010–2020). *Frontiers in Neuroergonomics*, 1, 606719.

[20] Hari, R., Baillet, S., Barnes, G., *et al.* (2018). IFCN-endorsed practical guidelines for clinical magnetoencephalography (MEG). *Clinical Neurophysiology*, 129(8), 1720–1747.

[21] Aumann, S., Donner, S., Fischer, J. and Müller, F. (2019). "Optical coherence tomography (OCT): principle and technical realization." In: Bille, J. (eds.). *High Resolution Imaging in Microscopy and Ophthalmology*. Cham: Springer. https://doi.org/10.1007/978-3-030-16638-0_3.

[22] Sharma, A. K., Nandal, A. and Dhaka, A. (2022). "Comparison between predictive model and transfer learning-based model for MRI classification." *RAMSACT 2022*, Springer, Manipal University Jaipur, India, May 2022.

[23] Fan, R., Chan, T. C., Prakash, G. and Jhanji, V. (2018). Applications of corneal topography and tomography: a review. *Clinical & Experimental Ophthalmology*, 46(2), 133–146.

[24] González Izard S., Juanes Méndez J. A., Ruisoto Palomera P. and García-Peñalvo F. J. (2019). Applications of virtual and augmented reality in biomedical imaging. *Journal of Medical Systems*, 43(4), 102. doi:10.1007/s10916-019-1239-z. PMID: 30874965.

[25] Robb, R. A., Hanson, D. P., Karwoski, R. A., Larson, A. G., Workman, E. L. and Stacy, M. C. (1989). "Analyze: a comprehensive, operator-interactive software package for multidimensional medical image display and analysis." *Computerized Medical Imaging and Graphics*, 13(6), 433–454.

[26] Shkirkova, K., Akam, E. Y., Huang, J., *et al.* (2017). Feasibility and utility of an integrated medical imaging and informatics smartphone system for management of acute stroke. *International Journal of Stroke*, 12(9), 953–960.

[27] Albahri, O. S., Zaidan, A. A., Albahri, A. S., *et al.* (2020). Systematic review of artificial intelligence techniques in the detection and classification of COVID-19 medical images in terms of evaluation and benchmarking: taxonomy analysis, challenges, future solutions and methodological aspects. *Journal of Infection and Public Health*, 13(10), 1381–1396.

[28] Zhou, L., Alenezi, F. S., Nandal, A., *et al.* (2023). "Fusion of overexposed and underexposed images using caputo differential operator for resolution and texture based enhancement." *Applied Intelligence*, 53, 15836–15854. https://doi.org/10.1007/s10489-022-04344-z.

Index

www.ingramcontent.com/pod-product-compliance
Lightning Source LLC
La Vergne TN
LVHW011837200926
848484LV00002B/365
* 9 7 8 1 8 3 9 5 3 5 9 3 2 *